Random Sets and Integral Geometry

Applied Probability and Statistics (Continued)

BARTHOLOMEW · Stochastic Models for Social Processes, *Second Edition*
BENNETT and FRANKLIN · Statistical Analysis in Chemistry and the Chemical Industry
BHAT · Elements of Applied Stochastic Processes
BOX and DRAPER · Evolutionary Operation: A Statistical Method for Process Improvement
BROWNLEE · Statistical Theory and Methodology in Science and Engineering, *Second Edition*
BURY · Statistical Models in Applied Science
CHERNOFF and MOSES · Elementary Decision Theory
CHIANG · Introduction to Stochastic Processes in Biostatistics
CHOW · Analysis and Control of Dynamic Economic Systems
CLELLAND, deCANI, BROWN, BURSK, and MURRAY · Basic Statistics with Business Applications, *Second Edition*
COCHRAN · Sampling Techniques, *Second Edition*
COCHRAN and COX · Experimental Designs, *Second Edition*
COX · Planning of Experiments
COX and MILLER · The Theory of Stochastic Processes, *Second Edition*
DANIEL and WOOD · Fitting Equations to Data
DAVID · Order Statistics
DEMING · Sample Design in Business Research
DODGE and ROMIG · Sampling Inspection Tables, *Second Edition*
DRAPER and SMITH · Applied Regression Analysis
DUNN and CLARK · Applied Statistics: Analysis of Variance and Regression
ELANDT-JOHNSON · Probability Models and Statistical Methods in Genetics
FLEISS · Statistical Methods for Rates and Proportions
GOLDBERGER · Econometric Theory
GROSS and HARRIS · Fundamentals of Queueing Theory
GUTTMAN, WILKS and HUNTER · Introductory Engineering Statistics, *Second Edition*
HAHN and SHAPIRO · Statistical Models in Engineering
HALD · Statistical Tables and Formulas
HALD · Statistical Theory with Engineering Applications
HARTIGAN · Clustering Algorithms
HOEL · Elementary Statistics, *Third Edition*
HOLLANDER and WOLFE · Nonparametric Statistical Methods
HUANG · Regression and Econometric Methods
JOHNSON and KOTZ · Distributions in Statistics
Discrete Distributions
Continuous Univariate Distributions-1
Continuous Univariate Distributions-2
Continuous Multivariate Distributions
JOHNSON and LEONE · Statistics and Experimental Design: In Engineering and the Physical Sciences, Volumes I and II
LANCASTER · The Chi Squared Distribution
LANCASTER · An Introduction to Medical Statistics

continued on back

Random Sets and Integral Geometry

G. MATHERON

John Wiley & Sons
New York · London · Sydney · Toronto

Library of Congress Cataloging in Publication Data:

Matheron, Georges.
Random sets and integral geometry.

(Wiley series in probability and mathematical statistics)
Bibliography: p.
1. Geometric probabilities. 2. Geometry, Integral.
3. Set theory. I. Title.

QA273.5.M37 519.2 74-13569
ISBN 0-471-57621-2

Printed in United States of America

10 9 8 7 6 5 4 3 2 1

Foreword

Modern statistics might well be defined as the application of computers and mathematics to data analysis. It must grow as new types of data are considered and as computing technology advances. With this growth a wider collection of mathematical tools will become useful. Since the computer also allows new types of data to be recorded, it appears to be ultimately responsible for the revolution in statistics.

Most statistics books suppose that each "individual" in the population of interest can be characterized by a single real number or perhaps by a vector. Hence the mathematical theory requires that one study real, or vector, valued random variables. Although at first the computer was used only for easy calculation of previously defined quantities (e.g., inverses, eigenvalues, and vectors of matrices), it is now used to generate informative data displays and statistical evaluations of quite new types.

With stochastic processes the mathematical picture is extended to random functions, but their statistical analysis simply was not possible with desk calculators. Much current research is still concerned with the joy of being able to do computations that were impossible 15 years ago. With the ability to record, store, and process more complex descriptions of "individuals," it is *certain* that statistical theory will be required to deal with more general random quantities.

For psychological and mathematical reasons, it is natural to think geometrically. Although many sciences have raised such problems, the statistics of geometrical objects has not, until recently, received much attention. The reason seems to be that both theory and practice needed radically new ideas. Ideas for dealing with theory or practice have been proffered, but only in Fontainebleau has the joint development been conscious and coordinated. At this research center of the Paris School of Mines, the necessity of analyzing spatial data from geology led Matheron and his colleagues to develop simultaneously the required computer technology and the matching probabilistic and statistical theory.

Random Sets and Integral Geometry deals with one aspect of this work and emphasizes that geometrical probability should be thought of as the theory of random sets. The treatment is forbiddingly mathematical but the basic ideas are simple and elegant—and important for this domain of statistical practice. However, much remains to be done; the most obvious need is for specific models for random sets.

If my vision of a *general* theory of statistics is correct, this will be a basic book for the future since the notion of a set is the cornerstone of mathematics. The only other book devoted to this kind of generalization is Grenander's *Probabilities on Algebraic Structures* (John Wiley & Sons, 1963). The possibilities are so vast that it seems safe to predict the next advance will again be suggested by a specific class of problems rather than by mathematical speculation. It is an exciting prospect.

G. S. WATSON

Princeton
May 1974

Preface

In the last ten years or so, image analysis techniques have developed rapidly. Various experimental devices now available combine an information recording unit with an information processing unit. The former usually involves scanning by a television camera so that tens of thousands of points in the sample may be investigated within a few minutes. The latter is a medium-sized computer, perhaps especially adapted to geometrical work (see, e.g., Moore, Wyman, and Joseph, 1968, or Klein and Serra, 1971). If we exclude the specific problems of pattern recognition, the aim of these techniques is, in general, the analysis of textures, that is, the objective characterization of an image with the help of some quantitative parameters such as surface, perimeter, number of connected components, size distribution, and curvature radii. But all this raises a number of theoretical problems of a geometrical or probabilistic nature, which pertain to what we can call "mathematical morphology" (see Haas, Matheron, and Serra, 1967).

In the first place, the parameters to be measured must have a rigorous geometrical definition (which is not always the case in the specialized literature when, for instance, we speak of size distribution). Once this parameter is defined, we must find a technique as simple as possible in order to make the corresponding measurements. Integral geometry proves to be quite useful by suggesting various possibilities for obtaining indirect measurements of a parameter that is difficult to measure otherwise. Examples of such parameters are the perimeter, the distribution of curvature radii, and the number of connected components. Also, problems of a *stereological* nature often arise. They involve reconstituting the geometrical properties of a tridimensional object, starting from observations made on planes or straight lines. Some of these problems are solvable; others are not. Here again, integral geometry provides indispensable guidance, especially with Steiner's and Crofton's formulae.

Lastly, the geometrical complexity of the objects studied (in fields like metallography, petrography, porous media, and biology) often necessitates a recourse to a statistical or probabilistic interpretation. But there are two quite different ways to consider geometrical probabilities. One may consider the object to be studied as entirely deterministic, and assume that the experimental observations are located at random with respect to that object. Integral geometry (which, since Deltheil, 1926, and Blaschke, 1936–37, has especially emphasized research on measures invariant under euclidean transformations) corresponds entirely to this first point of view. It is often necessary, however, to introduce randomness at a more fundamental level, by assuming that the object itself and its geometrical shape, and not only its location, are random. Then we are really thinking of the notion of *random set*, which is the main subject of this book.

This notion of a random set is strongly suggested, for instance, by porous media (see, e.g., Fara and Scheidegger, 1961) or by the phenomena that can be observed in metallography (see, e.g., Miles, 1972, and the Proceedings of the Symposium on Statistical and Probabilistic Problems in Metallurgy, published in *Advances in Applied Probability*, December 1972). It is this initial intuition that provides the present work with its starting point. From the mathematical point of view, the notion of a random closed set goes back to Choquet (1953–54), who gave it a complete mathematical characterization. A very wide generalization of this subject was recently made by Kendall (1973). Nevertheless, and despite the purely mathematical nature of the present treatise, the formulation and the very choice of problems for solution are directly inspired by experimental techniques of texture analysis that currently permit the thorough study of such media (see, e.g., Serra, 1969). Conversely, a large number of theoretical results that will be established herein lead to new and fruitful experimental procedures. For instance, the precise mathematical status given to the common but rather vague notion of particle size distribution leads to extremely interesting practical applications (see, e.g., Delfiner, 1970, 1971).

It is natural to begin by looking at the conditions under which a random set theory may be applied to physical reality. Consider, for instance, a porous medium. Clearly, it is in general not possible to examine one by one each of the "grains" that compose such a medium, or to study in detail their individual shapes and dimensions. Moreover, these grains are very commonly more or less connected, so that at the limit the notion of individual grains itself ceases to be realistic. Furthermore, the solid component of this medium is not necessarily the most interesting. The porous network itself, with its incredible complexity and the many possibilities it offers to the path of a fluid, is directly responsible for a property as important in practice as permeability. Therefore, in an earlier work (Matheron, 1967), I attempted a first interpretation of porous media in terms of random sets, with the aim of summing up the essential properties of their complex structures, using a small number of statistically representative and experimentally accessible parameters. Even at that time, it appeared that the morphological study of a porous medium required recourse to a certain number of geometrical probabilistic concepts that, in turn, could be linked to the central notion of a random set.

At the outset of "mathematical morphology," it is necessary to reflect on the notion of form, or of structure. In general, the structure of an object is defined as the set of relationships existing between elements or parts of the object. In order to experimentally determine this structure, we must try, one after the other, each of the possible relationships and examine whether or not it is verified. Of course, the image constructed by such a process will depend to the greatest extent on the choice made for the system $\mathfrak{R}$ of relationships considered as possible. Hence this choice plays a priori a constitutive role (in the Kantian meaning) and determines the relative worth of the concept of structure at which we will arrive.

In the case of a porous medium, let A be the solid component (union of grains), and A^c the porous network. In this medium, we shall move a figure B, called the structuring pattern, playing the role of a probe collecting information. This operation is experimentally attainable, and with a rather great facility, at least for planar sections of the medium. Of course, we cannot take any pattern B, but must choose it from a family $\mathfrak{B}$ that will be specified later on. The simplest relationships that can be envisaged to collect information on the structure of the porous medium are of the form $B \subset A$ (B is included in the grains) or $B \cap A \neq \varnothing$ (B hits the grains). Of course we will also have to concern ourselves with relationships that can be derived from the previous ones with the help of the logic operations "and," "or," and "no." In other words, the family of possible relationships on which we will have to work will be the σ-algebra generated by the simple relationships $B \subset A$, $B \cap A \neq \varnothing$, B scouring $\mathfrak{B}$. From a deterministic standpoint, we can consider the structure of the set A as known if, for any possible relationship $R \in \mathfrak{R}$, we know whether or not R is true for the set A. In the same way, from a probabilistic standpoint, we will consider that we know everything that is interesting, or possible in practice, to know if, for any $R \in \mathfrak{R}$, the probability $P(R)$ for this relationship to be verified is known. In other words, the random set A will be classically defined by the datum of a probability P on the σ-algebra $\mathfrak{R}$.

However, it becomes quickly apparent that the mathematical notion of sets is too rich to be applied as it stands to an experimental reality such as a porous medium. Indeed, if we define A as the set of points of space that belong to the grains, must A be considered as an open set, a closed set, or something topologically more complicated? In other words, does a point of the boundary of A belong to A or not? From an experimental point of view, this question is absolutely senseless, because no physical reality corresponds perfectly to the notion of a point belonging to the boundary of the grains. Indeed, for an experimentalist, there exist no true points but rather spots, with small but nonzero dimensions and rather poorly defined boundaries. To a first approximation, we can give a mathematical status to this experimental haziness by taking open sets for our structuring patterns B.

Thus the family $\mathfrak{B}$ from which we will generate our σ-algebra of relationships considered as possible will be (at most) the family of open sets. But if B is an open set, B is included in A if and only if it is included in the interior $\mathring{A}$ of A; and in the same way, B hits A if and only if it hits the closure $\bar{A}$ of A. Thus the logic of intersection based on the σ-algebra generated by events of the type $B \cap A \neq \varnothing$ naturally leads to the notion of a random closed set; and, likewise, the logic of inclusion is linked to the notion of a random open set. If the two logics are used simultaneously (as we generally do in the applications), we see that the relationships $R \in \mathfrak{R}$ express properties, not of A itself (which is a simple abstract mathematical concept to which corresponds no physical reality) but only of the pair $(\mathring{A}, \bar{A})$ made of the interior and the closure of the set A. In other words, when they have the same interior and the same closure, two sets A and A' are indistinguishable to the σ-algebra $\mathfrak{R}$, and must be considered as representing the same physical reality. This is really the minimum concession that physicists have the right to demand from mathematicians.

But it appears that these σ-algebras, adapted directly to experimental situations, are identified with the borelian tribes associated with topologies with which it is natural to provide the space $\mathcal{F}$ of closed sets, the space $\mathcal{G}$ of open sets, or the space $\mathcal{K}$ of compact sets. The initial probabilistic theme then bends toward *integral geometry* (expressed, e.g., after the manner of Hadwiger, 1957). It will be seen that many classical functionals of integral geometry (in particular, the well-known functionals of Minkowski) admit simple probabilistic interpretations in terms of random closed sets. This will be one of the major themes of this work, a theme that will be emphasized by the important role played by *convexity* in a certain number of questions that are apparently without link, but that arise about the euclidean space.

This book can be roughly divided into three subdivisions, each one consisting of three chapters. The first subdivision is devoted to definitions and general properties of random closed sets in a space E, which is not necessarily euclidean, but will always be assumed to be locally compact and separable (for the purpose of avoiding mathematical difficulties unrelated to our subject). Chapter 1 is devoted to the absolutely indispensable topological preliminaries that serve as a basis for what follows. The choice of definitions and specific problems, which may appear arbitrary at first sight, is, in fact, entirely justified by subsequent probabilistic applications. For instance, the rather exhaustive study of semicontinuity is justified by the fact that upper semicontinuous (u.s.c.) and lower semicontinuous (l.s.c.) mappings are automatically measurable, and therefore permit the construction of new random closed sets from an initial random closed set (RACS). This will be so for intersection, for union, and also, in the euclidean case, for Minkowski's addition and subtraction, as well as for the operation called opening A by a set B—an operation that leads to interesting notions concerning granulometries (i.e., the common notion of size distributions considered as set-valued mappings). Chapter 2 uses these topological results to define the general notion of a random closed set (RACS). The essential result is here the theorem due to Choquet (1953–54), according to which a function T on the space $\mathcal{K}$ can be related to a RACS according to the formula $T(K)=P(A\cap K\neq\emptyset)$ if and only if T is an alternating capacity of infinite order, verifying that $0\leqslant T\leqslant 1$. Certain important results concerning conditional RACS and regarding the space law (i.e., the maximum information concerning a RACS that can be collected from a countable infinity of points) will follow. In Chapter 3, the class of random closed sets infinitely divisible for the union is introduced; this class is identified with the Poisson processes on the space $\mathcal{F}'$ of nonempty closed sets, and also with a class of Choquet capacities.

Chapters 4 through 6 involve a study of the relationships between RACS in euclidean space and integral geometry. Here, convexity, above all, plays an essential role, especially in the very long Chapter 4 entirely devoted to this subject. After a short account of the classical functionals of Minkowski, this chapter identifies the almost certain convexity of a RACS with the C-additivity of its functionals T, and then examines classical questions concerning random intercepts and the convex cone $C_0(\mathcal{K})$ made of the compact convex sets containing the origin. This cone admits a subset $\mathcal{R}_1$, or *Steiner class*, made of the symmetric convex compact sets which are a finite Minkowski sum of segments or the limit of such finite sums. The Steiner class $\mathcal{R}_1$ is also a convex cone, and in fact even a *simplex*; this property leads to an important theorem of uniqueness. According to this theorem, a Poisson stationary process of hyperplanes, for instance, is determined when the densities induced on each line of space are known. Finally, we associate with the Minkowski functionals measures that are to represent the properties of curvature of the boundary of a compact convex set; their definition is generalized to different subspaces of $\mathcal{F}$ and $\mathcal{K}$ and, finally, to $\mathcal{F}$ itself. In the case of a RACS, the corresponding measures become random measures. It is in this sense, in particular, that we can speak of the specific surface area of a RACS. Chapter 4 gives a theory of semi-markovian RACS, and, in particular, a complete characterization of infinitely divisible and stationary semi-markovian RACS, two prototypes of which are Poisson linear varieties and boolean models with convex primary grains. As an application, Chapter 5 is a rather detailed study of Poisson hyperplanes. It will be noted that the notion of a Steiner compact set Λ associated with such a network allows the problem of calculating the characteristics of induced networks on linear varieties to be solved in a single step.

The third subdivision of the book is devoted to mappings valued in a space $\mathcal{F}$ or $\mathcal{K}$. We have not sought generality as such. It is here, on the contrary, that the experimental motivations of this book play the greatest role, and lead us to select a relatively small number of algebraic and topologic structures (the mathematical interest of which is not obvious at the start). Chapter 7, which is by far the most important for practical applications, is devoted to *granulometries*. It was necessary to give a precise axiomatic status to the usual notion (not always clearly defined) of grain size distribution, so that this notion could be applied to the pores as well as the grains of a porous medium, that is, to the case of a set that it is not necessarily possible or interesting to decompose into its connected components. Moreover, the set of axioms is simple, as is the purely algebraic study that follows. But if we want to apply these notions to random open or closed sets, we must make sure beforehand of the measurability of the corresponding mappings. This leads to a certain definition of "good" granulometries, the study of which necessitates rather deep topological investigations. Therefore (contrary to the logical sequence) we first study the simpler particular case of euclidean granulometries (the only ones useful in practical applications). Here, again, convexity intervenes in an unexpected way, in the sense that the best euclidean granulometries are shown to be, quite simply, the openings by the homothetic transforms of a convex compact set. The next chapter (Chapter 8) deals with the increasing mappings, particularly those which are compatible with translations in the euclidean space. Its purpose is essentially to clarify certain results observed in the case of granulometries. Finally, Chapter 9 is devoted to integrals and measures valued in a space $\mathcal{K}$ of compact sets or a space $C(\mathcal{K})$ of convex compact sets in an euclidean space.

G. Matheron

Fontainebleau, France
May 1974

Acknowledgments

I wish to thank G. S. Watson, who suggested that I write this book and encouraged me in the hazardous task, and N. Cressie, whose critical comments resulted in the elimination of many errors and obscurities. I am also greatly indebted to the team of the Mathematical Morphology Center of Fontainebleau, especially to P. Delfiner, J. Jacod, and J. Serra. Many of the ideas contained in this book were born during discussions with one or the other, and the whole work came from reflections on experimental methods of texture analysis. The support of the Ecole Nationale Supérieure des Mines de Paris is also gratefully acknowledged. Lastly, I thank Mrs. Kreyberg for the care she took in correcting this manuscript.

G. M.

Contents

Notations

$\cup$	Union
$\cap$	Intersection
$\varnothing$	Empty set
$\complement A$ or A^c	Complementary set of a set A
$A \setminus B$	Set difference (i.e., $A \cap B^c$)
$A_n \uparrow A$	$\{A_n\}$ is an increasing sequence of sets, and $\cup A_n = A$
$A_n \downarrow A$	$\{A_n\}$ is a decreasing sequence of sets, and $\cap A_n = A$
$\bar{A}$	Topological closure of a set A
$\mathring{A}$	Interior of a set A
∂A	Boundary of a set A (i.e., $\partial A = \bar{A} \cap \overline{A^c}$)
$\oplus$	Minkowski addition
$\ominus$	Minkowski subtraction
$\check{B}$	Symmetric set of B about the origin, that is, $\check{B} = \{-x, x \in B\}$
A_B	Opening of A by B, that is, $A_B = (A \ominus \check{B}) \oplus B$
A^B	Closing of A by B, that is, $A^B = (A \oplus \check{B}) \ominus B$
A_x	Translate of A by x, that is, $A_x = A \oplus \{x\}$
b_k	Volume of the unit ball in $\mathbf{R}^k$, that is, $b_k = \pi^{k/2}/\Gamma(1+k/2)$
$C(A)$	Convex hull of a set $A \in \mathbf{R}^d$
$C(\mathcal{K})$	Space of compact convex sets in $\mathbf{R}^d$
$C(\mathcal{K}_0)$	Space of compact convex sets containing the origin 0
$C(\mathcal{K}')$	Space of nonempty compact convex sets
$C(\mathcal{F})$	Space of closed convex sets
$C(\mathcal{F}')$	Space of nonempty closed convex sets
$C(\mathcal{G})$	Space of open convex sets
$\mathbf{C}(E)$ or $\mathbf{C}$	Space of continuous functions on E

$\mathbf{C}_{\mathcal{K}}(E)$ or $\mathbf{C}_{\mathcal{K}}$	Space of continuous functions on E with compact support
E	Generally a LCS space (i.e., locally compact, Hausdorff, and separable)
$\mathcal{F}(E)$ or $\mathcal{F}$	Space of all closed sets in E
$\mathcal{F}'$	Space of nonempty closed sets in E
$\mathcal{F}^A$	$=\{F, F\in\mathcal{F}, F\cap A=\varnothing\}$ $(A\subset E)$
$\mathcal{F}_B$	$=\{F, F\in\mathcal{F}, F\cap B\neq\varnothing\}$ $(B\subset E)$
$\mathcal{F}^A_{B_1,\ldots,B_n}$	$=\mathcal{F}^A\cap\mathcal{F}_{B_1}\cap\cdots\cap\mathcal{F}_{B_n}$
$\mathcal{F}_u(\mathcal{F})\subset\mathcal{F}$	Space of the $\cup$-hereditary families closed in $\mathcal{F}$
$\mathcal{F}_u(\mathcal{K})\subset\mathcal{K}$	Space of the $\cup$-hereditary families closed in $\mathcal{K}$
Φ_k	Space of the u.s.c. functions $E\to\mathbf{R}_+$ with compact support in E
Φ_g	Space of the l.s.c. functions $E\to\mathbf{R}_+$
Φ	Space of pseudo-integrable functions
$\mathcal{G}(E)$ or $\mathcal{G}$	Space of all open subsets of E
$\mathcal{G}_B$	$=\{G, G\in\mathcal{G}, G\supset B\}$
$\mathcal{G}^A$	$=\{G, G\in\mathcal{G}, G\not\supset A\}$
$\mathcal{H}$	Space of the pairs $\{\mathring{A},\bar{A}\}$; $A\in\mathcal{P}$, identified with $\{(G,F), G\in\mathcal{G}, F\in\mathcal{F}, G\subset F\}$
$\mathcal{T}$	Class of finite subsets of E
$\mathcal{T}'$	Class of nonempty finite subsets of E
Inf	Infimum
$\mathcal{K}(E)$ or $\mathcal{K}$	Space of compact subsets of E
$\mathcal{K}'$	Space of nonempty compact subsets of E
$\mathcal{K}_0$	Space of compact subsets of E containing the origin 0
$\mathcal{K}^A$	$=\{K, K\in\mathcal{K}, K\cap A=\varnothing\}$ $(A\subset E)$
$\mathcal{K}_B$	$=\{K, K\in\mathcal{K}, K\cap B\neq\varnothing\}$ $(B\subset E)$
$\mathcal{K}^A_{B_1,\ldots,B_n}$	$=\mathcal{K}^A\cap\mathcal{K}_{B_1}\cap\cdots\cap\mathcal{K}_{B_n}$
$\mathcal{P}(E)$ or $\mathcal{P}$	Space of all subsets of E
$\mathcal{P}_x$	$=\{A, A\in\mathcal{P}, x\in A\}$
Π_A	Projector on a closed convex set A
Π_0	Poisson polyhedron
$\mathbf{R}$	One-dimensional euclidean space
$\mathbf{R}^k$	k-Dimensional euclidean space
$\mathbf{R}_+$	$=[0,\infty[$ positive half straight line
$\bar{\mathbf{R}}_+$	$=[0,\infty]$ compactified half straight line
ρ	Hausdorff metric
$\mathcal{R}$	Convex cone of all supporting functions [identified with $C(\mathcal{K}_0)$]

$\mathcal{R}_s$	Class of nonempty symmetric compact convex sets
$\mathcal{R}_1$	Steiner class
$\mathcal{R}_\alpha$	See p. 000
S_0	(Generally) the unit sphere in $\mathbf{R}^d$
$\mathfrak{S}$	The convex ring
σ_f	Borelian tribe of $\mathcal{F}$
σ_g	Borelian tribe of $\mathcal{G}$
σ_h	Borelian tribe of $\mathcal{H}$
σ_k	Borelian tribe of $\mathcal{K}$
$\mathcal{S}_k$	Space of all k-dimensional subspaces in $\mathbf{R}^d$ $(k=0,1,\ldots,d)$
Sup	Supremum
Supp	Support (of a measure)
$\mathcal{T}_f$	Topology of $\mathcal{F}$
$\mathcal{T}_g$	Topology of $\mathcal{G}$
$\mathcal{T}_h$	Topology of $\mathcal{H}$
$\mathcal{T}_k$	Topology of $\mathcal{K}$ (myope topology)
W_k^d or W_k	Order k-Minkowski functional on $C(\mathcal{K}(\mathbf{R}^d))$
$W_k(dx)$ or W_k	Minkowski measure

Random Sets and
Integral Geometry

CHAPTER 1

The Topological Spaces $\mathcal{F}$, $\mathcal{G}$, $\mathcal{H}$, and $\mathcal{K}$

The spaces $\mathcal{F}$, $\mathcal{G}$, and $\mathcal{K}$ of the subsets respectively closed, open, and compact in a given topological space E have been topologized in various ways (see Michael, 1951). The particular topologies we have chosen seem to be most suitable for a probabilistic use. The myope topology on $\mathcal{K}$ is classical in integral geometry, and the compact topology on $\mathcal{F}$ may be deduced from the myope topology on the space $\mathcal{K}(\bar{E})$ of the compact subsets in the compactification $\bar{E}$ of the space E (see Choquet, 1953–54, and the remark following our Theorem 1-4-1.) Various continuous or semicontinuous operations are defined in the case of a euclidean space $E=\mathbf{R}^d$: Minkowski's addition $\oplus$ and subtraction $\ominus$, opening or closing of a set A by a set B, and so on. The case of convex compact sets is particularly interesting (cf. Chapter 4 and following). A theorem concerning compact sets infinitely divisible for $\oplus$ leads to a characterization of continuous semigroups on $C(\mathcal{K})$ and to the important notion of the granulometry of A in respect to a compact convex set K.

1-1. GENERAL HYPOTHESES AND NOTATIONS

Throughout the book, E will be a *locally compact*, *Hausdorff*, and *separable* (LCS) space (i.e., each point in E admits a compact neighborhood, and the topology of E admits a countable base). This implies the following properties, which will be used in the sequel without special reference.

a. Any compact set $K\subset E$ admits a countable fundamental system of open neighborhoods $G_1, G_2, \ldots$, and it is always possible to suppose that the sequence $\{G_n\}$ is decreasing in E. In other words, for any open set $G\supset K$, the inclusion $G_n\subset G$ is satisfied for n large enough; in particular, $K=\cap G_n$. If a closed set F and a compact set K are *disjoint* in E, that is,

$F \cap K = \emptyset$, there exist two disjoint open sets G and G' with $G \supset F$ and $G' \supset K$.

b. Let $G \subset E$ be an open set. Then there exists an increasing sequence of relatively compact and open sets $\{B_n\}$ (i.e., B_n is open, and its closure $\overline{B}_n$ is compact for each n) with $\overline{B}_n \subset B_{n+1}$ and $G = \cup B_n = \cup \overline{B}_{n+1}$. In particular, any compact set $K \subset G$ is included in B_n for n large enough.

c. There exists a countable family $\mathcal{B}$ of open sets in E such that each open set G is the union of a subfamily of $\mathcal{B}$ (i.e., $\mathcal{B}$ is a base for the topology of E). Moreover, it is possible to choose $\mathcal{B}$ in such a way that each $B \in \mathcal{B}$ is relatively compact (i.e., $\overline{B}$ is compact), and each open set G is the union of the sets $B \in \mathcal{B}$ satisfying $\overline{B} \subset G$.

If A is a subset of E, A^c or, sometimes, $\complement A$ denotes its *complementary set*, $\mathring{A}$ its interior, $\overline{A}$ its *closure* (or adherence), and $\partial A = \overline{A} \cap \complement \mathring{A}$ its *boundary*. If $B \subset A$, we put $A \backslash B = A \cap B^c$ (set difference). The relationship $A_n \uparrow A$ (resp. $A_n \downarrow A$) means that $\{A_n\}$ is an increasing (resp. decreasing) sequence of subsets in E satisfying $A = \cup A_n$ (resp. $A = \cap A_n$).

The notations $\mathcal{F}(E)$, $\mathcal{G}(E)$, and $\mathcal{K}(E)$, or, if there is no ambiguity simply $\mathcal{F}$, $\mathcal{G}$, and $\mathcal{K}$, denote the classes of the subsets respectively closed, open, and compact in E. In the same way, $\mathcal{P}(E)$, or $\mathcal{P}$, is the class of all the subsets of E. Generally speaking (and as far as possible), small italic letters $x, y\ldots$ denote elements or *points* in $E (x \in E)$, capital italic letters $A, F, K, \ldots$ denote *subsets* of $E (A \subset E)$, and cursive capitals $\mathcal{Q}$, $\mathcal{F}, \ldots$ denote *classes of subsets* in $E[\mathcal{Q} \subset \mathcal{P}(E)]$, so that we may write $a \in A \in \mathcal{Q}$.

Because we use intersection logics in the space $\mathcal{F}$, for each $B \in \mathcal{P}(E)$ we denote by $\mathcal{F}_B$ the class of the closed sets hitting B and by $\mathcal{F}^B$ the complementary class, that is, the class of the closed sets disjoint of B:

$$\mathcal{F}_B = \{F : F \in \mathcal{F}, \quad F \cap B \neq \emptyset\}$$

$$\mathcal{F}^B = \{F : F \in \mathcal{F}, \quad F \cap B = \emptyset\}$$

If $\{B_i, i \in I\}$ is a class of subsets in E, we obviously have the equalities

$$\cup \mathcal{F}_{B_i} = \mathcal{F}_{\cup B_i}; \qquad \cap \mathcal{F}^{B_i} = \mathcal{F}^{\cup B_i} \tag{1-1-1}$$

but only the inclusions

$$\cap \mathcal{F}_{B_i} \supset \mathcal{F}_{\cap B_i}; \qquad \cup \mathcal{F}^{B_i} \subset \mathcal{F}^{\cap B_i}$$

Nevertheless, if $K \in \mathcal{K}$ is *compact* and $\{G_n\}$ is a fundamental system of open neighborhoods of K, the equalities hold:

$$\cap \mathcal{F}_{G_n} = \mathcal{F}_K; \qquad \cup \mathcal{F}^{G_n} = \mathcal{F}^K \tag{1-1-2}$$

From the point of view of the inclusion logics in the space $\mathcal{G}$, we put, for any $B \in \mathcal{P}(E)$,

$$\mathcal{G}_B = \{G, G \in \mathcal{G}, G \supset B\}$$

$$\mathcal{G}^B = \{G, G \in \mathcal{G}, G \not\supset B\}$$

Obviously $G \in \mathcal{G}_B$ is equivalent to $G^c \in \mathcal{F}^B$, so that to each property of the space $\mathcal{F}$ corresponds the dual property of the space $\mathcal{G}$, and conversely. Generally, we shall give only the statements concerning the space $\mathcal{F}$ and leave to the reader their translation in terms of the space $\mathcal{G}$.

For the space $\mathcal{K}$, the notations are the same as for $\mathcal{F}$:

$$\mathcal{K}_B = \{K : K \in \mathcal{K}, K \cap B \neq \emptyset\}$$

$$\mathcal{K}^B = \{K : K \in \mathcal{K}, K \cap B = \emptyset\}$$

1-2. THE COMPACT SPACE $\mathcal{F}(E)$

Throughout the book, the space $\mathcal{F}$ is topologized by the topology $\mathcal{T}_f$, generated by the two families $\mathcal{F}^K$, $K \in \mathcal{K}$, and $\mathcal{F}_G$, $G \in \mathcal{G}$. By (1-1-1), the family $\mathcal{F}^K$, $K \in \mathcal{K}$, is closed for the finite union. Thus, by putting

$$\mathcal{F}^K_{G_1, G_2, \ldots, G_n} = \mathcal{F}^K \cap \mathcal{F}_{G_1} \cap \cdots \cap \mathcal{F}_{G_n} \tag{1-2-1}$$

for $K \in \mathcal{K}$, n integer $\geqslant 0$ and $G_1, \ldots, G_n \in \mathcal{G}$, the corresponding class of subsets of $\mathcal{F}$ is a *base for the topology* $\mathcal{T}_f$ on $\mathcal{F}$. Notice that $\mathcal{F}^{\emptyset}_G = \mathcal{F}_G$, and for $n = 0$ the set (1-2-1) is $\mathcal{F}^K$, so that the sets $\mathcal{F}_G$, $G \in \mathcal{G}$, and $\mathcal{F}^K$, $K \in \mathcal{K}$, belong to this base, as well as $\mathcal{F}^{\emptyset} = \mathcal{F}$ and $\mathcal{F}_{\emptyset} = \emptyset$. The sets $\mathcal{F}^K$, $K \in \mathcal{K}$, constitute a fundamental system of neighborhoods for the element $\emptyset \in \mathcal{F}$, and, in the same way, the sets $\mathcal{F}_{G_1, \ldots, G_n}$ (n integer > 0, $G_1, \ldots, G_n \in \mathcal{G}$) constitute a fundamental system of neighborhoods for $E \in \mathcal{F}$. Hence the empty set $\emptyset$ and the full set E are characterized by the poorness of the filters of their neighborhoods. Let us now examine the main properties of the space $\mathcal{F}$.

Theorem 1-2-1. *The space $\mathcal{F}$ is compact, Hausdorff, and separable.*

Proof. *a*. We show first that $\mathcal{F}$ is separable, that is, its topology $\mathcal{T}_f$ admits a countable base. Let $\mathcal{B}$ be a base of the topology $\mathcal{G}$ on E with property *c* of Section 1-1; that is, each $B \in \mathcal{B}$ is a relatively compact open set, and each $G \in \mathcal{G}$ is a union of the $B \in \mathcal{B}$ such that $\overline{B} \subset G$. Let $\mathcal{T}_b$ be the class of the subsets

$$\mathcal{F}_{B_1,\dots,B_n}^{\bar{B}'_1 \cup \cdots \cup \bar{B}'_k}$$

of $\mathcal{F}$, n and k integers $\geqslant 0$, $B_1,\dots,B_n$, $B'_1,\dots,B'_k \in \mathcal{B}$. Class $\mathcal{T}_b$ is countable, and it is sufficient to show that $\mathcal{T}_b$ is a base for $\mathcal{T}_f$. Let $F \in \mathcal{F}$ be a closed set, and $\mathcal{F}^K_{G_1,\dots,G_n}$ $(K \in \mathcal{K}, G_1,\dots,G_n \in \mathcal{G})$ an open neighborhood of F in $\mathcal{F}$. Then there exists a $\mathcal{V} \in \mathcal{T}_b$ with $F \in \mathcal{V} \subset \mathcal{F}^K_{G_1,\dots,G_n}$, for, if $F = \emptyset$, we have $n=0$, and we may choose $B \in \mathcal{B}$ with $B \supset K$, so that $\mathcal{V} = \mathcal{F}^{\bar{B}} \in \mathcal{T}_b$ verifies $F \in \mathcal{F}^{\bar{B}} \subset \mathcal{F}^K$. If $F \neq \emptyset$, for each $i = 1,2,\dots,n$, we may choose a point $x_i \in F \cap G_i$ and an open set $B_i \in \mathcal{B}$ with

$$x_i \in B_i \subset \bar{B}_i \subset G_i \cap K^c.$$

Then there exists a finite covering of the compact set K by open sets $B'_j \in \mathcal{B}$, $j = 1,2,\dots,k$, verifying for $j = 1,2,\dots,k$ the relationships $\bar{B}'_j \cap \bar{B}_i = \emptyset (i = 1,2,\dots,n)$ and $F \cap \bar{B}'_j = \emptyset$. Then, we have

$$F \in \mathcal{F}_{B_1,\dots,B_n}^{\bar{B}'_1 \cup \cdots \cup \bar{B}'_k} \subset \mathcal{F}^K_{G_1,\dots,G_n}$$

b. Let us now show that $\mathcal{F}$ is *separated* (or is a Hausdorff space) for the $\mathcal{T}_f$ topology. Let F, F' be the two closed sets with $F \neq F'$ and, for instance, x a point with $x \in F$ and $x \notin F'$. Thus there exists a relatively compact open set B such that $x \in B$ and $F' \cap \bar{B} = \emptyset$. Then we have $F \in \mathcal{F}_B$, $F' \in \mathcal{F}^{\bar{B}}$, and $\mathcal{F}_B \cap \mathcal{F}^{\bar{B}} = \emptyset$: $\mathcal{F}$ is separated.

c. It remains to show that $\mathcal{F}$ is compact. The class of the sets closed in $\mathcal{F}$ is the class closed for finite union and infinite intersection generated by the class $\mathbf{C} = \{\mathcal{F}_K, K \in \mathcal{K}\} \cup \{\mathcal{F}^G, G \in \mathcal{G}\}$. By a classical topological result, it is sufficient, in order to prove that $\mathcal{F}$ is compact, to show that the class $\mathbf{C}$ satisfies the "finite intersection" property. Thus let K_i, $i \in I$, be a family of compact sets, and G_j, $j \in J$, a family of open sets satisfying

$$\left(\bigcap_{i \in I} \mathcal{F}_{K_i}\right) \cap \left(\bigcap_{j \in J} \mathcal{F}^{G_j}\right) = \emptyset$$

Also, put $\Omega = \cup_{j \in J} G_j$. We have $\cap \mathcal{F}^{G_j} = \mathcal{F}^{\Omega}$, so that the above relationship may be rewritten as $\cap \mathcal{F}^{\Omega}_{K_i} = \emptyset$. But this intersection is empty if and only if there exists an index $i_0 \in I$ with $K_{i_0} \subset \Omega$. Otherwise, the closed set $\Omega^c \cap (\overline{\cup K_i})$ (which is empty if $I = \emptyset$) would be disjoint of Ω and would hit each K_i and thus belong to the intersection $\cap \mathcal{F}^{\Omega}_{K_i}$. Thus let $i_0 \in I$ be an index such that $K_{i_0} \subset \Omega = \cup \, G_j$. There exists a finite covering $G_{j_1}, \ldots, G_{j_n}$ of the compact set K_{i_0} for indices $j_1, \ldots j_n \in J$, and the intersection $\mathcal{F}_{K_{i0}} \cap \mathcal{F}^{G_{j1}} \cap \cdots \cap \mathcal{F}^{G_{jn}}$ is empty.

Remark. The space $\mathcal{F}' = \mathcal{F} \backslash \{\varnothing\}$ of the nonempty closed subsets of E is not necessarily compact, for, if $G_j, j \in J$, is a covering of E by a family of relatively compact open sets G_j, we have $\cap \, \mathcal{F}^{G_j} = \mathcal{F}^E = \{\varnothing\}$. But there exists a finite subfamily the intersection of which is $\{\varnothing\}$ only if E itself is compact (and not only locally compact). Conversely, if E is compact, $E \in \mathcal{K}$ implies that $\{\varnothing\} = \mathcal{F}^E$ is an open subset of $\mathcal{F}$, and $\mathcal{F}' = \mathcal{F} \backslash \{\varnothing\}$ is compact. The set $\{\varnothing\}$ is, at the same time, open and closed in $\mathcal{F}$, and $\varnothing$ is an isolated point in $\mathcal{F}$. Thus we may state:

Proposition 1-2-1. *The space* $\mathcal{F}' = \mathcal{F} \backslash \{\varnothing\}$ *is compact if and only if E itself is compact.*

The following proposition concerns the connectivity of $\mathcal{F}$ and will not be used in the sequel.

Proposition 1-2-2. *If E is connected and not compact, $\mathcal{F}(E)$ is connected. If E is connected and compact, $\mathcal{F}' = \mathcal{F} \backslash \{\varnothing\}$ is connected.*

Proof. Suppose that E is connected, and let $\mathcal{A}$ and $\mathcal{A}'$ be closed subsets of $\mathcal{F}$ satisfying $\mathcal{A} \cap \mathcal{A}' = \varnothing$ and $\mathcal{A} \cup \mathcal{A}' = \mathcal{F}$. The mapping $x \rightarrow \{x\}$ from E into $\mathcal{F}$ is continuous, so that the sets $A = \{x, \{x\} \in \mathcal{A}\}$ and $A' = \{x, \{x\} \in \mathcal{A}'\}$ are closed in E and satisfy $A \cap A' = \varnothing$ and $A \cup A' = E$. Thus one of them is empty, because E is connected; say, for instance that $A' = \varnothing$, so that every single element set $\{x\}$ belongs to $\mathcal{A}$. Let us show that every nonempty finite set belongs to $\mathcal{A}$: suppose, as a recurrence hypothesis, that every nonempty set containing at the most n distinct points is in $\mathcal{A}$, and show that the same is true for every set of the form $\{x_1, x_2, \ldots, x_n, x_{n+1}\}$. The mapping $y \rightarrow \{x_1, \ldots, x_n, y\}$ is continuous from E into $\mathcal{F}$, so that the sets $A_n = \{y : \{x_1, \ldots, x_n, y\} \in \mathcal{A}\}$ and $A_n' = \{y : \{x_1, \ldots, x_n, y\} \in \mathcal{A}'\}$ are closed in E and satisfy $A_n \cap A_n' = \varnothing$, $A_n \cup A_n' = E$. Thus we have $A_n = E$ and $A_n' = \varnothing$, because E is connected and $\{x_1, \ldots, x_n, x_n\} \in A_n$, and $\{x_1, \ldots, x_n, y\} \in \mathcal{A}$, for each $y \in E$. But the family of the nonempty finite sets is dense in $\mathcal{F}'$, by Corollary 2 of Theorem 1-2-2, if E is compact, and dense in $\mathcal{F}$ itself if E is not compact. Thus, $\mathcal{A}$ being closed, $\mathcal{A}'$ is empty if E is not compact, and $\mathcal{A}' = \varnothing$ or $\{\varnothing\}$ if E is compact, and the proposition is proved.

Convergence in $\mathcal{F}$

Let us now examine the convergence in the space $\mathcal{F}$. Because $\mathcal{F}$ admits a countable base, we may restrict ourselves to the convergence of sequences $n \to F_n$ in $\mathcal{F}$. By the definition of the topology $\mathcal{T}_f$ on $\mathcal{F}$, a sequence $\{F_n\}$ converges in $\mathcal{F}$ toward a limit F if and only if it satisfies two conditions:

1. If an open set G hits F, G hits all the F_n except, at the most, a finite number of them.
2. If a compact K is disjoint of F, it is disjoint of all the F_n except, at the most of a finite number of them.

The following theorem gives a more tractable criterion.

Theorem 1-2-2. *A sequence $\{F_n\}$ converges toward F in $\mathcal{F}$ if and only if it satisfies two conditions*:

1′. *For any $x \in F$, there exists for each integer n (except, at the most, for a finite number) a point $x_n \in F_n$ such that $\lim x_n = x$ in E.*
2′. *If $n_k \to F_{n_k}$ is a subsequence, every convergent sequence $n_k \to x_{n_k} \in F_{n_k}$ satisfies $\lim x_{n_k} \in F$.*

Moreover, conditions 1′ and 2′ are respectively equivalent to conditions 1 and 2 above.

Proof. It is obviously sufficient to prove the last statement. We show that condition 1 implies 1′. If F is empty, 1′ is trivially verified. If $F \neq \varnothing$, let x be a point in F and $G_1 = E \supset G_2 \supset \dots$ be a fundamental system of open neighborhoods of x. Each G_k hits F, and, by condition 1, there exists an integer N_k such that $F_n \cap G_k \neq \varnothing$ for $n \geqslant N_k$. Thus it is possible to form a sequence $\{x_n\}$, $n \geqslant N_1$, satisfying $x_p \in F_p \cap G_k$ for $p = N_k, N_k + 1, \dots, N_{k+1} - 1$, and this sequence clearly converges toward x.

Condition 1′ implies 1. If $F = \varnothing$, condition 1 is trivially satisfied. If $F \neq \varnothing$, let G be an open set with $F \cap G \neq \varnothing$ and $x \in F \cap G$. By condition 1′, there exists a sequence $\{x_n\}$, $n \geqslant n_0$, satisfying $x_n \in F_n$ and $\lim x_n = x$ in E. But G is a neighborhood of x, and we have $x_n \in G$ for n large enough, so that G hits F_n.

Condition 2 implies 2′. If $F = E$, condition 2′ is true. If $F \neq E$, let x be a point not belonging to F and K be a compact neighborhood of x disjoint of F. By condition 2, K is disjoint of F_n for n large enough, and x cannot be the limit of a sequence $n_k \to x_{n_k} \in F_{n_k}$. Thus condition 2′ is true.

Finally, if condition 2 is not true, there exists a compact set K disjoint of F, a subsequence $n_k \to F_{n_k}$, and, for each integer k, a point $x_{n_k} \in K \cap F_{n_k}$. Being included in the compact set K, the sequence $\{x_{n_k}\}$ admits a sub-

Clearly, a sequence $\{F_n\}$ converges in $\mathcal{F}$ if and only if $\overline{\lim}\, F_n = \underline{\lim}\, F_n = F$, and this common value is then $F = \lim F_n$. The following proposition characterizes these two sets, $\underline{\lim}\, F_n$ and $\overline{\lim}\, F_n$.

Proposition 1-2-3. *Let $\{F_n\}$ be a sequence in $\mathcal{F}$. Then:*

a. The set $\underline{\lim}\, F_n$ is the largest closed set $F \in \mathcal{F}$ satisfying the equivalent conditions 1 and 1′. In other words, $x \in \underline{\lim}\, F_n$ if and only if for any integer n large enough there exists $x_n \in F_n$ and $\lim x_n = x$ in E, or if and only if any neighborhood of x hits all the F_n, except for, at most, a finite number of them.

b. The set $\overline{\lim}\, F_n$ is closed, and it is the smallest closed set satisfying the quivalent conditions 2 and 2′. In other words, $x \in \overline{\lim}\, F_n$ if and only if there cists a subsequence $\{F_{n_k}\}$ and a sequence $n_k \to x_{n_k} \in F_{n_k}$ such that $x = \lim x_{n_k}$, , still, if and only if any neighborhood of x hits an infinite number of the F_n. oreover, the following relationship holds:

$$\overline{\lim}\, F_n = \bigcap_{N>0} \overline{\bigcup_{n \geq N} F_n}$$

'roof of a. Let F be the obviously closed set defined by $x \in F$ if and / if any neighborhood of x hits all the F_n, except, at most, a finite ıber of them. Any $x \in F$ is the limit of a sequence $\{x_n\}$ satisfying F_n for n large enough, so that $x \in \underline{\lim}\, F_n$ and so $F \subset \underline{\lim}\, F_n$. Con- ely, let y be a point with $y \notin F$. Then there exists a neighborhood V of ıd a subsequence $\{F_{n_k}\}$ satisfying $V \cap F_{n_k} = \varnothing$ for any k. But the ence $\{F_{n_k}\}$ in the compact space $\mathcal{F}$ admits a subsequence converging rd a closed set $A \supset \underline{\lim}\, F_n$. Thus $y \notin A$ implies $y \notin \underline{\lim}\, F_n$, and $\underline{\lim}\, F_n$.

ɔof of b. Let F' be the obviously closed set defined by $x \in F'$ if and f any neighborhood of x hits an infinite number of the F_n. Thus, if , there exist a subsequence $\{F_{n_k}\}$ and, for each k, a point $x_{n_k} \in F_{n_k}$, $= \lim x_{n_k}$ in E. If A is the limit in $\mathcal{F}$ of a subsequence of $\{F_{n_k}\}$, we $\in A \subset \overline{\lim}\, F_n$, and thus $F' \subset \overline{\lim}\, F_n$. Conversely, if $x \in \overline{\lim}\, F_n$, let ɔe a subsequence the limit A' of which in $\mathcal{F}$ satisfies $x \in A'$. By n 1′, there exists a sequence $n_k \to x_{n_k} \in F_{n_k}$ such that $x = \lim x_{n_k}$. Thus $x \in F'$ and $\overline{\lim}\, F_n \subset F'$.

ıd Lower Semicontinuities

ı 1-2-2. *Let Ω be a topological space, and ψ a mapping from Ω into ıy that ψ is upper semicontinuous (u.s.c.) if, for any $K \in \mathcal{K}$, the set* sequence converging toward a point $x \in K$. Hence, $x \notin F$, thu that condition 2′ is not true. The theorem is proved.

Corollary 1. *The mapping* $(F, F') \rightarrow F \cup F'$ *is continuous from space* $\mathfrak{F} \times \mathfrak{F}$ *onto* $\mathfrak{F}$.

Corollary 2. *Let* $\mathfrak{G}$ *be the class of the finite subsets in E, an the class of the nonempty finite subsets in E. Then* $\mathfrak{G}$ *is den dense in* $\mathfrak{F}' = \mathfrak{F} \setminus \{\varnothing\}$, *and, if E is not compact,* $\mathfrak{G}'$ *is dense*

Corollary 3. *Let* $\{F_n\}$ *and* $\{F'_n\}$ *be the two sequences in* $\mathfrak{F}$

a. $F_n \downarrow F$ *implies* $\lim F_n = F$ *in* $\mathfrak{F}$.

b. $F_n \uparrow A$ *implies* $\lim F_n = \bar{A}$ *in* $\mathfrak{F}$.

c. $F_n \downarrow F$ *and* $F'_n \downarrow F'$ *imply* $\lim(F_n \cap F'_n) = F \cap F'$ *and* lim *in* $\mathfrak{F}$.

d. $F_n \uparrow A$ *and* $F'_n \uparrow A$ *imply* $\lim(F_n \cup F'_n) = \overline{A \cup A'} = \bar{A} \cup$ $= (\overline{A \cap A'}) \subset \bar{A} \cap \bar{A}'$ *in* $\mathfrak{F}$.

e. $F_n \subset F'_n$ *for each n and* $\lim F_n = F$, $\lim F'_n = F'$ *impl*

Corollary 4. *For any* $B \in \mathfrak{P}(E)$, *the class of the clos in* $\mathfrak{F}$. *For any* $F_0 \in \mathfrak{F}(E)$, *the class of the closed sets*

Let us prove, for instance, Corollary 1. Let { sequences converging toward F and F', respective that $\lim(F_n \cup F'_n) = F \cup F'$. If x is a point in $F \cup F$ $x \in F$; thus (by criterion 1′ applied to the sequ sequence $n \rightarrow x_n \in F_n$ with $x = \lim x_n$. A fortiori criterion 1′ is verified by $\{F_n \cup F'_n\}$. We now sequence, and $n_k \rightarrow x_{n_k} \in F_{n_k} \cup F'_{n_k}$ a sequence Thus there exists a subsequence $n_{k_p} \rightarrow x_{n_{k_p}} \in F_{n_{k_p}}$ Criterion 2′ applied to $\{F_n\}$. A fortiori we ha satisfies 2′.

Note that the intersection operation is *not*

The Operations $\underline{\lim}$ and $\overline{\lim}$

Definition 1-2-1. *Let* $\{F_n\}$ *be a sequenc are respectively the intersection and th subsequences of* $\{F_n\}$ *converging in* $\mathfrak{F}$.

$\psi^{-1}(\mathcal{F}^K)$ is open in Ω. We say that ψ is lower semicontinuous (l.s.c.) if, for any $G \in \mathcal{G}$, the set $\psi^{-1}(\mathcal{F}_G)$ is open in Ω.

Clearly, the mapping ψ is continuous if and only if it is u.s.c. and l.s.c. If the topology of the space Ω admits a countable base, the following criterion holds.

Proposition 1-2-4. *Let Ω be a separable space and ψ a mapping from Ω into $\mathcal{F}$. Then:*

a. The mapping ψ is u.s.c. if and only if $\psi(\omega) \supset \overline{\lim}\, \psi(\omega_n)$ for any $\omega \in \Omega$ and any sequence $\{\omega_n\}$ converging toward ω in Ω.

b. The mapping ψ is l.s.c. if and only if $\psi(\omega) \subset \underline{\lim}\, \psi(\omega_n)$ for any $\omega \in \Omega$ and any sequence $\{\omega_n\}$ converging toward ω in Ω.

Proof. Since Ω is separable, $\psi^{-1}(\mathcal{F}_K)$ [respectively $\psi^{-1}(\mathcal{F}^G)$] is closed in Ω for any $K \in \mathcal{K}$ (resp. for any $G \in \mathcal{G}$) if and only if $\lim \omega_n = \omega$ in Ω and $\psi(\omega_n) \in \mathcal{F}_K$ (resp. $\in \mathcal{F}^G$) imply that $\psi(\omega) \in \mathcal{F}_K$ (resp. $\in \mathcal{F}^G$), that is, by criterion 2 (resp. by criterion 1) if and only if $\overline{\lim}\, \psi(\omega_n) \subset \psi(\omega)$ [resp. $\underline{\lim}\, \psi(\omega_n) \supset \psi(\omega)$].

Corollary 1. *The mapping $(F,F') \rightarrow (F \cap F')$ from $\mathcal{F} \times \mathcal{F}$ into $\mathcal{F}$ is u.s.c.*

Proof. Let $\{F_n\}$ and $\{F_n'\}$ be two sequences converging, respectively, toward F and F' in $\mathcal{F}$. If $\overline{\lim}\,(F_n \cap F_n')$ is not empty, let x be a point belonging to this set, and (Proposition 1-2-3) $\{x_{n_k}\}$ be a sequence converging toward x with $x_{n_k} \in F_{n_k} \cap F'_{n_k}$. By criterion 2′, $x_{n_k} \in F_{n_k}$ implies $x \in F$ and $x_{n_k} \in F'_{n_k}$ implies $x \in F'$. Thus $x \in F \cap F'$ and $\overline{\lim}\,(F_n \cap F_n) \subset F \cap F'$. Then, by the proposition, the mapping $(F,F') \rightarrow (F \cap F')$ is u.s.c.

Corollary 2. *The mapping $F \rightarrow \overline{F^c}$ from $\mathcal{F}$ into itself is l.s.c.*

Proof. $\overline{F^c} \cap G = \varnothing$ for $G \in \mathcal{G}$ is equivalent to $F^c \cap G = \varnothing$, and thus to $F \supset G$, and the set $\{F : F \in \mathcal{F}, F \supset G\}$ is closed by Theorem 1-2-2, Corollary 4. Thus $F \rightarrow \overline{F^c}$ is l.s.c.

Corollary 3. *If E is a locally connected space, the boundary mapping $F \rightarrow \partial F$ from $\mathcal{F}$ into itself is l.s.c.*

Proof. Let G be a connected open set. The boundary $\partial F = F \cap \overline{F^c}$ of a closed set F hits G if and only if G hits F and $\overline{F^c}$. Thus we have $\partial^{-1}(\mathcal{F}_G) = \mathcal{F}_G \cap \{F \not\supset G\}$, and this set is open in $\mathcal{F}$ by Theorem 1-2-2, corollary 4. If $G' \in \mathcal{G}$ is an open set, its connected components G_i, $i \in I$, are open, because E is locally connected. Then we have $\partial^{-1}(\mathcal{F}_G) = \partial^{-1}(\cup \mathcal{F}_{G_i}) = \cup \partial^{-1}(\mathcal{F}_{G_i})$, and this set is open in $\mathcal{F}$.

Corollary 4. *Let E and E' be two LCS spaces and ψ an increasing mapping from $\mathcal{F}(E)$ into $\mathcal{F}(E')$ [i.e., $F_1 \subset F_2$ in $\mathcal{F}(E)$ implies $\psi(F_1) \subset \psi(F_2)$]. Then ψ is u.s.c. if and only if $F_n \downarrow F$ in $\mathcal{F}(E)$ implies $\psi(F_n) \downarrow \psi(F)$.*

Proof. Suppose that ψ is u.s.c. If $F_n \downarrow F$, we have $F = \lim F_n$ in $\mathcal{F}(E)$, by Theorem 1-2-2, Corollary 2, and thus $\overline{\lim}\, \psi(F_n) \subset \psi(F)$. On the other hand, $F_n \supset F$ implies $\underline{\lim}\, \psi(F_n) \supset \psi(F)$. Hence $\psi(F) = \lim \psi(F_n)$ in $\mathcal{F}(E')$. But ψ is increasing, and we have also $\psi(F_n) \downarrow \cap \psi(F_n)$. Thus $\cap \psi(F_n) = \psi(F)$ by Theorem 1-2-2, Corollary 3, that is, $\psi(F_n) \downarrow \psi(F)$.

Conversely, suppose that $\psi(F_n) \downarrow \psi(F)$ if $F_n \downarrow F$. Let $\{F_n\}$ be a sequence in $\mathcal{F}(E)$ converging toward a limit F. By relationship (*b*), Proposition 1-2-3, we have $F = \cap A_N$ with $A_N = \overline{\cup_{n \geqslant N} F_n}$. But the sequence $\{A_N\}$ is decreasing, and thus $\psi(A_N) \downarrow \psi(F)$. By $F_N \subset A_N$, we have also $\overline{\lim}\, \psi(F_N) \subset \overline{\lim}\, \psi(A_N) = \lim \psi(A_N) = \psi(F)$, and ψ is u.s.c.

Corollary 5. *Let T be an increasing mapping from $\mathcal{F}(E)$ into the extended real line $[-\infty, +\infty]$. Then T is u.s.c. if and only if $F_n \downarrow F$ in $\mathcal{F}$ implies $T(F_n) \downarrow T(F)$.*

This is an immediate consequence of Corollary 4, when identifying each point $x \in [-\infty, +\infty]$ with the closed set $\{y : y \leqslant x\}$.

The Case of a Metric Space E

We suppose now that E is a locally compact and separable metric space, the topology of which is defined by a distance $d: E \times E \to \mathbf{R}_+$. Let $B_\epsilon(x)$ [respectively $\overline{B}_\epsilon(x)$] be the open (resp., closed) ball with center x and radius $\epsilon > 0$. Then the topology of $\mathcal{F}(E)$ is generated by the $\mathcal{F}_{B_\epsilon(x)}$ and the $\mathcal{F}^{\overline{B}_\epsilon(x)}$, $x \in E, \epsilon > 0$. For any $F \in \mathcal{F}$ and $x \in E$, if $F \neq \varnothing$, put

$$d(x, F) = Inf\{d(x,y), y \in F\},$$

and $d(x, \varnothing) = \infty$. Clearly, $F \in \mathcal{F}_{B_\epsilon(x)}$ is equivalent to $d(x, F) < \epsilon$, and $F \in \mathcal{F}^{\overline{B}_\epsilon(x)}$ is equivalent to $d(x, F) > \epsilon$. Hence a sequence $\{F_n\}$ converges toward F if and only if for any $x \in E$ the sequence $\{d(x, F_n)\}$ converges toward $d(x, F)$ for the topology of the compactified half straight line $\overline{\mathbf{R}}_+$. More precisely:

Proposition 1-2-5. *Suppose that the topology of E is defined by a metric d. Then a sequence $\{F_n\}$ converges in $\mathcal{F}$ if and only if for any $x \in E$ the sequence $\{d(x, F_n)\}$ converges in $\overline{\mathbf{R}}_+$.*

The "only if" part is true by the above consideration. Conversely, suppose that for any $x \in E$ the sequence $\{d(x, F_n)\}$ admits a limit $f(x) \in [0, +\infty]$. There exists a subsequence $\{F_{n_k}\}$ converging toward a limit F in the compact space $\mathcal{F}$, and we have $d(x, F) = \lim d(x, F_{n_k}) = f(x)$, by the "only if" part of the proof. Thus every subsequence converges toward the closed set $F = \{x : f(x) = 0\}$, and this implies $F = \lim F_n$.

1-3. THE SPACES $\mathcal{G}$ AND $\mathcal{K}$

The topology $\mathcal{T}_g$ on $\mathcal{G}$ is generated by the classes $\mathcal{G}_K$, $K \in \mathcal{K}$, and $\mathcal{G}^G$, $G \in \mathcal{G}$. The mapping $\complement : F \to F^c$ is obviously a homeomorphism from $\mathcal{F}$ onto $\mathcal{G}$, so that all the results of the preceding section may be translated by duality. For instance, the convergence criterion of Theorem 1-2-2 becomes the following:

Theorem 1-3-1. *A sequence $\{G_n\}$ converges toward G in $\mathcal{G}$ if and only if two conditions are satisfied:*

1′. *For any point $x \not\in G$, there exists a sequence $\{x_n\}$ with $x = \lim x_n$ in E and $x_n \not\in G_n$ for n large enough.*

2′. *For any subsequence $\{G_{n_k}\}$ and any sequence $n_k \to x_{n_k} \not\in G_{n_k}$ converging toward a limit x in E, we have $x \not\in G$.*

Let ψ be a mapping from a topological space Ω into $\mathcal{G}$. Then we say that ψ is l.s.c. (resp. u.s.c.) if $\psi^{-1}(\mathcal{G}_K)$ [resp. $\psi^{-1}(\mathcal{G}^G)$] is open in Ω for any $K \in \mathcal{K}$ (resp. for any $G \in \mathcal{G}$). In other words:

Proposition 1-3-1. *A mapping ψ from a topological space Ω into $\mathcal{G}$ is l.s.c. (resp. u.s.c.) if and only if the mapping $\complement \circ \psi$ from Ω into $\mathcal{F}$ is u.s.c. (resp. l.s.c.).*

Let us now examine a new space, called $\mathcal{K}$.

Definition 1-3-1. *The space $\mathcal{K}$ is the range of the mapping $A \to (\mathring{A}, \overline{A})$ from $\mathcal{P}(E)$ into the product space $\mathcal{G} \times \mathcal{F}$.*

The space $\mathcal{K}$ is obviously included in the subspace $\{(G, F); G \subset F\}$ of $\mathcal{G} \times \mathcal{F}$. If $G \in \mathcal{G}$ and $F \in \mathcal{F}$, there does not necessarily exist a set

$A \in \mathcal{P}(E)$ such that $\mathring{A} = G$ and $\overline{A} = F$. Nevertheless, it is always the case if the following axiom is satisfied by E.

Axiom 1-3. *The space E admits a subset D dense in E, as well as its complementary set D^c.*

If $G \in \mathcal{G}$, $F \in \mathcal{F}$, and $G \subset F$, the set

$$A = G \cup \left((\overline{G})^c \cap \mathring{F} \cap D \right) \cup \partial F$$

satisfies $\mathring{A} = G$ and $\overline{A} = F$. Axiom 1-3-1 is verified, in particular, if the open sets in E are not countable, as is obviously the case if E is a euclidean space.

Proposition 1-3-2. *Suppose that Axiom* 1-3-1 *is satisfied. Then $\mathcal{K}$ is compact and separable for the relative topology $\mathcal{T}_h$ induced by the product topology $\mathcal{T}_f \otimes \mathcal{T}_g$.*

Proof. $\mathcal{G} \times \mathcal{F}$ is compact and separable, and it remains to prove that $\mathcal{K}$ is closed in $\mathcal{G} \times \mathcal{F}$. By Axiom 1-3-1, it is sufficient to show that for any sequences $\{G_n\}$ and $\{F_n\}$ converging toward limits G and F in $\mathcal{G}$ and $\mathcal{F}$, respectively, and satisfying $G_n \subset F_n$ for any n we have $G \subset F$. If $G = \varnothing$, this result is trivial. If $G \neq \varnothing$, let x be a point belonging to G. The set $\mathcal{G}_{\{x\}}$ is open in $\mathcal{G}$, so that $x \in G_n \subset F_n$ for n large enough. But $\mathcal{F}_{\{x\}}$ is closed in $\mathcal{F}$, and $x \in F_n$ implies $x \in F$. Thus $G \subset F$.

1-4. THE SPACE $\mathcal{K}(E)$ AND THE MYOPE TOPOLOGY

The myope topology $\mathcal{T}_k$ on $\mathcal{K}(E)$ is generated by the classes $\mathcal{K}^F$, $F \in \mathcal{F}$, and $\mathcal{K}_G$, $G \in \mathcal{G}$. If the space E itself is compact, the topological spaces $\mathcal{F}$ and $\mathcal{K}$ are identical. But if E is not compact, the myope topology is strictly finer than the relative topology induced by $\mathcal{T}_f$ on $\mathcal{K} = \mathcal{K} \cap \mathcal{F}$, so that $\mathcal{K}$ is not a topological subspace of $\mathcal{F}$. In fact, the $\mathcal{T}_f$ relative topology on $\mathcal{K}$ is generated by the classes $\mathcal{F}^K \cap \mathcal{K} = \mathcal{K}^K$, $K \in \mathcal{K}$, and $\mathcal{F}_G \cap \mathcal{K} = \mathcal{K}_G$, $G \in \mathcal{G}$, which are open for $\mathcal{T}_k$; but if F is a noncompact closed set, $\mathcal{K}^F$ is not open for relative $\mathcal{T}_f$-topology. The one-to-one mapping $K \to K$ from $\mathcal{K}$ into $\mathcal{F}$ is continuous, so that any compact subset $\mathcal{V} \subset \mathcal{K}(E)$ is compact (and thus closed) in $\mathcal{F}$, and then any subset $\mathcal{V}' \subset \mathcal{V}$ closed (and thus compact) for the myope topology is also closed in $\mathcal{F}$. Hence the myope topology and the $\mathcal{T}_f$ topology are identical on the compact subsets of $\mathcal{K}(E)$. Conversely, if $\mathcal{V} \subset \mathcal{K}$ is closed (and thus compact in $\mathcal{F}$ for the $\mathcal{T}_f$ topology, and if the relative myope and $\mathcal{T}_f$ topologies are identical on $\mathcal{V}$, $\mathcal{V}$ is also compact for $\mathcal{T}_k$. This condition is obviously verified by every $\mathcal{K}^{K^c}$, $K \in \mathcal{K}$ (i.e., the class of the compact sets included in a given $K \in \mathcal{K}$). On the other hand, if $\mathcal{B}$ is a base of the topology of E such that $\overline{B} \in \mathcal{K}$ for any $B \in \mathcal{B}$, the $\mathcal{K}^{B^c}$, $B \in \mathcal{B}$, constitute an open covering of the space $\mathcal{K}(E)$. Thus, if $\mathcal{V}$ is compact in $\mathcal{K}$ for the myope topology, there exists a relatively compact set B such that $\mathcal{V} \subset \mathcal{K}^{B^c}$ and (a fortiori) $\mathcal{V} \subset \mathcal{K}^{K^c}$ (with $K = \overline{B} \in \mathcal{K}$). In other words, a subset $\mathcal{V} \subset \mathcal{K}$ is compact for the myope topology if and only if it is closed in $\mathcal{F}$ and included in $\mathcal{K}^{K^c}$ for a compact set $K \in \mathcal{K}$. It follows without difficulty that $\mathcal{K}$ is LCS for the myope topology. Let us state these results as follows.

Proposition 1-4-1. *The myope topology is finer than the relative topology induced on $\mathcal{K}$ by $\mathcal{F}$, and strictly finer if E is not compact. Nevertheless, these two topologies are equivalent on the subsets $\mathcal{V} \subset \mathcal{K}$ compact for the myope topology. A subset $\mathcal{V} \subset \mathcal{K}$ is compact for the myope topology if and only if it is closed in $\mathcal{F}$ and there exists a $K_0 \in \mathcal{K}$ satisfying $K_0 \supset K$ for every $K \in \mathcal{V}$.*

Corollary . *$\mathcal{K}$ is LCS.*

The following convergence criterion in $\mathcal{K}$ is also an obvious consequence of Proposition 1-4-1.

Theorem 1-4-1. *A sequence $\{K_n\}$ converges in $\mathcal{K}$ for the myope topology if and only if two conditions are satisfied:*

1. *There exists $K_0 \in \mathcal{K}$ with $K_0 \supset K_n$ for any n.*
2. *The sequence $\{K_n\}$ converges in $\mathcal{F}$.*

Corollary 1. *The empty set $\varnothing$ is an isolated point in $\mathcal{K}$.*

Corollary 2. *The mappings $(K, K') \to K \cup K'$ from $\mathcal{K} \times \mathcal{K}$ into $\mathcal{K}$ and $(K, F) \to (K \cup F)$ from $\mathcal{K} \times \mathcal{F}$ into $\mathcal{F}$ are continuous.*

Corollary 3. *Let $\{K_n\}$ be a sequence in $\mathcal{K}$. Then $K_n \downarrow K$ implies $K = \lim K_n$ in $\mathcal{K}$. $K_n \uparrow A$ implies $\lim K_n = \overline{A}$ in $\mathcal{K}$ if and only if $\overline{A}$ is compact.*

Corollary 4. *The set $\mathcal{G}' = \mathcal{G} \setminus \{\varnothing\}$ (i.e., the class of the nonempty finite sets in E) is dense in $\mathcal{K}' = \mathcal{K} \setminus \{\varnothing\}$.*

Remark. If E is not compact, the preceding results become clearer when we consider the compact space $\bar{E}=E\cup\{\omega\}$, obtained by adding to E a point at infinity ω (see Choquet, 1953–54). The open neighborhoods of ω are the complementary sets (in $\bar{E}$) of the $K\in\mathcal{K}(E)$, and the compact sets $K\in\mathcal{K}(E)$ are identical with the closed sets of $\mathcal{F}(\bar{E})$ not including ω, so that $\mathcal{K}(E)=\mathcal{F}^{\{\omega\}}(\bar{E})$ is an open subset of $\mathcal{F}(\bar{E})$. It is easy to verify that the myope topology is identical with the relative topology induced by $\mathcal{F}(\bar{E})$ on the open subset $\mathcal{K}(E)$. In other words, $\mathcal{K}(E)$ *is a topological subspace of* $\mathcal{F}(\bar{E})$, *open in* $\mathcal{F}(\bar{E})$. On the other hand, the complementary set of $\mathcal{K}(E)$ in $\mathcal{F}(\bar{E})$ is the closed set $\mathcal{F}_{\{\omega\}}(\bar{E})$, and the mapping $F\to F\cup\{\omega\}$ is a homeomorphism from $\mathcal{F}(E)$ onto $\mathcal{F}_{\{\omega\}}(\bar{E})$. Hence $\mathcal{F}(\bar{E})$ is the union of two disjoint topological subspaces $\mathcal{F}^{\{\omega\}}$ and $\mathcal{F}_{\{\omega\}}$, homeomorphic, respectively, to $\mathcal{K}(E)$ and $\mathcal{F}(E)$.

By Theorem 1-4-1, the mappings defined or valued on $\mathcal{K}$ will satisfy properties analogous to those encountered for the space $\mathcal{F}$. If Ω is a topological space, a mapping $\psi:\Omega\to\mathcal{K}$ is said to be u.s.c. (resp. l.s.c.) if $\psi^{-1}(\mathcal{K}^F)$ [resp. $\psi^{-1}(\mathcal{K}_G)$] is open in Ω for any $F\in\mathcal{F}$ (resp. for any $G\in\mathcal{G}$). Let us examine only u.s.c. and increasing mappings from $\mathcal{K}$ into the extended line $[-\infty,+\infty]$.

Proposition 1-4-2. *Let T be an increasing mapping from $\mathcal{K}$ into the extended line [i.e., $K\subset K'$ in $\mathcal{K}$ implies $T(K)\leqslant T(K')$]. Then T is u.s.c. if and only if $K_n\downarrow K$ in $\mathcal{K}$ implies $T(K_n)\downarrow T(K)$.*

By Theorem 1-4-1, this is an immediate consequence of Proposition 1-2-4, Corollary 5.

Proposition 1-4-3. *a. Let T be an increasing mapping from $\mathcal{K}$ into the extended line, and put $T_g(G)=\mathrm{Sup}\{T(K),K\in\mathcal{K},K\subset G\}$ for any $G\in\mathcal{G}$, and $T_k(K)=\mathrm{Inf}\{T_g(G),G\in\mathcal{G},G\supset K\}$ for any $K\in\mathcal{K}$. Then T_g is l.s.c. on $\mathcal{G}$ and T_k u.s.c. on $\mathcal{K}$. More precisely, T_k is the smallest u.s.c. upper bound of T on $\mathcal{K}$, and $T=T_k$ if and only if T itself is u.s.c. Moreover, the converse formula, $T_g(G)=\mathrm{Sup}\{T_k(K),K\in\mathcal{K},K\subset G\}$, holds for any $G\in\mathcal{G}$.*

b. Let T' be an increasing mapping from $\mathcal{G}$ into the extended line, and put $T_k'(K)=\mathrm{Inf}\{T'(G),G\in\mathcal{G},G\supset K\}$ for any $K\in\mathcal{K}$ and $T_g'(G)=\mathrm{Sup}\{T_k'(K),K\in\mathcal{K},K\subset G\}$ for any $G\in\mathcal{G}$. Then T_k' is u.s.c. on $\mathcal{K}$, T_g' is the largest l.s.c. lower bound of T' on $\mathcal{G}$, and $T'=T_g'$ if and only if T' itself is l.s.c. Moreover, the converse formula, $T_k'=\mathrm{Inf}\{T_g'(G),G\in\mathcal{G},G\supset K\}$, holds for any $K\in\mathcal{K}$.

Proof. Let us prove, for instance, the first statement. Let G be an open set and ϵ a real number >0. There exists a $K_0 \in \mathcal{K}$ such that $K_0 \subset G$ and $T_g(G) \leqslant T(K_0) + \epsilon$. Then, for any $G' \in \mathcal{G}_{K_0}$ (i.e., $G' \supset K_0$), we have $T_g(G') \geqslant T(K_0) \geqslant T_g(G) - \epsilon$, and thus T_g is l.s.c. Now let K be a compact set. There exists $G_0 \in \mathcal{G}$ such that $G_0 \supset K$ and $T_k(K) \geqslant T_g(G_0) - \epsilon$. For any $K' \in \mathcal{K}^{G_0^c}$ (i.e., $K' \subset G_0$), we have $T_k(K') \leqslant T_g(G_0) \leqslant T_k(K) + \epsilon$, and thus T_k is u.s.c.

Obviously, $T_k \geqslant T$ on $\mathcal{K}$. If T is u.s.c., we have the equality $T_k = T$. In fact, let K be a compact set, and $\{G_n\}$ a fundamental system of open neighborhoods of K satisfying $G_n \downarrow K$ and $T_g(G_n) \downarrow T_k(K)$. For a given $\epsilon > 0$ and any integer n, there exists $K_n \in \mathcal{K}$ with $K \subset K_n \subset G_n$ and $T(K_n) \geqslant T_g(G_n) - \epsilon$. But the sequence $\{K_n\}$ converges toward K in $\mathcal{K}$, as it is easy to verify by Theorem 1-4-1, and thus we have $\overline{\lim}\, T(K_n) \leqslant T(K)$, for T is u.s.c. But T is increasing, and this implies $\underline{\lim}\, T(K_n) \geqslant T(K)$. Hence $T(K) = \lim T(K_n) \geqslant \lim T_g(G_n) - \epsilon = T_k(K) - \epsilon$. Thus $T(K) = T_k(K)$.

If T is not u.s.c., let T' be an u.s.c. upper bound of T. Then the same procedure will yield $T'_g \geqslant T_g$ and $T'_k \geqslant T_k$. But $T'_k = T'$ by the upper semicontinuity of T'. Thus $T' \geqslant T_k$, and T_k is the smallest u.s.c. upper bound of T. Finally, by putting $\overline{T}_g(G) = \operatorname{Sup}\{T_k(K), K \in \mathcal{K}, K \subset G\}$, we have $T(K) \leqslant T_k(K) \leqslant T_g(G)$ for any compact set $K \subset G$, and, taking the supremum, $T_g(G) \leqslant \overline{T}_g(G) \leqslant T_g(G)$. Thus $\overline{T}_g = T_g$.

The Hausdorff Metric

Let us now suppose that the topology of the LCS space E is defined by a metric $d: E \times E \to \mathbf{R}_+$. Then we may define a metric ρ on $\mathcal{K}' = \mathcal{K} \setminus \{\varnothing\}$ by putting

$$\rho(K, K') = \operatorname{Max}\left\{ \operatorname*{Sup}_{x \in K} d(x, K'), \quad \operatorname*{Sup}_{x' \in K'} d(x', K) \right\}$$

and ρ is called the *Hausdorff metric* on $\mathcal{K}'$.

Proposition 1-4-4. *The topology defined by the Hausdorff metric ρ is equivalent to the relative myope topology on $\mathcal{K}' = \mathcal{K} \setminus \{\varnothing\}$.*

Proof. For a given $K \in \mathcal{K}'$, the function $x \to d(x, K)$ is continuous, and thus for any $K' \in \mathcal{K}'$ there exists $x' \in K'$ such that $d(x', K) = \operatorname{Sup}\{d(y, K), y \in K'\}$. If $B_\epsilon(y)$ is the open ball with center y and radius

$\epsilon > 0$, it follows that

$$\operatorname*{Sup}_{x \in K'} d(x,K) < \epsilon \qquad \text{if and only if } K' \subset \bigcup_{y \in K} B_\epsilon(y)$$

$$\operatorname*{Sup}_{y \in K} d(y,K') < \epsilon \qquad \text{if and only if } K' \in \bigcap_{y \in K} \mathcal{K}_{B_\epsilon(y)}$$

Let $\mathcal{B}_\epsilon(K) = \{K', K' \in \mathcal{K}', \rho(K,K') < \epsilon\}$ be the open ball with center K and radius $\epsilon > 0$ in $\mathcal{K}'$. The compact set K admits a finite covering by open balls $B_1, B_2, \ldots, B_n$ with radii $\leqslant \epsilon/2$, and thus we have $\mathcal{K}_{B_1,\ldots,B_n} \subset \cap \{\mathcal{K}_{B_\epsilon(y)}, y \in K\}$. If $F \in \mathcal{F}$ is the complementary set of the open set $\cup \{B_\epsilon(y), y \in K\}$, it follows that $\mathcal{K}^F_{B_1,\ldots,B_n}$ is a myope open neighborhood of K included in the open ball $\mathcal{B}_\epsilon(K)$, so that this ball is open for the myope topology.

Conversely, let F be a closed set disjoint of $K \in \mathcal{K}$, and $\epsilon = \operatorname{Inf}\{d(x,F), x \in K\}$. Then $\rho(K',K) < \epsilon$ implies $K' \subset \cup \{B_\epsilon(y), y \in K\}$ and thus $K' \cap F = \varnothing$. It follows that $\mathcal{K}^F$ is open for the Hausdorff topology on $\mathcal{K}'$. If an open set G hits $K \in \mathcal{K}$, there exists $y \in K \cap G$ and $\epsilon > 0$ so that $B_\epsilon(y) \subset G$. Then $\rho(K,K') < \epsilon$ implies $K' \in \mathcal{K}_{B_\epsilon(y)} \subset \mathcal{K}_G$. Hence $\mathcal{K}_G$ is open for the Hausdorff topology, and the proposition is proved.

Remark. If E is a *euclidean space*, we define the Minkowski sum in $\mathcal{P}(E)$ by putting $A \oplus B = \{a+b, a \in A, b \in B\}$ (see below). If B_ϵ is the ball with center O and radius $\epsilon \geqslant 0$, the Hausdorff distance $\rho(K,K'), K, K' \in \mathcal{K}'$, is given by the very useful formula

$$\rho(K,K') = \operatorname{Inf}\{\epsilon : K \subset K' \oplus B_\epsilon, K' \subset K \oplus B_\epsilon\} \tag{1-4-1}$$

1-5. CASE OF THE EUCLIDEAN SPACE $E = \mathbf{R}^d$

In this section, we examine the case $E = \mathbf{R}^d$. By the vector structure of E, it is possible to define new operations on $\mathcal{P}(E)$. In what follows, A, B,... are subsets of $E = \mathbf{R}^d$.

Minkowski's Addition and Subtraction

The *Minkowski addition* $\oplus$ is defined in $\mathcal{P}(E)$ by putting $A \oplus B = \{a+b, a \in A, b \in B\}$. It is an associative and commutative operation, by which $\mathcal{P}(E)$ is an abelian semigroup. Note that we have $A \oplus \{o\} = A$ and $A \oplus \varnothing = \varnothing$ by definition, so that the semigroup $\mathcal{P}(E)$ admits the unit element $\{o\}$ but is not a group. For $x \in \mathbf{R}^d$, $A \oplus \{x\}$ is the *translate* of A by the translation x. We shall often write A_x instead of $A\oplus\{x\}$. For any $B\in\mathcal{P}(E)$, we put $\check{B}=\{-x, x\in B\}$, that is, $\check{B}$ is the *symmetrical set* of B with respect to the origin O. Clearly, $y\in A\oplus B$ is equivalent to $(\check{B})_y\cap A\neq\varnothing$. Hence we have the following formulae:

$$A\oplus B=\bigcup_{y\in A} B_y=\bigcup_{x\in B} A_x=\{z: A\cap(\check{B})_z\neq\varnothing\} \tag{1-5-1}$$

The set $\{z: A\cap B_z\neq\varnothing\}$ of the points z such that A hits the translate B_z is called the *dilatation* of A by B and is equal to $A\oplus\check{B}$.

The operation dual to the Minkowski addition is the *Minkowski subtraction* $\ominus$, defined by putting $A\ominus B=(A^c\oplus B)^c$. By (1-5-1), we obtain

$$A\ominus B=(A^c\oplus B)^c=\bigcap_{x\in B} A_x=\{z:(\check{B})_z\subset A\} \tag{1-5-2}$$

The operation dual to dilatation is called *erosion*. The erosion of A by B is the set $\{z: B_z\subset A\}$ of the points z such that the translate B_z is included in A and, by (1-5-2), is equal to $A\ominus\check{B}$.

The following relationships are obvious:

$$\begin{aligned}&(A\ominus C)\oplus B\subset(A\oplus B)\ominus C\\&(A\ominus B)\ominus C=(A\ominus C)\ominus B=A\ominus(B\oplus C)\end{aligned} \tag{1-5-3}$$

In the same way, $B\subset C$ implies $A\oplus B\subset A\oplus C$, $A\ominus B\supset A\ominus C$, and $B\ominus A\subset C\ominus A$.

Note that the inclusion $\{o\}\subset A\ominus\check{A}$ is satisfied by any $A\in\mathcal{P}$, but not necessarily the equality. However, if a point $x\neq 0$ belongs to $A\ominus\check{A}$, we have $A_x\subset A$ and thus also $A_{nx}\subset A$ for any integer $n\geqslant 0$; the unbounded sequence $\{nx\}$ is included in A. It follows that for any *bounded* set A we have the equality

$$A\ominus\check{A}=\{o\} \tag{1-5-4}$$

With respect to the union, we may write

$$A\oplus(B\cup C)=\bigcup_{x\in A}(B_x\cup C_x)=\left(\bigcup_{x\in A} B_x\right)\cup\left(\bigcup_{x\in A} C_x\right)$$

and thus we obtain the following formulae:

$$\begin{aligned}&A\oplus(B\cup C)=(A\oplus B)\cup(A\oplus C)\\&A\ominus(B\cup C)=(A\ominus B)\cap(A\ominus C)\\&(B\cap C)\ominus A=(B\ominus A)\cap(C\ominus A)\end{aligned} \tag{1-5-5}$$

On the other hand, the following inclusions hold:

$$A\oplus(B\cap C)\subset(A\oplus B)\cap(A\oplus C)$$

$$A\ominus(B\cap C)\supset(A\ominus B)\cup(A\ominus C)$$

$$(B\cup C)\ominus A\supset(B\ominus A)\cup(C\ominus A)$$

but generally not the corresponding equalities.

If A is *convex*, $A\ominus B$ is convex for any $B\in\mathcal{P}$. If A and B are convex, $A\oplus B$ is convex. This is so because $A\ominus B=\cap\{A_x; x\in B\}$ is the intersection of convex sets, and the second statement is obvious.

If A and B are *connected*, $A\oplus B$ is connected.

Proof. For $b\in B$, A_b is connected and included in $A\oplus B$, and thus included in a connected component C of $A\oplus B$, say $A_b\subset C$. Thus, for any $x\in A$, we have $x+b\in C$. But $(x+b)\in A_b\cap B_x$ implies that $A_b\cup B_x$ is connected, as it is the union of two nondisjoint connected sets. It follows that $B_x\subset C$ for any $x\in A$, that is, $A\oplus B\subset C$.

Let B be connected and C_i, $i\in I$, be the connected components of A. Then we have

$$A\ominus B=\bigcup_{i\in I} C_i\ominus B \tag{1-5-6}$$

since $C_i\subset A$ implies $C_i\ominus B\subset A\ominus B$ and $\cup(C_i\ominus B)\subset A\ominus B$. Conversely, $z\in A\ominus B$, that is, $(\check{B})_z\subset A$ implies $(\check{B})_z\subset C_i$ for an $i\in I$, that is, $z\in C_i\ominus B$ and $A\ominus B\subset\cup(C_i\ominus B)$.

Opening and Closing A by B

By combining erosion and dilatation, we obtain two other operations called opening and closing; the *opening* A_B of A by B and the *closing* A^B of A by B are defined as follows:

$$A_B=(A\ominus\check{B})\oplus B;\qquad A^B=(A\oplus\check{B})\ominus B \tag{1-5-7}$$

For a given $B\in\mathcal{P}$, the mappings $A\to A_B$ and $A\to A^B$ are *increasing* (i.e., $A\subset A'$ implies $A_B\subset A'_B$ and $A^B\subset A'^B$), *idempotent* $[(A_B)_B=A_B$ and $(A^B)^B=A^B]$, and dual to each other $[(A^c)_B=(A^B)^c$ and $(A^c)^B=(A_B)^c]$. Moreover, the closing is *extensive* and the opening is *antiextensive*, that is, $A_B\subset A\subset A^B$.

We say that a set A is *open with respect to* B if $A_B=A$, and *closed with respect to* B if $A^B=A$. The set A is open (resp. closed) with respect to B if and only if there exists $C\in\mathcal{P}$ such that $A=C\oplus B$ (resp. $A=C\ominus B$).

The opening of A by B is the union of the translates B_y included in A, that is,

$$A_B = \cup \{ B_y : y \in E, B_y \subset A \} \tag{1-5-8}$$

Proof. Observe that $x \in (A \ominus \check{B}) \oplus B$ is equivalent to $(\check{B})_x \cap (A \ominus \check{B}) \neq \varnothing$, and thus to {there exists $y \in E$ such that $B_y \subset A$ and $x \in B_y$}.

If A and B are *convex*, A_B and A^B are convex. If B is *connected*, and if the connected components of A are C_i, $i \in I$, we have

$$A_B = \bigcup_{i \in I} (C_i)_B$$

This is an immediate consequence of the analogous results proved in the case of the operations $\oplus$ and $\ominus$.

In the sequel, we shall chiefly consider dilatations, erosions, openings, or closings by a *nonempty compact set.* For any $K \in \mathcal{K}'$, the following rules hold:

$F \in \mathcal{F}$ implies $F \oplus K$, $F \ominus K$, F_K, and $F^K \in \mathcal{F}$.
$G \in \mathcal{G}$ implies $G \oplus K$, $G \ominus K$, G_K, and $G^K \in \mathcal{G}$.
$K' \in \mathcal{K}$ implies $K' \oplus K$, $K' \ominus K$, K'_K, and $K'^K \in \mathcal{K}$.

Let us prove, for instance, the rules concerning the space $\mathcal{F}$. Let $\{x_n\} \subset F \oplus K$ be a sequence converging toward x in E. For each n, we have $x_n = f_n + k_n$ for a $f_n \in F$ and a $k_n \in K$. But K is compact, and there exists a subsequence $n_p \to k_{n_p} \in K$ converging toward a point $k \in K$. Thus we have $\lim f_{n_p} = x - k = f \in F$, and $x = k + f \in F \oplus K$. Hence $F \oplus K \in \mathcal{F}$. If G is an open set, $G \oplus K = \cup_{x \in K} G_x$ is open. Thus $F \ominus K = (F^c \oplus K)^c$ is closed. Finally, $F \oplus \check{K}$ and $F \ominus \check{K}$ being closed, so also are $F_K = (F \ominus \check{K}) \oplus K$ and $F^K = (F \oplus \check{K}) \ominus K$.

Concerning the *continuity* of these operations, we have the following results.

Proposition 1-5-1. *The following mappings are continuous*:
$(F,K) \to F \oplus K$ *from* $\mathcal{F} \times \mathcal{K}$ *onto* $\mathcal{F}$.
$(K,K') \to F \oplus K'$ *from* $\mathcal{K} \times \mathcal{K}$ *onto* $\mathcal{K}$.
$F \to \check{F}$ *from* $\mathcal{F}$ *onto* $\mathcal{F}$, *and* $K \to \check{K}$ *from* $\mathcal{K}$ *onto* $\mathcal{K}$.
$(\lambda, F) \to \lambda F$ *from* $\mathbf{R}_+ \times \mathcal{F}$ *onto* $\mathcal{F}$, *and* $(\lambda, K) \to \lambda K$ *from* $\mathbf{R}_+ \times \mathcal{K}$ *onto* $\mathcal{K}$.

We recall the definition $\check{A} = \{-x, x \in A\}$, $\lambda A = \{\lambda x, x \in A\}$. The mapping $A \to \lambda A$ (for a given $\lambda \geqslant 0$) is called a *positive homothetic*.

Proof. Let us prove, for instance, the continuity of $(F,K)\to F\oplus K$. Let $\{F_n\}$ and $\{K_n\}$ be two sequences converging toward F and K, respectively, in $\mathcal{F}$ and $\mathcal{K}$. If $K=\varnothing$, we have $K_n=\varnothing$ for n large enough by Theorem 1-4-1, Corollary 1, and thus $F_n\oplus K_n=\varnothing$, so that the sequence $\{F_n\oplus K_n\}$ converges toward $F\oplus K=\varnothing$ in $\mathcal{F}$. Now we suppose $K\neq\varnothing$, and show that $\{F_n\oplus K_n\}$ converges toward $F\oplus K$ by verifying criteria 1′ and 2′ of Theorem 1-2-2.

Criterion 1′. If $F\oplus K\neq\varnothing$, let x be a point with $x=f+k\in F\oplus K, f\in F$ and $k\in K$. By criterion 1′ applied to the sequences $\{F_n\}$ and $\{K_n\}$, there exist for each n (except, at most, a finite number) points $f_n\in F_n$ and $k_n\in K_n$ such that $f=\lim f_n$ and $k=\lim k_n$ in E. Thus we have $x=\lim x_n$ with $x_n=f_n+k_n\in F_n\oplus K_n$, and the sequence $\{F_n\oplus K_n\}$ satisfies criterion 1′.

Criterion 2′. Let $n_p\to x_{n_p}=f_{n_p}+k_{n_p}$ be a sequence coverging toward $x\in E$ with $f_{n_p}\in F_{n_p}$ and $k_{n_p}\in K_{n_p}$. By Theorem 1-4-1, the compact sets K_{n_p} are included in a fixed $K_0\in\mathcal{K}$, and the sequence $\{k_{n_p}\}$ admits a subsequence converging toward a point $k\in K$. Then the point $f=x-k$ is also an adherent point for the sequence $\{f_{n_p}\}$, so that $f\in F$ by criterion 2′ applied to $\{F_n\}$. Thus $x=f+k\in F\oplus K$.

Proposition 1-5-2. *The mappings $(A,K)\to A\ominus K$, $(A,K)\to A_K$, and $(A,K)\to A^K$ from $\mathcal{F}\times\mathcal{K}'$ (resp. $\mathcal{K}\times\mathcal{K}'$) into $\mathcal{F}$ (resp. $\mathcal{K}$) are u.s.c.*

Proof. Let $\{A_n\}$ and $\{K_n\}$ be two sequences converging toward A and K, respectively, in $\mathcal{F}$ and in $\mathcal{K}'$. By Proposition 1-2-4, the mapping $(A,K)\to A\ominus K$ is u.s.c. if $A\ominus K\supset\overline{\lim}\,(A_n\ominus K_n)$. Let $n_k\to x_{n_k}\in A_{n_k}\ominus K_{n_k}$ be a sequence converging toward a point $x\in E$. We have $\check{K}_{n_k}\oplus\{x_{n_k}\}\subset A_{n_k}$, and thus $\check{K}\oplus\{x\}\subset A$, by Proposition 1-5-1 and Theorem 1-2-2, Corollary 3, that is, $x\in A\ominus K$. By Proposition 1-2-3, it follows that $\overline{\lim}\,(A_n\ominus K_n)\subset A\ominus K$. The proof is analogous for A_K and A^K.

Corollary. *Let A be a closed (resp. a compact) set, and $\{K_n\}$ an increasing sequence in $\mathcal{K}'$. If $\overline{\cup K_n}=K\in\mathcal{K}$, we have $A\ominus K_n\downarrow A\ominus K$, and* $\lim(A\ominus K_n)=A\ominus K$ *in $\mathcal{F}$ (resp. in $\mathcal{K}$).*

Proof. By Theorem 1-4-1, Corollary 3, $\lim K_n=K$ in $\mathcal{K}$, and the upper semicontinuity gives $A\ominus K\supset\overline{\lim}\,(A\ominus K_n)$. But $A\ominus K_n\supset A\ominus K$ implies $A\ominus K\subset\underline{\lim}(A\ominus K_n)$, and thus $A\ominus K=\lim(A\ominus K_n)$, and also $A\ominus K_n\downarrow A\ominus K$, because the sequence $\{A\ominus K_n\}$ is decreasing.

Convexity

The properties related to convexity in the euclidean space $E=\mathbf{R}^d$ will play a decisive role in the sequel, and specifically in Chapters 4, 5, and 6. Let us now indicate a few preliminary results. If $A\in\mathcal{P}(E)$ is a subset of $E=\mathbf{R}^d$, the *convex hull* of A is the set of the points $x=\sum_{i=1}^n\lambda_i x_i$, n integer >0, λ_i real with $0\leqslant\lambda_i\leqslant 1$, and $\sum\lambda_i=1$ and $x_i\in A$ for $i=1,2,\ldots,n$. This convex hull will be denoted by $C(A)$. If F is closed (resp. compact), $C(F)$ is closed (resp. compact). Also note that the following relationship holds for any A, $A'\in\mathcal{P}$:

$$C(A\oplus A')=C(A)\oplus C(A') \tag{1-5-9}$$

Proposition 1-5-3. *Let F be a closed set and B a bounded subset of E. Then $F^B\subset C(F)$. Moreover, if F is convex, F is closed with respect to B. If F and F' are closed and convex and satisfy $F\oplus B=F'\oplus B$, then $F=F'$.*

Proof. If $F\in\mathcal{F}$, $C(F)$ is the intersection of the half spaces containing F. If $x\not\in C(F)$, there exists a closed half space $H\supset F$ such that $x\not\in H$, and a translate B_y of B disjoint of H and including the point x. By (1-5-8), this implies $x\in(F^c)_B=(F^B)^c$, so that we have $F^B\subset C(F)$. If F is convex, we also have $C(F)=F\subset F^B$, and thus $F=F^B$. If F and F' are closed and convex, $F\oplus B=F'\oplus B$ implies $F^{\check{B}}=F'^{\check{B}}$, by the definition of the closing, and thus $F=F'$ by the preceding results.

Proposition 1-5-4. *The mapping $K\to C(K)$ is continuous on $\mathcal{K}$, and the mapping $F\to C(F)$ is l.s.c. on $\mathcal{F}$.*

Proof. Let $\{F_n\}$ be a sequence converging toward F in $\mathcal{F}$, x a point belonging to the convex hull $C(F)$, and $x=\sum_{i=1}^k\lambda_i x_i$ a representation of x with $x_i\in F$, $\sum\lambda_i=1$, and $\lambda_i\geqslant 0$. By the convergence criterion $1'$ (Theorem 1-2-2) applied to the sequence $\{F_n\}$, for each $i=1,2,\ldots,k$, the point x_i is the limit of a sequence $n\to x_n(i)\in F_n$. Putting $x_n=\sum\lambda_i x_n(i)$, we have $x_n\in C(F_n)$ and $x=\lim x_n$. Thus $x\in\underline{\lim}\,C(F_n)$, by Proposition 1-2-3, and $\underline{\lim}\,C(F_n)\supset C(F)$. By criterion b, Proposition 1-2-4, the mapping $F\to C(F)$ is l.s.c. on $\mathcal{F}$.

Now let $\{K_n\}$ be a sequence converging toward K in $\mathcal{K}$, so that there exists $K_0\in\mathcal{K}$ such that $K_n\subset K_0$ for any $n>0$, and also $C(K_n)\subset C(K_0)$. The sequence $\{C(K_n)\}$ satisfies the first condition of Theorem 1-4-1. By the first part of the proof, we have $\underline{\lim}\,C(K_n)\supset C(K)$, and it remains to show that $\overline{\lim}\,C(K_n)\subset C(K)$. Let H be an open half space including K, so that

$K_n \subset H$ for n large enough by the definition of the myope topology, and also $C(K_n) \subset H$. Thus, if x is an adherent point for a sequence $n_k \to x_{n_k} \in C(K_{n_k})$, we have $x \in \overline{H}$. But $C(K)$ is the intersection of the closed half spaces $\overline{H}$ such that $H \supset K$. This implies $x \in C(K)$ and $\overline{\lim}\, C(K_n) \subset C(K)$ by criterion b, Proposition 1-2-3. Thus the mapping C is continuous on $\mathcal{K}$.

Corollary. *The space $C(\mathcal{F})$ of the closed convex sets is closed in $\mathcal{F}$. The space $C(\mathcal{K})$ of the compact convex sets is closed in $\mathcal{K}$.*

Proof. Let $\{F_n\}$ be a sequence converging toward F in $\mathcal{F}$ and satisfying $F_n \in C(\mathcal{F})$, that is, $F_n = C(F_n)$. The mapping C being l.s.c. on $\mathcal{F}$, we have $C(F) \subset \underline{\lim}\, C(F_n) = \lim F_n = F$, and thus $F = C(F)$, that is, $F \in C(\mathcal{F})$. Thus the space $C(\mathcal{F})$ is closed. The proof is the same for the space $C(\mathcal{K})$.

Compact Sets Infinitely Divisible for $\oplus$

For any $\lambda \geqslant 0$, the homothetic $(\lambda, B) \to \lambda B$ obviously satisfies the relationship $\lambda B \oplus \mu B \supset (\lambda + \mu) B$. If the set B is convex, this inclusion becomes an equality, so that the family $B_\lambda = \lambda B$ is a one-parameter semigroup in $\mathcal{P}(E)$, that is, $B_\lambda \oplus B_\mu = B_{\lambda+\mu}$, $\lambda, \mu \geqslant 0$. Moreover, if B is a compact set, this semigroup is continuous in $\mathcal{K}$ by Proposition 1-5-1. We shall prove that the converse is also true. For this purpose, we denote by $A^{\oplus n}$ the Minkowski sum of n terms; each of them is equal to A, and we say that a set A is *infinitely divisible for the Minkowski sum* $\oplus$ if for any integer $n > 0$ there exists a set B_n such that $A = B_n^{\oplus n}$.

Theorem 1-5-1. *A compact set $A \in \mathcal{K}$ is infinitely divisible for the Minkowski sum if and only if it is convex.*

Proof. If $A \in C(\mathcal{K})$, by taking $B_n = (1/n)A$ we have $A = B_n^{\oplus n}$, and A is infinitely divisible. Conversely, suppose that $A \in \mathcal{K}$ and for any $n > 0$ there exists a set B_n such that $A = B_n^{\oplus n}$. Because A is compact, B_n is bounded, and by substituting $\overline{B}_n$ for B_n we may suppose that $B_n \in \mathcal{K}$. (This is a consequence of the relationship $\overline{B \oplus B'} = \overline{B} \oplus \overline{B}'$, which holds for any bounded sets B and B'.) Obviously, $nB_n \subset A \in \mathcal{K}$, so that, by Theorem 1-4-1, there exists a subsequence $\{n_k B_{n_k}\}$ converging in $\mathcal{K}$ toward a limit B.

Let us show that $C(A)=C(B)$. For n and k integers >0, we have $nB_n \subset A \subset B_k^{\oplus k} \subset kC(B_k)$, and thus $nC(B_n) \subset kC(B_k)$. This implies $nC(B_n) = kC(B_k) = C(A)$. By Proposition 1-5-4, it follows that $C(B) = \lim n_k C(B_{n_k}) = C(A)$.

Let us now show that $C(B) \subset A$. Let x be a point in $C(B)$, and

$$x = \sum_{i=1}^{r} \lambda_i x_i, \qquad x_i \in B, \qquad \lambda_i \geq 0, \qquad \sum \lambda_i = 1$$

a representation of x. By the convergence criterion 1′ applied to the sequence $\{n_k B_{n_k}\}$, there exists for each $i=1,2,\ldots,r$ a sequence $n_k \to y_{n_k}(i) \in n_k B_{n_k}$ satisfying $x_i = \lim y_{n_k}(i)$. Choosing integers $N(i,k) \geq 0$ such that

$$\sum_{i=1}^{r} N(i,k) = n_k, \qquad \lim_{k\to\infty} \frac{N(i,k)}{n_k} = \lambda_i$$

and putting $x_k = (1/n_k)\Sigma_{i=1}^r N(i,k) y_{n_k}(i)$, we have $x_k \in B_{n_k}^{\oplus n_k} = A$ and $x = \lim x_k$. Thus $x \in A$ and $C(B) \subset A$.

The relationships $C(B) \subset A$ and $C(A) = C(B)$ clearly imply $A = C(A)$, and thus A is convex.

Corollary. *A family B_λ, $\lambda \geq 0$, in $\mathcal{K}'$ is a one-parameter continuous semigroup (i.e., $B_\lambda \oplus B_\mu = B_{\lambda+\mu}$, $\lambda, \mu \geq 0$) if and only if $B_\lambda = \lambda B$ for $B \in C(\mathcal{K})$.*

Proof. The "if" part has already been shown. Conversely, if a family B_λ, $\lambda \geq 0$, is a semigroup in $\mathcal{K}'$, we have $B_\lambda = (B_{\lambda/n})^{\oplus n}$ for any integer $n>0$, and so $B_\lambda \in C(\mathcal{K})$ by the theorem. By the semigroup relation we obtain $B_\lambda = nB_{\lambda/n} = (n/k)B_{\lambda k/n}$ for n, k integers >0, and thus $B_\lambda = \lambda B_1$ if the mapping $\lambda \to B_\lambda$ is continuous.

It is possible to improve this result by assuming a hypothesis weaker than the continuity of the semigroup B_λ (see Section 9-1). The following result will also be useful in Chapter 9:

Proposition 1-5-5. *For any $A \in \mathcal{K}'$, the sequence $\{(1/n)A^{\oplus n}\}$ converges in $\mathcal{K}$ toward $C(A)$.*

Proof. By $(1/n)A^{\oplus n} \subset C(A)$ and Theorem 1-4-1, we only have to show the convergence in $\mathcal{F}$. Obviously, $\overline{\lim}\,(1/n)A^{\oplus n} \subset C(A)$, and it remains to prove that $\underline{\lim}(1/n)A^{\oplus n} \supset C(A)$. Let $x = \Sigma_{i=1}^r \lambda_i x_i$ be a representation of a point $x \in C(A)$, with $x_i \in A$, $\lambda_i \geq 0$, $\Sigma\lambda_i = 1$. For any $n \geq r$, there exist integers $N(n,i)$ satisfying

$$\sum_{i=1}^{r} N(n,i)=n, \qquad \lim_{n\to\infty} \frac{N(n,i)}{n}=\lambda_i$$

By putting $y_n=\Sigma(N(n,i)/n)x_i$, we get $y_n\in(1/n)A^{\oplus n}$ and $x=\lim y_n$, and thus $x\in \underline{\lim}(1/n)A\oplus^n$. Hence $C(A)\subset \underline{\lim}(1/n)A^{\oplus n}$, and the proposition is proved.

Granulometries

Let $B_\lambda, \lambda\geq 0$, be a one-parameter continuous semigroup on $\mathcal{K}'$; that is, by the corollary of Theorem 1-5-1, $B_\lambda=\lambda B$ for a given $B\in C(\mathcal{K}')$. For any $A\in\mathcal{P}(E)$ and $\lambda>0$, we put

$$\psi_\lambda(A)=A_{\lambda B}=\cup\{\lambda B_x, x\in E, \lambda B_x\subset A\}$$

and we say that the mapping $\lambda\to\psi_\lambda(A)$ is the *granulometry* of the set A with respect to the compact convex set B. A more general theory of granulometries will be given in Chapter 7. The mapping ψ_λ satisfies the following conditions:

1. $\psi_0(A)=A$, and $\lambda\geq\mu$ implies $\psi_\lambda(A)\subset\psi_\mu(A)$.
2. For any $\lambda\geq 0$ and $A\subset A'$, $\psi_\lambda(A)\subset\psi_\lambda(A')$.
3. For any $\lambda\geq\mu\geq 0$, $\psi_\lambda \circ \psi_\mu=\psi_\mu \circ \psi_\lambda=\psi_\lambda$.

Proof. Condition 1 is a consequence of the convexity of B, and 2 is true because the opening by λB is an increasing operation. It remains to prove condition 3. Put $\lambda=\mu+\nu$, $\nu\geq 0$. We have $\psi_\lambda(A)=C\oplus\mu B$ with $C=(A\ominus\lambda\check{B})\oplus\nu B$, and thus $\psi_\lambda(A)$ is open with respect to μB, so that $\psi_\mu \circ \psi_\lambda=\psi_\lambda$. But the opening is idempotent, and it follows that $\psi_\lambda \circ \psi_\mu \circ \psi_\lambda=\psi_\lambda$. By the antiextensivity of the opening, we get

$$\psi_\lambda \circ \psi_\mu(A)\supset\psi_\lambda \circ \psi_\mu \circ \psi_\lambda(A)=\psi_\lambda(A),$$

and also $\psi_\lambda \circ \psi_\mu(A)\subset\psi_\lambda(A)$ because the opening is an increasing operation. This implies $\psi_\lambda \circ \psi_\mu=\psi_\lambda$ and $\psi_\lambda\subset\psi_\mu$.

With respect to the translations $A\to A_x$ and the positive homothetics, the granulometry ψ_λ satisfies the following conditions (which will lead us in Chapter 7 to the general definition of the *euclidean granulometries*):

4. $\psi_\lambda(A_x)=\psi_\lambda(A)\oplus\{x\}$.
5. $\psi_\lambda(\mu A)=\mu\psi_{\lambda/\mu}(A)$.

By Proposition 1-5-2 and the continuity of the homothetics, the mapping $(\lambda, F)\to\psi_\lambda(F)$ from $\mathbf{R}_+\times\mathcal{F}$ into $\mathcal{F}$ is *u.s.c.* In the same way, the mapping $(\lambda, G)\to\psi_\lambda(G)$ from $\mathbf{R}_+\times\mathcal{G}$ into $\mathcal{G}$ is *l.s.c.* Moreover, ψ_λ is decreasing in λ, and this implies the following properties:

6. $F\in\mathcal{F}$, and $\lambda\uparrow\mu$ implies $\psi_\lambda(F)\downarrow\psi_\mu(F)$ and $\psi_\mu(F)=\lim\psi_\lambda(F)$.
7. $G\in\mathcal{G}$, and $\lambda\downarrow\mu$ implies $\psi_\lambda(G)\uparrow\psi_\mu(G)$ and $\psi_\mu(G)=\lim\psi_\lambda(G)$.

Now let V be a positive measure on E, and suppose that $V(F)<\infty$ or $V(G)<\infty$. For instance, V may be the Lebesgue measure if the closed set F or the open set G has a finite volume. The mapping $V:\lambda\to V(\psi_\lambda(F))$ is decreasing and continuous from the left if $F\in\mathcal{F}$, and $\lambda\to V(\psi_\lambda(G))$ is decreasing and continuous from the right if $G\in\mathcal{G}$. By the definition, $V(\psi_\lambda(F))$ is the volume occupied by the translates of λB included into F, so that the function $\lambda\to V(\psi_\lambda(F))$ represents in a certain sense the "size distribution" (or granulometry) of the closed set F.

It is also possible to give a local meaning to the granulometry of A with respect to $B\in C(\mathcal{K}')$ by putting, for any $x\in A$,

$$\Lambda_A(x)=\operatorname{Sup}\{\lambda:\exists y\in E, x\in(\lambda B)_y\subset A\} \tag{1-5-10}$$

[and $\Lambda_A(x)=0$ if $x\notin A$]. Then $\Lambda(x)$ represents the size of the set A as measured at point x in respect to standard B. If $A\in\mathcal{F}$, the supremum is reached [i.e., $x\in\psi_{\Lambda_A(x)}(A)$], and there actually exists a translate of $\Lambda_A(x)B$ including x and included in A. On the contrary, if $A\in\mathcal{G}$, the supremum is not reached. More precisely:

Proposition 1-5-6. *The mapping $(F,x)\to\Lambda_F(x)$ is u.s.c. from $\mathcal{F}\times E$ into $\mathbf{R}_+$, and $\psi_\lambda(F)=\{x:\Lambda_F(x)\geq\lambda\}$ for any $F\in\mathcal{F}$. The mapping $(G,x)\to\Lambda_G(x)$ is l.s.c. from $\mathcal{G}\times E$ into $\mathbf{R}_+$, and $\psi_\lambda(G)=\{x:\Lambda_G(x)>\lambda\}$ for any $G\in\mathcal{G}$.*

Proof. The supremum being reached in (1-5-10), the relationship $\psi_\lambda(F)=\{x:\Lambda_F(x)\geq\lambda\}$ follows. The set $\{(F,x), \Lambda_F(x)\geq\lambda\}=\{(F,x), x\in\psi_\lambda(F)\}$ is closed in $\mathcal{F}\times E$, because the mapping $F\to\psi_\lambda(F)$ is u.s.c. on $\mathcal{F}$, and thus $(F,x)\to\Lambda_F(x)$ is u.s.c. on $\mathcal{F}\times E$. The proof is analogous in the case of the space $\mathcal{G}$.

Finally, let us note the following result, which will be used in Chapter 6:

Proposition 1-5-7. *Let K be a convex compact set, the interior $\mathring{K}$ of which is nonempty, and $F\in\mathcal{F}$. Then we have*

$$C(F)=\overline{\bigcup_{\rho>0} F^{\rho K}}=\lim_{\rho\to\infty} F^{\rho K}$$

The limit is taken in $\mathcal{F}$, or in $\mathcal{K}$ if F is compact.

Proof. By Proposition 1-5-3, we already have $F^{\rho K} \subset C(F)$. If we put $F' = \cup F^{\rho K}$, that is, the limit in $\mathcal{F}$ of the increasing family $F^{\rho K}$, $F' \subset C(F)$ follows. If $F \in \mathcal{K}$, $C(F)$ is compact, and F' is also the limit in $\mathcal{K}$ of the increasing family $F^{\rho K}$. It remains to show that $F' \supset C(F)$. Let x be a point not belonging to F', and V a compact neighborhood of x disjoint of F'. We may obviously suppose that $0 \in \mathring{K}$, and take for V a translate $(\epsilon K)_x$ of a homothetic of K. It follows that $(\epsilon K)_x \subset (F^c \ominus \rho \check{K}) \oplus \rho K$ for any $\rho > 0$, that is, $x \in (F^c \ominus \rho \check{K}) \oplus (\rho - \epsilon) K$ if $\rho \geq \epsilon$, and thus, by substituting $\rho + \epsilon$ for ρ, $x \in (F^c \ominus (\rho + \epsilon)\check{K}) \oplus \rho K = (F^c \ominus \epsilon \check{K})_{\rho K}$. Let $\{\rho_n\}$ be a sequence in $\mathbf{R}_+$ such that $\rho_n \uparrow \infty$. For any $n > 0$, there exists a point $y_n \in E$ such that $x \in (\rho_n K)_{y_n} \subset F^c \ominus \epsilon \check{K}$, and this implies $x \in (\epsilon \mathring{\check{K}})_x \subset ((\rho_n + \epsilon)\mathring{\check{K}})_{y_n} \subset F^c$. Thus the sequence $n \to \{(\rho_n + \epsilon)\mathring{\check{K}}\}$ admits in $\mathcal{G}$ an adherent point satisfying $x \in (\epsilon \mathring{\check{K}})_x \subset G \subset F^c$. But G is necessarily an open half space H, as it is easy to verify, and this implies $x \in H$ and $F \subset H^c$, that is, $C(F) \subset H^c$ and $x \notin C(F)$. Thus $F' \supset C(F)$.

Corollary. *Let K be a nonempty compact set and B_r a ball with center O and radius $r \geq 0$. Then, for any $\epsilon > 0$, we have $K \oplus B_{r+\epsilon} \supset C(K) \oplus B_r$ for r large enough.*

Proof. By Proposition 1-5-7, $C(K) = \lim K^{B_r}$ in $\mathcal{K}$ for $r \uparrow \infty$. By (1-4-1), it follows that $C(K) \subset K^{B_r} \oplus B_\epsilon$ for r large enough, and $K^{B_r} \oplus B_\epsilon \subset (K \oplus B_{r+\epsilon}) \ominus B_r$ implies $C(K) \oplus B_r \subset (K \oplus B_{r+\epsilon})_{B_r} \subset K \oplus B_{r+\epsilon}$.

CHAPTER 2

The Random Closed Sets (RACS)

The definition of a RACS given in Section 2-1 is due to Choquet, (1953–54), as well as the characterization of the important functional T, which is, for a RACS, analogous to the distribution function of a random variable: the class of these functionals T is a class of alternating capacities of infinite order. For a more general notion of a random set, as well as a generalized version of the Choquet theorem, see Kendall (1973). The notion of a conditional RACS is given in Section 2-3, and questions concerning separability, P-continuity and a.s. continuity, and measurability are examined in Sections 2-4 and 2-5. Special topics important for practical applications (the functionals P and Q, the covariance, the linear granulometries) are treated in the last section.

2-1. THE FUNCTIONAL T ASSOCIATED WITH A RACS

Let E be a LCS space and $\mathcal{F} = \mathcal{F}(E)$ the space of the closed subsets of E topologized as indicated in Chapter 1. With this topology $\mathcal{T}_f$ on $\mathcal{F}$, we shall associate its borelian tribe σ_f, that is, the σ-algebra generated by classes $\mathcal{F}^K$, $K \in \mathcal{K}$, and $\mathcal{F}_G$, $G \in \mathcal{G}$. Actually, σ_f is generated by the single class $\mathcal{F}^K$, $K \in \mathcal{K}$, as well as by the single class $\mathcal{F}_G$, $G \in \mathcal{G}$, for, if $\{K_n\}$ is a sequence in $\mathcal{K}$ such that $K_n \uparrow G \in \mathcal{G}$, we also have $\mathcal{F}_{K_n} \uparrow \mathcal{F}_G$ in $\mathcal{P}(\mathcal{F})$. Conversely, if $\{G_n\}$ is a fundamental countable system of open neighborhoods of $K \in \mathcal{K}$ such that $G_n \downarrow K$, we have $\mathcal{F}_{G_n} \downarrow \mathcal{F}_K$ by (1-1-2).

Then we may define a *random closed set* (or RACS) as a measurable mapping A from a probabilized space $(\Omega, \mathcal{Q}, P')$ into the measurable space $(\mathcal{F}, \sigma_f)$. Let P be the probability induced on σ_f, that is, $P(\mathcal{V}) = P'(A^{-1}(\mathcal{V}))$ for any $\mathcal{V} \in \sigma_f$. It is also possible to define directly the RACS A by providing the measurable space $(\mathcal{F}, \sigma_f)$ with this probability P: A is the identical mapping from $\mathcal{F}$ onto itself, that is, $A(F) = F$, and in this case we shall often say that A is the canonical RACS on the probability space ($\mathcal{F}$, σ_f, P). The compactness of the space $\mathcal{F}$ grants us that there actually exist probabilities on σ_f. In the same way, we may define a family A_i, $i \in I$, of RACS on the corresponding product space. In Section 2-4, we shall examine the point of view of the space law, that is, the possible relationships between the notion of a RACS and the more general notion of a random set (not necessarily closed) in $\mathcal{P}(E)$.

If Ω is a topological space, $\mathcal{Q}$ its borelian tribe, and α an u.s.c. or a l.s.c. mapping from Ω into $\mathcal{F}$, α is measurable, and defines a RACS if a probability P' is given on $\mathcal{Q}$, for σ_f is generated by the single class $\mathcal{F}^K$, $K \in \mathcal{K}$, or by the single class $\mathcal{F}_G$, $G \in \mathcal{G}$, as indicated above. Hence, by the results of Chapter 1, if A and A' are RACS, so are $A \cup A'$, $A \cap A'$, $\complement \mathring{A}$, and $\bar{\mathring{A}}$, and also $\partial A = A \cap \complement \mathring{A}$. If E is a euclidean space and K a nonempty compact set, $A \oplus K$, $A \ominus K$, A^K, A_K, and so forth are also RACS.

It is well known that the probability law of an ordinary random variable is entirely determined if the corresponding distribution function is given. In the case of a RACS, there exists a very similar notion. Let A be the canonical RACS associated with a probability P on σ_f. For any $K \in \mathcal{K}$ and any $G \in \mathcal{G}$, let us put

$$T(K) = P(\mathcal{F}_K) = P(A \cap K \neq \varnothing)$$

$$T(G) = P(\mathcal{F}_G) = P(A \cap G \neq \varnothing)$$

If the function T defined in this way is given on $\mathcal{K}$ (or on $\mathcal{G}$), then the probability P on σ_f is entirely determined. To give a straightforward proof of this result would not be difficult, but it is simpler to consider it as an immediate corollary of the Choquet theorem which will be proved in the next section.

Any function on $\mathcal{K}$ or $\mathcal{G}$ may generally not be associated with a RACS. Obviously, the functional of a RACS A must be *increasing* and must satisfy $T(\varnothing) = 0$, for no closed set hits the empty set, and also $0 \leqslant T \leqslant 1$. If $G \in \mathcal{G}$ and $K \in \mathcal{K}$, $K \subset G$ implies $T(K) \leqslant T(G)$. If $K_n \uparrow G$, we also have $\mathcal{F}_{K_n} \uparrow \mathcal{F}_G$ in $\mathcal{P}(\mathcal{F})$, and thus $T(K_n) \uparrow T(G)$ (sequential monotone continuity). In the same way, if $\{G_n\}$ is a countable fundamental system of open neighborhoods of a compact set K, $G_n \downarrow K$ implies $\mathcal{F}_{Gn} \downarrow \mathcal{F}_K$, as we have seen above, and thus also $T(G_n) \downarrow T(K)$. It follows that the functional T must verify two conditions:

$$\begin{aligned} T(G) &= \operatorname{Sup}\{T(K), K \in \mathcal{K}, K \subset G\} \qquad (G \in \mathcal{G}) \\ T(K) &= \operatorname{Inf}\{T(G), G \in \mathcal{G}, G \supset K\} \qquad (K \in \mathcal{K}) \end{aligned} \tag{2-1-1}$$

By Proposition 1-4-3, this implies that T is u.s.c. on $\mathcal{K}$ and l.s.c. on $\mathcal{G}$. By Proposition 1-2-4, Corollary 5, Proposition 1-3-1, and Proposition 1-4-2, these upper and lower semicontinuity conditions are respectively equivalent to the following conditions of sequential monotone continuity:

$$\begin{aligned} &K_n \downarrow K \text{ in } \mathcal{K} \quad \text{implies} \quad T(K_n)\downarrow T(K) \\ &G_n \uparrow G \text{ in } \mathcal{G} \quad \text{implies} \quad T(G_n)\uparrow T(G) \end{aligned} \tag{2-1-2}$$

Now let $K, K_1, K_2, \ldots$ be compact sets, and let us define by recurrence procedure the functions $S_1, S_2, \ldots$ by putting

$$S_1(K; K_1) = T(K \cup K_1) - T(K)$$

. . . .

$$S_n(K; K_1, \ldots, K_n) = S_{n-1}(K; K_1, \ldots, K_{n-1}) - S_n(K \cup K_n; K_1, \ldots, K_{n-1}) \tag{2-1-3}$$

Clearly, $S_n(K; K_1, \ldots, K_n)$ is the probability $P(\mathcal{F}^K_{K_1}, \ldots, K_n)$ for the RACS A to hit $K_1, K_2, \ldots, K_n$ and not to hit K. Thus the functions S_n must be $\geqslant 0$ for any $K, K_1, \ldots, K_n \in \mathcal{K}$, as well as the analogous functions we can obtain by starting from open sets $G, G_1, G_2, \ldots$.

We will show that, by these conditions, T is a capacity. Let us first recall the general definition (see, e.g., Choquet, 1953–54, or Meyer, 1966). If E is a locally compact space, a capacity on E is a mapping T from $\mathcal{P}(E)$ into the extended line $[-\infty, +\infty]$ satisfying three conditions:

1. $A, A' \in \mathcal{P}(E)$, and $A \subset A'$ implies $T(A) \leqslant T(A')$.
2. $A_n \uparrow A$ in $\mathcal{P}(E)$ implies $T(A_n) \uparrow T(A)$.
3. $K_n \downarrow K$ in $\mathcal{K}(E)$ implies $T(K_n) \downarrow T(K)$.

If E is a LCS space, as is always assumed in this book, condition 3 is equivalent to the upper semicontinuity of T on $\mathcal{K}$.

A subset $B \subset E$ is said to be *capacitable* if the following approximation formula is satisfied:

$$T(B) = \text{Sup}\,\{T(K), K \in \mathcal{K}, K \subset B\} \tag{2-1-4}$$

and it can be shown (see, e.g., Meyer, 1966) that any borelian set in E is capacitable if the space E is LCS.

Now let T be an u.s.c. function on $\mathcal{K}$ such that the functions S_n defined by (2-1-3) are $\geqslant 0$ for any integer $n > 0$, and put $T^*(G) = \text{Sup}\,\{T(K),$

$K \in \mathcal{K}$, $K \subset G\}$ for any $G \in \mathcal{G}$, and $T^*(A) = \text{Inf}\ \{T^*(G),\ G \in \mathcal{G},\ G \supset A\}$ for any $A \in \mathcal{P}(E)$. Then it can be shown that the function T^* on $\mathcal{P}(E)$ satisfies conditions 1, 2, and 3 above. Moreover, we have $T^*(K) = T(K)$ for $K \in \mathcal{K}$, by Proposition 1-4-3, so that T^* is a *capacity* by the above definition. Because $T^* = T$ on $\mathcal{K}$, we shall write $T(A)$ instead of $T^*(A)$ for any $A \in \mathcal{P}(E)$. It can be verified that the functions $S_n(A;\ A_1, \ldots, A_n)$ defined according to (2-1-3) are still $\geqslant 0$ for any $A, A_1, \ldots \in \mathcal{P}(E)$. Such a capacity T is said to be an *alternating Choquet capacity of infinite order* or, for brevity, simply a *Choquet capacity*.

We obtain a very similar construction when starting from a function T l.s.c. on $\mathcal{G}$ and such that the S_n functions in (2-1-3) are $\geqslant 0$. The functions S_n remain $\geqslant 0$ under transformations (2-1-1), so that these two points of view are strictly equivalent.

Now let B be a borelian or, more generally, a T-capacitable subset in E. It is not at all obvious that $\mathcal{F}_B$ belongs to σ_f, and the relationship $T(B) = P(\mathcal{F}_B)$ is questionable. However, by the definition of a Choquet capacity, there exists a decreasing sequence $\{G_n\} \subset \mathcal{G}$ satisfying $G_n \supset B$ and $T(G_n) \downarrow T(B)$, and, by (2-1-4), also an increasing sequence $\{K_n\} \subset \mathcal{K}$ with $K_n \subset B$ and $T(K_n) \uparrow T(B)$. By putting $\mathcal{V} = \cup \mathcal{F}_{K_n} = \mathcal{F}_{\cup K_n}$ and $\mathcal{V}' = \cap \mathcal{F}_{G_n}$, we get $\mathcal{V} \in \sigma_f$, $\mathcal{V}' \in \sigma_f$, $\mathcal{V} \subset \mathcal{F}_B \subset \mathcal{V}'$, and $P(\mathcal{V}) = T(B) = P(\mathcal{V}')$. Hence $\mathcal{F}_B$ is measurable for the completion $(\mathcal{F}, \tilde{\sigma}_f, \tilde{P})$ of the probability space $(\mathcal{F}, \sigma_f, P)$ and satisfies $\tilde{P}(\mathcal{F}_B) = T(B)$.

By the preceding considerations, the functional T associated with a RACS is a Choquet capacity satisfying $T(\varnothing) = 0$ and $0 \leqslant T \leqslant 1$. We shall see that the converse is also true.

2-2. THE CHOQUET THEOREM

The following theorem was proved by Choquet by means of the theory of integral representations on convex cones. The proof we shall give is purely probabilistic.

Theorem 2-2-1(Choquet). *Let E be a LCS space and T a function defined on $\mathcal{K}(E)$ [resp. on $\mathcal{G}(E)$]. Then there exists a (necessarily unique) probability P on σ_f satisfying $P(\mathcal{F}_K) = T(K)$ for $K \in \mathcal{K}$ [resp. $P(\mathcal{F}_G) = T(G)$ for $G \in \mathcal{G}$] if and only if T is an alternating Choquet capacity of infinite order such that $0 \leqslant T \leqslant 1$ and $T(\varnothing) = 0$.*

We recall that the statement "T is a Choquet capacity" means that two conditions are satisfied:

1. The functions S_n of (2-1) are $\geqslant 0$ for any $K, K_1, \ldots \in \mathcal{K}$ ($G, G_1, \ldots \in \mathcal{G}$).

2. T is u.s.c. on $\mathcal{K}$ (l.s.c. on $\mathcal{G}$), or, which is the same, $K_n \downarrow K$ in $\mathcal{K}$ implies $T(K_n) \downarrow T(K)$ [resp. $G_n \uparrow G$ in $\mathcal{G}$ implies $T(G_n) \uparrow T(G)$].

The "only if" part of the theorem was proved in the preceding section. In regard to the "if" part, it is sufficient to prove the statement concerning the space $\mathcal{K}$, because by transformation (2-1-1) u.s.c. functions on $\mathcal{K}$ are changed into l.s.c. functions on $\mathcal{G}$, and the functions S_n remain $\geqslant 0$. We begin with two lemmas.

Lemma 2-2-1. *Let $\mathcal{V}$ be a class of subsets in E including $\varnothing$ and closed for finite union, and $\mathcal{S}$ be the class closed for the finite intersection generated in $\mathcal{P}(\mathcal{F})$ by the classes $\mathcal{F}_V$, $V \in \mathcal{V}$, and $\mathcal{F}^V$, $V \in \mathcal{V}$. Then $\mathcal{S}$ is a semialgebra, and any nonempty $S \in \mathcal{S}$ admits a representation $S = \mathcal{F}^V_{V_1,\ldots,V_n}$ for an integer $n \geqslant 0$ and subsets $V, V_1, \ldots, V_n \in \mathcal{V}$ such that $V_i \not\subset V \cup V_j$ if $i \neq j$. Such a representation is called a reduced representation. If $\mathcal{F}^V_{V_1,\ldots,V_k'}$ is another reduced representation of S, then $V = V'$ and there exists a permutation $(i_1, i_2, \ldots, i_n)$ of the n first integers such that $V_j \cup V = V'_{i_j} \cup V$ for $j = 1, 2, \ldots, n$.*

Proof. We recall $\mathcal{S}$ is a semialgebra if $\varnothing \in \mathcal{S}$, $\mathcal{S}$ is closed for the finite intersection, and the complementary set of any $S \in \mathcal{S}$ is a finite union of disjoint sets in $\mathcal{S}$. By its definition, $\mathcal{S}$ is closed for the finite intersection. We have $\varnothing = \mathcal{F}_\varnothing \in \mathcal{S}$, since $\varnothing \in \mathcal{V}$. If $S = \mathcal{F}^V_{V_1,\ldots,V_n}$, the complementary set is

$$S^c = \mathcal{F}_V \cup \mathcal{F}^{V \cup V_1} \cup \mathcal{F}^{V \cup V_2}_{V_1} \cup \cdots \cup \mathcal{F}^{V \cup V_n}_{V_1, V_2, \ldots, V_{n-1}}$$

Thus $\mathcal{S}$ is a semialgebra.

Let $S = \mathcal{F}^V_{V_1,\ldots,V_n}$ be a nonempty element in $\mathcal{S}$. If $V_i \subset V \cup V_j$ for two indices $i \neq j$, V_j is superfluous in the representation $\mathcal{F}^V_{V_1,\ldots,V_n}$ and may be canceled. Hence there exists a reduced representation.

Let $\mathcal{F}^V_{V_1,\ldots,V_n}$ and $\mathcal{F}^{V'}_{V_1',\ldots,V_k'}$ be two reduced representations of $S \in \mathcal{S}$, $S \neq \varnothing$, and prove that $V = V'$. Suppose, for instance, that there exists $x \in V' \cap V^c$. Because S is nonempty, we have $V_j' \not\subset V'$ for $j = 1, 2, \ldots, k$, and there exist k points $x_1, \ldots, x_k$ such that $x_j \in V_j'$, $x_j \notin V'$, and

$$\{x_1, \ldots, x_k\} \in \mathcal{F}^{V'}_{V_1',\ldots,V_k'} = S = \mathcal{F}^V_{V_1,\ldots,V_n}$$

But $x \notin V$ implies $\{x, x_1, \ldots, x_k\} \in \mathcal{F}^V_{V_1,\ldots,V_n}$, and $x \in V'$ yields $\{x, x_1, \ldots, x_k\} \notin \mathcal{F}^{V'}_{V_1',\ldots,V_k'}$. From this contradiction, we conclude that $V = V'$.

For $i = 1, 2, \ldots, n-1$, we have $V_i \not\subset V_n \cup V$, because the representations are reduced. Then let $y_1, y_2, \ldots, y_{n-1}$ be $n-1$ points such that $y_i \in V_i \cap V_n^c \cap V^c$, and $y \in V_n \cap V^c$. By $\{y_1, \ldots, y_{n-1}\} \notin S$ and $\{y, y_1, \ldots, y_{n-1}\} \in S$, there exists an index $j_n (0 < j_n \leqslant k)$ such that $y \in V_{j_n}'$ and $y_i \notin V_{j_n}'$, $i = 1, 2, \ldots, n-1$. Let y' be another point in $V_n \cap V^c$. In the same way, we obtain $y' \in V'_{j_n}$. Hence $V_n \cap V^c \subset V'_{j_n}$, that is, $V_n \subset V \cup V'_{j_n}$. In the same way, we see that there exists an index i_n $(0 < i_n \leqslant n)$ such that $V'_{j_n} \cap V'^c = V'_{j_n} \cap V^c \subset V_{i_n}$. But $i_n \neq n$ implies $V_n \subset V_{i_n} \cup V$. It follows that $i_n = n$ and $V_n \cap V^c = V'_{j_n} \cap V^c$. By repeating this proof for each index i, the lemma follows.

Lemma 2-2-2. *With the same notations as in Lemma 2-2-1, let T be a function on $\mathcal{V}$ such that $T(\varnothing)=0$, and the functions S_n be defined by putting*

$$S_0(V) = 1 - T(V)$$

(*a*) $\quad\ldots\ldots\ldots\ldots\ldots\ldots\ldots\ldots\ldots\ldots\ldots\ldots\ldots\ldots\ldots\ldots\ldots\ldots\ldots\ldots$

$$S_n(V; V_1,\ldots,V_n) = S_{n-1}(V; V_1,\ldots,V_{n-1}) - S_{n-1}(V \cup V_n; V_1,\ldots,V_{n-1})$$

are $\geqslant 0$ for any integer $n \geqslant 0$ and $V, V_1, V_2, \ldots \in \mathcal{V}$. Then there exists a unique additive mapping from $\mathcal{S}$ into [0, 1] *such that $P(\varnothing)=0$, $P(\mathcal{F}_V) = T(V)$ for $V \in \mathcal{V}$, and*

(*b*) $$P\left(\mathcal{F}^V_{V_1,\ldots,V_n}\right) = S_n(V; V_1,\ldots,V_n)$$

Proof. First, we must verify that relationship (*b*) actually defines a function on $\mathcal{S}$, that is, $\mathcal{F}^V_{V_1,\ldots,V_n} = \mathcal{F}^{V'}_{V'_1,\ldots,V'_i,\ldots,V'_k}$ implies $S_n(V; V_1,\ldots,V_n) = S_k(V'; V'_1,\ldots,V'_k)$. Let $S \in \mathcal{S}$ be the element admitting these two representations. If $S=\varnothing$, we have $V_i \subset V$ and $V'_j \subset V'$ for an index i and an index $j (0 < i \leqslant n,\ 0 < j \leqslant k)$. But it can easily be shown by a recurrence procedure that $V_i \subset V$ implies $S_n(V; V_1,\ldots,V_n)=0$, and we have the required equality $S_n(V; V_1,\ldots,V_n) = S_k(V'; V'_1,\ldots,V'_k)$.

Suppose now that $S \neq \varnothing$. It is possible to change $\mathcal{F}^V_{V_1,\ldots,V_n}$ into one of its reduced forms without changing the value of $S_n(V; V_1,\ldots,V_n)$, for if, for instance, $V_1 \subset V \cup V_n$, that is, $\mathcal{F}^V_{V_1,\ldots,V_n} = \mathcal{F}^V_{V_1,\ldots,V_{n-1}}$, relationships (*a*) also imply $S_n(V; V_1,\ldots,V_n) = S_{n-1}(V; V_1,\ldots,V_{n-1})$. Thus we may suppose that the two representations of S are reduced. By Lemma 2-2-1, we have $V = V'$ and $k = n$, and, by permuting indices j, for instance, which does not change the value of S_n, we may also suppose that $V_i \cup V = V'_i \cup V$, $i = 1, 2, \ldots, n$. It is easy to verify that relationships (*a*) imply $S_n(V; V_1,\ldots,V_n) = S_n(V; V \cup V_1,\ldots,V \cup V_n)$. Thus we have $S_n(V; V_1,\ldots,V_n) = S_n(V; V'_1,\ldots,V'_n)$, and relationship (*b*) defines a function P on $\mathcal{S}$.

This function P is $\geqslant 0$ on $\mathcal{S}$, for the S_n are $\geqslant 0$, and satisfies $P(\varnothing) = P(\mathcal{F}_\varnothing) = T(\varnothing) = 0$. By (*a*) and $S_n \geqslant 0$, we also obtain

$$0 \leqslant P\left(\mathcal{F}^V_{V_1,\ldots,V_n}\right) \leqslant P\left(\mathcal{F}^V_{V_1,\ldots,V_{n-1}}\right) \leqslant \cdots \leqslant P(\mathcal{F}^V) \leqslant 1$$

Thus P is a mapping from $\mathcal{S}$ into [0, 1].

It remains to prove that P is *additive* on $\mathbb{S}$. Let $\mathcal{A}$ and $\mathcal{A}'$ be two nonempty elements in $\mathbb{S}$ admitting the reduced representations

$$\mathcal{A} = \mathfrak{F}^V_{V_1,\dots,V_n}; \qquad \mathcal{A}' = \mathfrak{F}^{V'}_{V'_1,\dots,V'_k}$$

We must show that $\mathcal{A} \cap \mathcal{A}' = \varnothing$ and $\mathcal{A} \cup \mathcal{A}' \in \mathbb{S}$ imply $P(\mathcal{A} \cup \mathcal{A}') = P(\mathcal{A}) + P(\mathcal{A}')$. Obviously, we have

$$\mathcal{A} \cap \mathcal{A}' = \mathfrak{F}^{V \cup V'}_{V_1,\dots,V_n,\, V'_1,\dots,V'_n}$$

and this intersection is empty if and only if one of the sets V_i or V'_j is included in $V \cup V'$. Suppose, for instance, that

$$(c) \qquad V_n \subset V \cup V'$$

If $\mathcal{A} \cup \mathcal{A}' \in \mathbb{S}$, this union admits a representation of the form

$$\mathcal{A} \cup \mathcal{A}' = \mathfrak{F}^{V''}_{V''_1,\dots,V''_p}$$

If $V^c = \varnothing$, we have $\mathcal{A} = \{\varnothing\}$ or $\mathcal{A} = \varnothing$, and the additivity relationship is trivial. Suppose that $V^c \neq \varnothing$, and choose $x \in V^c$, $x_i \in V_i \cap V^c$ $(i = 1, 2, \dots, n)$, so that $F = \{x, x_1, \dots, x_n\} \in \mathcal{A}$. Thus $F \in \mathcal{A} \cup \mathcal{A}'$ implies $F \cap V'' = \varnothing$, that is, $x \notin V''$. It follows that $V'' \subset V$, and in the same way $V'' \subset V'$. In other words,

$$(d) \qquad V'' \subset V \cap V'$$

Then let us show that $V = V''$. Suppose that there exist two points x and x' such that $x \in V \cap \complement V''$ and $x' \in V' \cap \complement V''$, so that $\{x, x'\}$ and V'' are disjoint. Choose points $x''_i \in V''_i \cap \complement V''$, $i = 1, 2, \dots, p$. Then we have $\{x, x', x''_1, \dots, x''_p\} \in \mathfrak{F}^{V''}_{V''_1,\dots,V''_p} = \mathcal{A} \cup \mathcal{A}'$. But this implies either $\{x, x'\} \cap V = \varnothing$ or $\{x, x'\} \cap V' = \varnothing$ and yields a contradiction. We conclude that one of the relationships $V \subset V''$ and $V' \subset V''$ is true. If $V' \subset V''$ is true, we obtain $V' = V''$ by (d), but this is impossible because it implies $V' \subset V$, and then, by (c), $V_n \subset V$ and $\mathcal{A} = \varnothing$, and contradicts our hypothesis. It follows that $V \subset V''$ and, by (d) and (c),

$$(e) \qquad V = V''; \qquad V \subset V'; \qquad V_n \subset V'$$

Now let F be an element in $\mathcal{A} \cup \mathcal{A}'$. Then $F \in \mathfrak{F}_{V_n}$ implies $F \notin \mathcal{A}'$, because $V_n \subset V'$, and thus $F \in \mathcal{A}$. On the other hand, $F \in \mathfrak{F}^{V_n}$ implies $F \notin \mathcal{A}$ and thus $F \in \mathcal{A}'$. It follows that

$$\mathcal{A} = (\mathcal{A} \cup \mathcal{A}') \cap \mathfrak{F}_{V_n} = \mathfrak{F}^V_{V''_1,\dots,V''_p,\, V_n}$$

$$\mathcal{A}' = (\mathcal{A} \cup \mathcal{A}') \cap \mathfrak{F}^{V_n} = \mathfrak{F}^{V \cup V_n}_{V''_1,\dots,V''_p}$$

By the recurrence formula $S_{p+1}(V; V_1'',\ldots,V_p'', V_n) = S_p(V; V_1'',\ldots,V_p'') - S_p(V \cup V_n; V_1'',\ldots,V_p'')$, we obtain $P(\mathcal{Q}) = P(\mathcal{Q} \cup \mathcal{Q}') - P(\mathcal{Q}')$, and so P is additive.

Proof of Theorem 2-2-1. We are now able to prove the "if" part of the Choquet theorem. Let T be a function on $\mathcal{K}$ satisfying the stated conditions, $\mathcal{V}$ be the class of the sets $V = G \cup K$, $G \in \mathcal{G}$, $K \in \mathcal{K}$, and $\mathcal{S}$ be the semialgebra on $\mathcal{F}$, the elements of which are the $\mathcal{F}^V_{V_1,\ldots,V_n}$, n an integer $\geqslant 0$, $V, V_1,\ldots,V_n \in \mathcal{V}$. The Choquet capacity T may be extended on $\mathcal{V}$, and the corresponding functions S_n are still $\geqslant 0$. By the two preceding lemmas, the relationship

$$P\left(\mathcal{F}^V_{V_1,\ldots,V_n}\right) = S_n(V; V_1,\ldots,V_n)$$

defines an additive mapping P from $\mathcal{S}$ into $[0, 1]$ satisfying $P(\varnothing) = 0$. On the other hand, the σ-algebra generated by $\mathcal{S}$ is σ_f. By the classical theorems (see, e.g., Neveu, 1964, Proposition I-6-2), P is σ-additive on $\mathcal{S}$ and admits on σ_f an extension by a necessarily unique probability if there exists in $\mathcal{S}$ a compact class $\mathbf{C}$ such that, for any $\mathcal{Q} \in \mathcal{S}$,

$$P(\mathcal{Q}) = \operatorname{Sup}\{P(C), C \in \mathbf{C}, C \subset \mathcal{Q}\} \tag{2-2-1}$$

Let $\mathbf{C}$ be the class $\{\mathcal{F}^G_{K_1,\ldots,K_n}$, n an integer $\geqslant 0$, $G \in \mathcal{G}$, $K_1,\ldots,K_n \in \mathcal{K}\}$. Class $\mathbf{C}$ is closed for $\cap$ and its elements are compact in $\mathcal{F}$, so that $\mathbf{C}$ is actually a compact class in $\mathcal{S}$. It remains to prove (2-2-1), and this will complete the proof.

Let $\mathcal{Q} = \mathcal{F}^V_{V_1,\ldots,V_k}$ be an element of $\mathcal{S}$, and $V = G_0 \cup K_0$ for an open set G_0 and a compact set K_0. There exists a sequence $\{G_n\} \subset \mathcal{G}$ satisfying $G_n \supset \overline{G}_{n+1} \in \mathcal{K}$ and $G_n \downarrow K_0$. It follows that $\mathcal{F}^{G_n} \uparrow \mathcal{F}^{K_0}$ and $\mathcal{F}^{G_0 \cup G_n} \uparrow \mathcal{F}^V$ by (1-1-2). In the same way, for each $i = 1, 2,\ldots,k$, V_i is the union of an open and a compact set, and there exists a sequence $\{K_n(i)\} \subset \mathcal{K}$ satisfying $K_n(i) \uparrow V_i$ and thus $\mathcal{F}_{K_n(i)} \uparrow \mathcal{F}_{V_i}$. If we put

$$\mathcal{Q}_n = \mathcal{F}^{G_0 \cup G_n}_{K_n(1),\ldots,K_n(k)}$$

then $\mathcal{Q}_n \in \mathbf{C}$ and $\mathcal{Q}_n \uparrow \mathcal{Q}$. It remains to prove that $P(\mathcal{Q}_n) \uparrow P(\mathcal{Q})$. By elementary calculus, we obtain

$$P(\mathcal{Q}) = -T(V) + \sum_i T(V \cup V_i) - \sum_{i_1 < i_2} T(V \cup V_{i_1} \cup V_{i_2}) + \cdots$$

$$P(\mathcal{Q}_n) = -T(G_0 \cup G_n) + \sum_i T(G_0 \cup G_n \cup K_n(i))$$

$$- \sum_{i_1 < i_2} T(G_0 \cup G_n \cup K_n(i_1) \cup K_n(i_2)) + \cdots$$

These sums being finite, it is sufficient to verify that each term appearing in the expression of $P(\mathcal{A}_n)$ converges toward the corresponding term in the expression of $P(\mathcal{A})$. But this is an immediate consequence of the following lemma.

Lemma 2-2-3. *Let T be a Choquet capacity, G and G_0 two open sets, K a compact set, $\{K_n\} \subset \mathcal{K}$ a sequence such that $K_n \uparrow G$, and $\{G_n\} \subset \mathcal{G}$ a sequence such that $G_n \supset \overline{G}_{n+1} \in \mathcal{K}$ and $G_n \downarrow K$. Then $T(G_0 \cup K \cup G) = \lim T(G_0 \cup G_n \cup K_n)$.*

Proof. The fact that T is increasing implies

$$(f) \qquad T(G_0 \cup G_n \cup G) \geqslant T(G_0 \cup G_n \cup K_n) \geqslant T(G_0 \cup K \cup K_n)$$

Let us prove that $T(G_0 \cup G_n \cup G) \downarrow T(G_0 \cup G \cup K)$. T is a capacity, and this implies $T(G_0 \cup G \cup K) = \text{Inf } \{T(G'),\ G' \in \mathcal{G},\ G' \supset G_0 \cup G \cup K\}$. But if $G' \in \mathcal{G}$ and $G' \supset G_0 \cup G \cup K$, we have $G_n \subset G'$ for n large enough, and $T(G') \geqslant T(G_0 \cup G \cup G_n)$. Hence the stated convergence is true. In the same way, $K \cup K_n \cup G_0 \uparrow K \cup G \cup G_0$ implies $T(K \cup K_n \cup G_0) \uparrow T(K \cup G \cup G_0)$, by Axiom 2 of capacities. These two convergences and inequalities (f) imply $T(G_0 \cup K \cup G) = \lim T(G_0 \cup G_n \cup K_n)$.

Remark. The functionals defined on $\mathcal{K}$ are of great importance in integral geometry (see, e.g., the book by Hadwiger, 1957). Among these functionals, the Choquet theorem characterizes the ones that admit a probabilistic interpretation, and in the same way indicates interesting connections between integral geometry and the RACS theory. We shall give many examples in the following chapters.

2-3. CONDITIONAL RACS

The following proposition is a simple corollary to the Choquet theorem.

Proposition 2-3-1. *Let $\mathcal{B}$ be a countable base of the topology $\mathcal{G}$ on E such that $\overline{B} \in \mathcal{K}$ for any $B \in \mathcal{B}$, closed for finite union, and such that, for any $K \in \mathcal{K}$ and any $G \in \mathcal{G}$,*

$$(a) \qquad K = \cap \{\overline{B},\ B \in \mathcal{B},\ B \supset K\}; \qquad G = \cup \{B;\ B \in \mathcal{B},\ \overline{B} \subset G\}$$

Let $\mathcal{B}' = \{\overline{B},\ B \in \mathcal{B}\}$, and T be a function defined on $\mathcal{B}'$. Then there exists a necessarily unique probability P on σ_f such that $P(\mathcal{F}_{B'}) = T(B')$ for any $B' \in \mathcal{B}'$ if and only if three conditions are fulfilled:

1. $T(\emptyset)=0$ and $0 \leqslant T \leqslant 1$.
2. The functions S_n defined in (2-1-3) are $\geqslant 0$ on $\mathcal{B}'$.
3. T is u.s.c. for the relative myope topology on $\mathcal{B}'$.

Proof. The "only if" part is obvious. Conversely, suppose that T satisfies these three conditions. Put $T^*(G)=\text{Sup}\{T(B'), B' \in \mathcal{B}', B' \subset G\}$ for any $G \in \mathcal{G}$, and $T^*(K)=\text{Inf}\{T^*(G), G \in \mathcal{G}, G \supset K\}$ for any $K \in \mathcal{K}$. By relationship (a), it is possible to prove that T^* is u.s.c. on $\mathcal{K}$ and l.s.c. on $\mathcal{G}$ and that $T=T^*$ on $\mathcal{B}'$ (the proof is the same as for Proposition 1-4-3). Thus we shall write T instead of T^*. Obviously, $T(\emptyset)=0$ and $0 \leqslant T \leqslant 1$ on $\mathcal{G}$. It remains to prove that the functions S_n are $\geqslant 0$ on $\mathcal{G}$—in other words, that the following inequalities hold:

$$(b) \qquad \sum_{i=1}^{k} T(G \cup G_i) + \sum_{i_1<i_2<i_3} T(G \cup G_{i_1} \cup G_{i_2} \cup G_{i_3}) + \cdots$$
$$\geqslant T(G) + \sum_{i_1<i_2} T(G \cup G_{i_1} \cup G_{i_2}) + \cdots$$

for $G, G_1, \ldots, G_k \in \mathcal{G}$. There exist in $\mathcal{B}$ sequences $\{B_n\}$ and $\{B_n(i)\}$ such that $\bar{B}_n \subset B_{n+1}$, $\bar{B}_n(i) \subset B_{n+1}(i)$, $\bar{B}_n \uparrow G$, and $\bar{B}_n(i) \uparrow G_i$, $i=1, 2, \ldots, k$. On the other hand, by condition 2, we have

$$\sum_i T(\bar{B}_n \cup \bar{B}_n(i)) + \ldots \geqslant T(\bar{B}_n) + \sum_{i_1<i_2} T(\bar{B}_n \cup \bar{B}_n(i_1) \cup \bar{B}_n(i_2)) + \cdots$$

and inequalities (b) hold by the definition of the extension of T on $\mathcal{G}$.

Thus the conditions of the Choquet theorem are satisfied by T, and there exists a probability P on σ_f such that $P(\mathcal{F}_{B'}) = T(B')$. The uniqueness follows from relationship (a): if $K \in \mathcal{K}$, by (a) there exists a sequence $\{B'_n\} \subset \mathcal{B}'$ such that $B'_n \downarrow K$. Hence $T(B'_n) \downarrow P(\mathcal{F}_K) = T(K)$, and P is unique.

The space E being LCS, it is always possible to find a countable base $\mathcal{B}$ of its topology satisfying the conditions stated in Proposition 2-3-1. For instance, if E is a euclidean space, the class of the finite unions of open balls with rational centers and radii will be suitable.

Then it is possible to define the *notion* of a *RACS conditional with respect to a random variable u.* Let $A=(\mathcal{F}, \sigma_f, P)$ be a RACS, u a measurable mapping from $\mathcal{F}$ into a measurable space $(\Omega, \mathcal{H})$, and F the probability induced on $\mathcal{H}$ by P, that is, $F(H)=P(u^{-1}(H))$ for any $H \in \mathcal{H}$. Also let $\mathcal{B}$ be a base of $\mathcal{G}$ satisfying the conditions of Proposition 2-3-1, and

obtain, for any $\mathcal{V} \in \sigma_f$ and $H \in \mathcal{H}$,

$$P'(\mathcal{V} \cap u^{-1}(H)) = P[\varphi^{-1}(\mathcal{V}) \cap u^{-1}(\gamma^{-1}(H))]$$

$$= \int_{\gamma^{-1}(H)} P_u(\varphi^{-1}(\mathcal{V});\omega)F'(d\omega)$$

$$= \int_H P_u(\varphi^{-1}(\mathcal{V});\gamma^{-1}(\omega))F(d\omega)$$

And thus

$$P'_u(\mathcal{V};\omega) = P_u(\varphi^{-1}(\mathcal{V});\gamma^{-1}(\omega)) \quad (F\text{-a.s.}) \tag{2-3-3}$$

In other words, the *conditional RACS* $A'_u(\omega)$ and $\varphi[A_u(\gamma^{-1}(\omega))]$ *are equivalent*. As a particular case, the following result will be used in the sequel:

Proposition 2-3-2. *Let* $A = (\mathcal{F}, \sigma_f, P)$ *be a RACS,* φ *a measurable mapping from* $\mathcal{F}$ *into itself under which the probability* P *remains invariant,* u *a measurable mapping from* $\mathcal{F}$ *into a measurable space* $(\Omega, \mathcal{H})$, *and* F *the probability on* $\mathcal{H}$ *associated with the random variable* $v = u \bigcirc \varphi$. *If there exists a one-to-one mapping* γ *from* Ω *onto itself, such that* γ, *as well as its reciprocal mapping* γ^{-1}, *is measurable, and such that* $u \bigcirc \varphi = \gamma \bigcirc u$ *P-a.s., then, for F-almost every* $\omega \in \Omega$, *the conditional RACS* $A_u(\omega)$ *and* $\varphi[A_u(\gamma^{-1}(\omega))]$ *are equivalent. In other words, for any* $\mathcal{V} \in \sigma_f$,

$$P_u(\mathcal{V};\omega) = P_u(\varphi^{-1}(\mathcal{V});\gamma^{-1}(\omega)) \quad (F\text{-a.s.}) \tag{2-3-4}$$

Proof. The RACS A and $A' = \varphi(A)$ are equivalent, so that also $P'_u(\mathcal{V};\omega) = P_u(\mathcal{V};\omega)$, and (2-3-4) follows from (2-3-3).

Example. Let E be a *euclidean* space, and φ be the *translation* $F \to \varphi(F) = F_h = F \oplus \{h\}$ for a given $h \in E$. Let K be a compact set, and u be the mapping $F \to u(F) = F \oplus K$ (or $F \ominus K$, F_k, F^K, etc.). These mappings are compatible with the translation, that is $u(F_h) = (u(F))_h$. It follows that $v(F) = u(F_h) = (u(F))_h$, and γ is the same translation, $F \to \gamma(F) = F_h$. If the RACS A is *stationary* (i.e., its probability is invariant under the translations) (2-3-4) holds, and the RACS $A_u(F)$ [i.e., the RACS A taken conditionally if $u(A) = F$] is equivalent to $(A_u(F_{-h}))_h$, that is, the translate of A taken conditionally if $u(A) = F_{-h}$.

Independent RACS.

There is no difficulty in defining a pair (A_1, A_2) of RACS by a probability P on the product σ-algebra $\sigma_f \otimes \sigma_f$. We may define the independence of two RACS A_1 and A_2, that is, $P(\mathcal{V} \times \mathcal{V}') = P(\mathcal{V} \times \mathcal{F})P(\mathcal{F} \times \mathcal{V}')$ for any $\mathcal{V}$, $\mathcal{V}' \in \sigma_f$. This independence holds if and only if, for any $K, K' \in \mathcal{K}$,

$$P(\mathcal{F}_K \times \mathcal{F}_{K'}) = T_1(K)T_2(K') \tag{2-3-5}$$

with $T_1(K) = P(\mathcal{F}_K \times \mathcal{F})$ and $T_2(K') = P(\mathcal{F} \times \mathcal{F}_{K'})$.

Inductive Limit of RACS.

Let $\{B_n\} \subset \mathcal{G}$ be a sequence such that $B_{n+1} \supset \overline{B}_n$, $\overline{B}_n \in \mathcal{K}$ and $B_n \uparrow E$, and, for each integer $n > 0$, A_n a RACS a.s. included into $\overline{B}_n$. Then A_n is defined by the functional T_n on $\mathcal{K}$ such that $T_n(K) = P(A_n \in \mathcal{F}_K)$, $K \in \mathcal{K}(\overline{B}_n)$, and $T_n(K) = T_n(K \cap \overline{B}_n)$ if $K \in \mathcal{K}$, $K \not\subset \overline{B}_n$. We suppose that the following condition is fulfilled:

$$T_{n+m}(K) = T_n(K) \qquad \left(n, m > 0, \quad K \in \mathcal{K}, \quad K \subset \overline{B}_n\right) \tag{2-3-6}$$

Then the RACS $A_{n+m} \cap \overline{B}_n$ and A_n are equivalent. Under this condition it is possible to define a RACS A on $(\mathcal{F}, \sigma_f)$ such that, for any $n > 0$, $A \cap \overline{B}_n$ is equivalent to A_n. We shall say that the RACS A is the *inductive limit* of the sequence $\{A_n\}$ constituted by the RACS A_n, the functionals of which verify (2-3-6).

In order to prove the existence of A, let K be a compact set. By $B_n \uparrow E$, we have $K \subset \overline{B}_n$ for any n large enough, say $n \geqslant n_0$, and also $T_n(K) = T_{n_0}(K)$ by (2-3-6). Thus we may define on $\mathcal{K}$ a functional T by putting $T(K) = T_{n0}(K)$. Obviously, $T(\varnothing) = 0$ and $0 \leqslant T \leqslant 1$. T is u.s.c. on $\mathcal{K}$, for, if $K_n \downarrow K$ in $\mathcal{K}$, the compact sets K_n remain included in a fixed $\overline{B}_{n_0}$, and T_{n_0} is u.s.c. For the same reason, the functions S_n associated with T are $\geqslant 0$, because T_{n_0} is a Choquet capacity. Hence by Theorem 2-2-1, there exists a RACS A, the probability P of which satisfies $P(\mathcal{F}_K) = T(K)$ for any $K \in \mathcal{K}$. Moreover, the RACS $A \cap \overline{B}_n$ admits the functional $K \to T(K \cap \overline{B}_n) = T_n(K)$, and thus is equivalent to the RACS A_n.

2-4. THE POINT OF VIEW OF THE SPACE LAW

Let us now examine a more general notion of a *random set* (no longer closed). A set $A' \in \mathcal{P}(E)$ is entirely determined if its *indicator* $1_{A'}$ is given

[$1_{A'}(x)=1$ if $x \in A'$ and $1_{A'}(x)=0$ if $x \notin A'$]. If this indicator is changed into a random functional valued in $\{0, 1\}$, we obtain a random set A' on $\mathcal{P}(E)=\{0,1\}^E$—but obviously A' is not a RACS. We shall denote as $\mathcal{I}$ *the class of the finite subsets in E* and put, for any $I, I' \in \mathcal{I}$,

$$M_I^{I'} = \{A' : A' \in \mathcal{P}(E), I \subset A', I' \cap A' = \varnothing\}$$

Let $\mathcal{M}$ be the σ-algebra generated by the class $M_I^{I'}$, $I, I' \in \mathcal{I}$, with P' a probability on $\mathcal{M}$, and put

$$P'(I, I') = P'(M_I^{I'})$$

$$T'(I) = 1 - P'(M^I) = P'(\{A' \cap I \neq \varnothing\})$$

By the obvious recurrence formulae:

$$\begin{aligned} P'(\varnothing, I') &= 1 - T'(I') \\ P'(I \cup \{x\}, I') &= P'(I, I') - P'(I, I' \cup \{x\}) \end{aligned} \tag{2-4-1}$$

the function P' on $\mathcal{I} \times \mathcal{I}$ is determined if T' is given on $\mathcal{I}$. The two following conditions are obviously fulfilled by T':

1. $T'(\varnothing)=0$, and $0 \leqslant T' \leqslant 1$.
2. The function P' defined on $\mathcal{I} \times \mathcal{I}$ by formulae (2-4-1) is $\geqslant 0$.

Conversely, if a function T' on $\mathcal{I}$ verifies these two conditions, there exists a unique probability P' on $\mathcal{M}$ such that $P'(M^I) = 1 - T'(I)$ for any $I \in \mathcal{I}$. It follows from the classical Kolmogorov theorem applied to the particular case of a random indicator. Alternatively, we may remark that the class $M_I^{I'}$, $I, I' \in \mathcal{I}$ is a semialgebra $\mathcal{S}$ generating $\mathcal{M}$. By conditions 1 and 2, the function P' defined by (2-4-1) is additive and $0 \leqslant P' \leqslant 1$ on $\mathcal{S}$. Finally, by

$$\bigcap_{j \in J} M_{I_j}^{I'_j} = M_{\cup I_j}^{\cup I'_j}$$

$\mathcal{S}$ is also a compact class in $\mathcal{M}$. Thus, by the classical theorems, there exists a unique extension of P' by a probability on $\mathcal{M}$.

We shall say that a function T' on $\mathcal{I}$ satisfying conditions 1 and 2 above is a *space law*, and state:

Proposition 2-4-1. *If T' is a space law, there exists a unique probability P' on $\mathcal{M}$ such that the associated random set A' satisfies $P'(\{A' \cap I \neq \varnothing\}) = T'(I)$ for any $I \in \mathcal{I}$.*

Now let $A=(\mathcal{F}, \sigma_f, P)$ be a RACS and T be its functional, defined by $T(K)=P(\mathcal{F}_K)$, $K\in\mathcal{K}$. Clearly, the restriction of T to $\mathcal{I}$ satisfies conditions 1 and 2, and thus is a space law, with which is associated a probability P' on $\mathfrak{M}$ such that $P'(M_I{}^{I'})=P(\{I\subset A\}\cap\mathcal{F}^{I'})$. If φ is the one-to-one mapping $A\to\varphi(A)=A$ from $\mathcal{F}$ into $\mathcal{P}(E)$, the random set $\varphi(A)$ is clearly equivalent to $A'=(\mathcal{P}(E), \mathfrak{M}, P')$.

Then the following question arises: Under what conditions is it possible to associate a RACS A with a given space law T' in such a way that $T'(I)=P(A\in\mathcal{F}_I)$ for any $I\in\mathcal{I}$? Let us first give a partial answer:

Proposition 2-4-2. *Let T' be a space law. Then there exists a RACS A the space law of which is identical with T' if and only if T' is u.s.c. for the relative myope topology on $\mathcal{I}$.*

Proof. The "only if" part is obvious. Conversely, suppose T' to be u.s.c. on $\mathcal{I}$ and put $T(G)=\text{Sup}\ \{T'(I),\ I\in\mathcal{I},\ I\subset G\}$ for any $G\in\mathcal{G}$ and $T(K)=\text{Inf}\ \{T(G),\ G\in\mathcal{G},\ G\supset K\}$ for any $K\in\mathcal{K}$. It is easy to verify that T is l.s.c. on $\mathcal{G}$ and u.s.c. on $\mathcal{K}$ and satisfies the condition stated in the Choquet theorem Theorem 2-2-1. It remains to show that $T(I)=T'(I)$ for any $I\in\mathcal{I}$. Obviously, $T(I)\geqslant T'(I)$. Conversely, let $\{G_n\}\subset\mathcal{G}$ be a sequence such that $G_n\downarrow I$ and $T(G_n)\downarrow T(I)$, and $\epsilon>0$. For each $n>0$, there exists $I_n\in\mathcal{I}$ such that $I\subset I_n\subset G_n$ and $T'(I_n)+\epsilon\geqslant T(G_n)$. If $n\uparrow\infty$, the sequence I_n converges toward I in $\mathcal{K}$, so that $\overline{\lim}\ T'(I_n)\leqslant T'(I)$, because T' is u.s.c. on $\mathcal{I}$. On the other hand, $T'(I_n)\geqslant T'(I)$ yields $\underline{\lim}\ T'(I_n)\geqslant T'(I)$, and thus $T'(I)=\lim T'(I_n)\geqslant\lim\,(T(G_n)-\epsilon)=T(I)-\epsilon$. It follows that $T'(I)=T(I)$, and the proposition is proved.

Remark. The RACS A the space law of which is T' is never unique. For, if A' is a Poisson point process independent of A, the RACS $A\cup A'$ admits the same space law T'. We shall now improve Proposition 2-4-2, and show that there exists a unique RACS (up to an equivalence) which admits the space law T' and is separable in the following sense:

Definition 2-4-1. *We say that a RACS $A=(\mathcal{F}, \sigma_f, P)$ is separable if there exists a countable subset $D\subset E$ dense in E such that $A=\overline{A\cap D}$ a.s. The set D is called a separating subset for A.*

In order to justify this definition, we must show that the set $\{A=\overline{A\cap D}\}$ is measurable. For any $G\in\mathcal{G}$, $\overline{A\cap D}\cap G\neq\varnothing$ is equivalent to $A\cap D\cap G\neq\varnothing$. Thus, if $\mathfrak{B}$ is a countable base of $\mathcal{G}$, we have

$$\{A=\overline{A\cap D}\}=\bigcap_{B\in\mathfrak{B}}(\mathcal{F}^B\cup\mathcal{F}_{B\cap D})\in\sigma_f$$

Note also that A is separable and D is a separating subset for A if and only if $\mathcal{F}_G = \mathcal{F}_{G \cap D}$ a.s. for each $G \in \mathcal{G}$ (or each $B \in \mathcal{B}$), and that, if D is a separating subset for A, so is ever countable set $D' \supset D$.

Theorem 2-4-1. *Let T' be a space law, and T the function defined on $\mathcal{G}$ by putting $T(G) = \sup \{T'(I), I \in \mathcal{I}, I \subset G\}$, $G \in \mathcal{G}$, and on $\mathcal{K}$ by putting $T(K) = \operatorname{Inf} \{T(G), G \in \mathcal{G}, G \supset K\}$, $K \in \mathcal{K}$. Then:*

a. There exists a RACS $A = (\mathcal{F}, \sigma_f, P)$ such that $P(\mathcal{F}_K) = T(K)$ for any $K \in \mathcal{K}$, and T is the smallest function on $\mathcal{K}$ which satisfies the conditions of the Choquet theorem and $T' \leqslant T$ on $\mathcal{I}$.

b. This RACS A is separable. If D is a separating subset for A, and if $A' = (\mathcal{P}(E), \mathcal{M}, P')$ is the random set associated with the space law T', then A is equivalent to $\overline{A' \cap D}$.

c. There exists a RACS the space law of which is T' if and only if the following relationship holds for any $x \in E$:

$$(c) \qquad T'(\{x\}) = T(\{x\})$$

If so, $\overline{A' \cap D}$ is (up to an equivalence) the unique separable RACS the space law of which is T'.

Proof. *a.* It is easy to verify that the conditions of the Choquet theorem are fulfilled by T and $T \geqslant T'$ on $\mathcal{I}$. Let T'' be another capacity satisfying these conditions and such that $T'' \geqslant T'$ on $\mathcal{I}$. Then, by $T''(G) = \operatorname{Sup} \{T''(K), K \in \mathcal{K}, K \subset G\}$, we obtain $T'' \geqslant T$ on $\mathcal{G}$ and thus also on $\mathcal{K}$.

b. Let $\mathcal{B} \subset \mathcal{G}$ be a countable base of the topology $\mathcal{G}$ on E closed under finite union. For each $B \in \mathcal{B}$, let $\{I_n(B)\} \subset \mathcal{I}$ be an increasing sequence such that $I_n(B) \subset B$ and $T'(I_n(B)) \uparrow T(B)$. The set $D(B) = \cup_n I_n(B)$ is countable. $I_n(B) \uparrow D(B)$ implies

$$T'(I_n(B)) \uparrow P'(A' \cap D(B) \neq \varnothing)$$

so that $T(B) = P'(A' \cap D(B) \neq \varnothing)$. Then let us put $D = \cup \{D(B), B \in \mathcal{B}\}$. The set D is countable, and by $D \cap B \supset D(B)$ we obtain $T(B) \leqslant P'(A' \cap D \cap B \neq \varnothing)$ and thus

$$(b) \qquad T(B) = P'(A' \cap D \cap B \neq \varnothing) \qquad (B \in \mathcal{B})$$

for the converse inequality always holds by the definition of $T(B)$. For any $G \in \mathcal{G}$, there exists a sequence $\{B_n\} \subset \mathcal{B}$ such that $B_n \uparrow G$, because $\mathcal{B}$ is closed under finite union. But T is l.s.c. on $\mathcal{G}$, and it follows that $T(B_n) \uparrow T(G)$, by Proposition 1-2-4, Corollary 5. On the other hand, $B_n \uparrow G$ implies $\{A' \cap D \cap B_n \neq \varnothing\} \uparrow \{A' \cap D \cap G \neq \varnothing\}$, and thus $P'(A' \cap D \cap B_n$

$\neq\varnothing)\uparrow P'(A'\cap D\cap G\neq\varnothing)$. Then, by ($b$), we obtain

$$(b')\qquad T(G)=P'(A'\cap D\cap G\neq\varnothing)\qquad (G\in\mathcal{G})$$

The mapping α: $A'\to\alpha(A')=(\overline{A'\cap D})$ from $(\mathcal{P}(E), \mathfrak{M})$ into $(\mathcal{F}, \sigma_f)$ is measurable, for $\overline{A'\cap D}\cap G\neq\varnothing$ is equivalent to $A'\cap D\cap G\neq\varnothing$, $G\in\mathcal{G}$, and $\alpha^{-1}(\mathcal{F}_G)=\{A'\cap D\cap G\neq\varnothing\}\in\mathfrak{M}$. Thus the RACS $\alpha(A')$ exists and satisfies $P(\alpha(A')\in\mathcal{F}_G)=T(G)$, that is, is equivalent to A. Moreover, $\alpha(A')$ is separable and admits D as a separating subset, because $\alpha(A')\cap D\supset A'\cap D$ implies $\overline{\alpha(A')\cap D}\supset\alpha(A')$, that is, $\overline{\alpha(A')\cap D}=\alpha(A')$.

It remains to show that, for another separating subset D', $\overline{A'\cap D'}$ is still equivalent to A, that is, to prove the relationship

$$(b'')\qquad \overline{A'\cap D'}=\overline{A'\cap D}\qquad (P'\text{-a.s.})$$

Notice at first that $D''=D\cup D'$ is still a separating subset for A. This implies, for any $G\in\mathcal{G}$,

$$T(G)=P'(A'\cap G\cap D''\neq\varnothing)=P'(A'\cap G\cap D'\neq\varnothing)=P'(A'\cap G\cap D\neq\varnothing)$$

Thus, by the inclusions $\{A'\cap G\cap D'\neq\varnothing\}\subset\{A'\cap G\cap D''\neq\varnothing\}$ and $\{A'\cap G\cap D\neq\varnothing\}\subset\{A'\cap G\cap D''\neq\varnothing\}$, we get the a.s. equalities $\{A'\cap G\cap D''\}=\{A'\cap G\cap D'\}=\{A'\cap G\cap D\}$. It follows that ($b''$) holds.

c. By part a, T' is the space law of a RACS if and only if $T=T'$ on $\mathcal{I}$. This condition obviously implies relationship (c). Conversely, suppose that (c) holds. Let x be a point in E, and $D'=D\cup\{x\}$ be a separating subset for A including x. By (b''), we may write

$$\{x\in A'\}\subset\{x\in\overline{A'\cap D'}\}\overset{\text{a.s}}{=}\{x\in\overline{A'\cap D}\}$$

But, by condition (c), $P'(x\in A')=P'(x\in\overline{A'\cap D})$. It follows that $\{x\in A'\}=\{x\in\overline{A'\cap D}\}$ P'-a.s., and this implies that the space law of $A'\cap D$ is T'.

Finally, let $A''=(\mathcal{F}, \sigma_f, P'')$ be another separable RACS, the space law of which is T', and D'' be a separating subset for A. For any $G\in\mathcal{G}$, we get $P''(\mathcal{F}_G)=P''(\mathcal{F}_{G\cap D''})=\text{Sup}\ \{T'(I),\ I\in\mathcal{I},\ I\subset G\cap D''\}$. Hence $P''(\mathcal{F}_G)\leqslant T(G)$. But a strict inequality is impossible, by part a, and so $P''(\mathcal{F}_G)=T(G)$, that is, $P''=P$, so that A is unique up to an equivalence.

Remark. Let $A=(\mathcal{F}, \sigma_f, P)$ be a nonseparable RACS, $A_1=(\mathcal{F}, \sigma_f, P_1)$ be the separable RACS admitting the same space law as A, and D be a separating subset for A_1. The mapping α: $A\to\alpha(A)=\overline{A\cap D}$ from $(\mathcal{F}, \sigma_f)$ into itself is measurable and thus defines a RACS α. This RACS is clearly equivalent to A_1. By Section 2-3, for P_1-almost every α, there exists a conditional RACS $A_\alpha = (\mathcal{F}, \sigma_f, P_\alpha)$ such that

$$\int_H P_\alpha(\mathcal{V})\, P_1(d\alpha) = P(\mathcal{V} \cap \{\alpha(A) \in H\})$$

for any H and $\mathcal{V}$ in σ_f. A_α is a version of the RACS A conditional in $\alpha = (\overline{A \cap D})$. In other words, the probability P_α represents the residual indetermination that remains concerning A once we have all the information we can obtain by a countable set of observations. Obviously, $P_\alpha(A_\alpha \supset \alpha) = 1$ and $P_\alpha(x \in A_\alpha,\ x \notin A) = 0$ for any $x \in E$, (P_1-a.s. for every α).

2-5. A.S. CONTINUITY, P-CONTINUITY AND MEASURABILITY

Definition 2-5-1. *We say that a RACS $A = (\mathcal{F}, \sigma_f, P)$ is P-continuous at a point $x \in E$ if* $\lim P(\{x \in A\} \cap \{y \notin A\}) = \lim P(\{x \notin A\} \cap \{y \in A\}) = 0$ *for y converging toward x in E. We say that A is P-continuous if it is P-continuous at every $x \in E$.*

This definition depends only on the space law of A. Thus, more generally, we may define the P-continuity of a space law without referring to a RACS.

Proposition 2-5-1. *a. A space law T' is P-continuous at $x \in E$ if and only if* $\lim x_n = x$ *in E implies $T'(\{x\}) = \lim T'(\{x_n\})$ and $T'(\{x\}) = \lim T'(\{x, x_n\})$.*

b. Let A be a RACS and T its space law. Then A is P-continuous at $x \in E$ if and only if $x = \lim x_n$ in E implies $T(\{x\}) = \lim T(\{x_n\})$

c. Let $A' = (\mathcal{P}(E), \mathfrak{M}, P')$ be a P-continuous random set. Then $\overline{A' \cap D} = \overline{A' \cap D'}$ P'-a.s. for any countable subsets D, D' dense in E. Moreover, the separable RACS A of Theorem 2-4-1 admits every countable subset D dense in E as a separating subset, and A is equivalent to $\overline{A' \cap D}$.

d. If a RACS is P-continuous, it admits every countable subset dense in E as a separating subset.

Proof. Part *a* follows from $P(\{x \in A\} \cap \{y \notin A\}) = T'(\{x, y\}) - T'\{y\}$. If T is the space law of a RACS, $x = \lim x_n$ always implies $T(\{x\}) = \lim T(\{x, x_n\})$, because T is u.s.c. on $\mathcal{K}$, and part *b* follows.

In order to prove part c, we note first that the P-continuity implies $M_{\{x\}}^{G \cap D} = \emptyset$a.s.; thus $M^{G \cap D} \subset M^{G \cap D'}$ for any $x \in G \in \mathcal{G}$, and D, D'

countable subsets dense in E. In the same way, $M_{G\cap D}\subset M_{G\cap D'}$ a.s., so that $M^{G\cap D'}=M^{G\cap D}$ a.s. This implies $\overline{A'\cap D}=\overline{A'\cap D'}$ P'-a.s. The separable RACS A of Theorem 2-4-1 is equivalent to $\overline{A'\cap D}$ for a countable dense subset D, and thus also to $\overline{A'\cap D'}$ for any other countable dense subset D', provided we can verify that D' is a separating subset. But this is easy.
Finally, part d is a simple corollary of part c.

Remark. If $E=\mathbf{R}^d$ is a euclidean space, we say that a RACS A is *order* 2 (or *weakly*) *stationary* if its space law T satisfies $T(\{x,y\})=T(\{x\})+T(\{y\})-C(x-y)$ for a function C on $E=\mathbf{R}^d$ such that $C(x-y)=P(\mathcal{F}_{\{x\},\{y\}})$. By the preceding proposition, we have the following:

Corollary. *Any weakly stationary RACS A is P-continuous, and its random indicator* 1 *is continuous in quadratic mean (i.e.,* $\lim E[(1_A(x_n)-1_A(x))^2]=0$ *for any sequence* $\{x_n\}$ *converging toward a point* $x\in E$*).*

Proof. By the stationarity of the RACS A, $T(\{x\})=C(0)$ is constant, and thus A is P-continuous by condition b of the proposition. On the other hand, the random indicator 1_A of the RACS A is quadratic-mean continuous if and only if the covariance $C(x-y)=E[1_A(x)1_A(y)]$ is continuous, and this condition is always verified by the continuity of $(x, y)\to T(\{x,y\})$.

By Proposition 1-2-4, Corollary 3, the boundary mapping $A\to\partial A$ is l.s.c., and thus measurable, if E is locally connected. Hence the following:

Definition 2-5-2. *If E is a locally connected LCS space, we say that a RACS A on E is a.s. continuous at* $x\in E$ *if* $P(\{x\in\partial A\})=0$, *and A is a.s. continuous if it is a.s. continuous at any* $x\in E$.

The a.s. continuity is really only a property of the space law, as shown by the following:

Proposition 2-5-2. *A RACS A is a.s. continuous if and only if it is P-continuous and* $\overline{\mathring{A}\cap D}$ *is a.s. continuous for a countable dense subset* $D\subset E$. *If so, we have a.s.* $\overline{\mathring{A}}=\overline{\mathring{A}\cap D'}$ *for any countable subset D′ dense in E, and* $\overline{\mathring{A}}$ *is equivalent to the RACS obtained from the space law T of A as in Theorem* 2-4-1.

Proof. Clearly, the a.s. continuity implies the P-continuity. Thus we may suppose that A is P-continuous. If D is a countable subset dense in E, we get $\mathring{A} = \overline{\mathring{A} \cap D}$ a.s., and thus

$$\partial(\overline{A \cap D}) \subset \partial A \subset \partial(\overline{A \cap D}) \cup \left((\overline{A \cap D})^c \cap A \right)$$

Because A is P-continuous, Proposition 2-5-1 and Theorem 2-4-1 imply that $\overline{A \cap D}$ and A have the same space law. Then by $\overline{A \cap D} \subset A$, we obtain $P(x \in A \cap (\overline{A \cap D})^c) = 0$ for any $x \in E$, and thus

$$P(x \in \partial A) = P\left(x \in \partial(\overline{A \cap D})\right)$$

It follows that A is a.s. continuous if and only if $\overline{A \cap D}$ is so. If A is a.s. continuous, we have $A \cap D \subset \mathring{A}$ a.s. for any countable subset D dense in E, and thus $\overline{A \cap D} \subset \overline{\mathring{A}}$ a.s. But the inclusion $\overline{\mathring{A}} \subset \overline{A \cap D}$ always holds, so that $\overline{\mathring{A}} = \overline{A \cap D}$ a.s.

Let us now examine the *measurability* of a RACS, which will yield a more tractable criterion for the a.s. continuity.

Theorem 2-5-1. *Any RACS A is measurable in the following sense: the mapping k: $E \times \mathcal{F} \to \{0,1\}$, defined by putting $k(x,A) = 1$ if $x \in A$ and $k(x,A) = 0$ if $x \notin A$, is measurable for the product σ-algebra $\sigma(\mathcal{G}) \otimes \sigma_f$.*

Proof. Let $\mathcal{B}$ be a countable base of the topology $\mathcal{G}$ on E. Then $x \notin A$ is equivalent to {there exists $B \in \mathcal{B}$ such that $x \in B$ and $A \cap B = \varnothing$}. Thus

$$k^{-1}(0) = \bigcup_{B \in \mathcal{B}} B \times \mathcal{F}^B \in \sigma(\mathcal{G}) \otimes \sigma_f$$

Corollary 1. *Let μ be a positive measure on $(E, \sigma(\mathcal{G}))$. Then the mapping $A \to \mu(A) = \int k(x,A)\mu(dx)$ from $\mathcal{F}$ into $\mathbf{R}_+$ is a random variable $\geqslant 0$, the expectation of which is*

$$E(\mu(A)) = \int P(\{x \in A\})\mu(dx)$$

In the same way, if α is a measurable mapping from $\mathcal{F}$ into itself, $\mu(\alpha(A))$ is a random variable, the expectation of which is

$$E[\mu(\alpha(A))] = \int P(\{x \in \alpha(A)\})\mu(dx)$$

This corollary holds for all the measurable mappings we met in Chapter 1.

Let us, for instance, consider the boundary mapping $A \to \partial A$ (which is measurable, by $\partial A = A \cap \overline{A^c}$ and Proposition 1-2-4, Corollaries 1 and 2). By the preceding corollary, we conclude that *a RACS A is a.s. continuous if and only if* $\mu(\partial A)=0$ *a.s. for any measure* $\mu \geqslant 0$. *In particular*:

Corollary 2. *Let* $E=\mathbf{R}^d$ *be a euclidean space, and* μ *the Lebesgue measure on E. If the mapping* $x \to P(x \in \partial A)$ *is continuous on E, and, in particular, if A is stationary (i.e., P is invariant under the translations), then A is a.s. continuous if and only if* $\mu(\partial A)=0$ *a.s.*

Proof. Let μ be the Lebesgue measure on $E=\mathbf{R}^d$. The $E(\mu(\partial A))=0$ if and only if $P(x \in \partial A)=0$ for μ-almost every $x \in E$, that is, for every $x \in E$ if $x \to P(x \in \partial A)$ is continuous on E. In the stationary case, $P(x \in \partial A)$ is a constant, and this condition is automatically satisfied.

2-6. OPEN RANDOM SETS, AND RANDOM SETS ON $\mathcal{K}$

It is not difficult to define the notion of an open random set: let σ_g be the borelian tribe on $\mathcal{G}$, which is generated by the family $\mathcal{G}_K$, $K \in \mathcal{K}$, as well as by the family $\mathcal{G}^G$, $G \in \mathcal{G}$. Then an open random set is a measurable mapping from a probabilized space $(\Omega, \mathcal{Q}, P)$ into $(\mathcal{G}, \sigma_g)$. In the canonical form, an open random set is simply defined by a probability P on σ_g. The homeomorphism $F \to F^c$ from $\mathcal{F}$ onto $\mathcal{G}$ is obviously also an isomorphism for the σ-algebras σ_f and σ_g, so that the results obtained in the preceding section may immediately be translated by duality.

Let us now examine the case of the space $\mathcal{K}$ identified with the subspace $\{G \subset F\} \subset \mathcal{G} \times \mathcal{F}$, and suppose that Axiom 1-3-1 is satisfied (see Section 1-3). An element of $\mathcal{K}$ is the pair $(\mathring{A}, \overline{A})$, where $\mathring{A}$ is the interior and $\overline{A}$ the closure of a set $A \in \mathcal{P}(E)$. We may define an equivalence in $\mathcal{P}(E)$ by putting $A \equiv A'$ if $\mathring{A} = \mathring{A}'$ and $\overline{A} = \overline{A}'$, and identify the space $\mathcal{K}$ with the quotient space $\mathcal{P}(E)/\equiv$. From this point of view, two subsets A and A' in E are regarded as undistinguishable if they have the same interior and the same closure, a criterion which seems very suitable to the experimental conditions involved by the applications of the theory.

It is also possible to define on $\mathcal{P}(E)$ itself the σ-algebra σ_p generated by the class $M_G{}^{G'} = \{A,\ A \in \mathcal{P}(E),\ A \supset G,\ A \cap G' = \varnothing\}$, G', $G \in \mathcal{G}$. Because $A \supset G$ is equivalent to $\mathring{A} \supset G$, and $A \cap G' = \varnothing$ to $\overline{A}' \cap G = \varnothing$, we may also write

$$M_G{}^{G'} = \{A\colon (\mathring{A}, \overline{A}) \in \mathcal{G}_G \times \mathcal{F}^{G'}\}$$

so that the σ-algebra σ_p is compatible with the above equivalence relationship$\equiv$, and the quotient σ-algebra $\alpha_p/\equiv$ may be identified with the σ-algebra σ_h induced on $\mathcal{H}$ by the product σ-algebra $\sigma_g \otimes \sigma_f$. In other words, $(\mathcal{P}, \sigma_p)/\equiv = (\mathcal{H}, \sigma_h)$.

The results we have obtained in the spaces $\mathcal{F}$ and $\mathcal{G}$ may easily be translated in terms of space $\mathcal{H}$. In particular, A is separable if the RACS $\bar{A}$ and $\complement \mathring{A}$ are separable, and it is easy to prove the following:

Proposition 2-6-1. *If a random set $A = (\mathcal{H}, \sigma_h, P)$ is separable and a.s. continuous, we have $\bar{A} = \bar{\mathring{A}}$ and $\mathring{A} = \mathring{\bar{A}}$ a.s.*

These relationships characterize well-constructed sets, such that neither $\bar{A}$ nor $\complement \mathring{A}$ has isolated points. A model of this kind is very often suitable to practical applications. For instance, we may think that a porous medium will be described satisfactorily as a realization of a separable a.s. continuous random set on $(\mathcal{H}, \sigma_h)$.

2-7. EXAMPLES

Let A be a RACS. In practical applications, the following notations are often useful:

$$P(B) = P(\{B \subset A\}); \qquad Q(B) = P(\{B \cap A = \varnothing\}) = 1 - T(B)$$

($B \in \mathcal{G}$ or $B \in \mathcal{K}$). The set $\{F\colon F \in \mathcal{F},\ F \supset B\}$ is closed in $\mathcal{F}$ for any subset $B \subset E$, so that the probability $P(B)$ always exists.

It is generally difficult (or practically impossible) to form an explicit expression of P knowing Q, or conversely. However, in the particular case where $B \in \mathcal{G}$ is a *finite set*, we may write

$$Q(\{x_1, \ldots, x_n\}) = E\left(\prod_{i=1}^{n} 1_{A^c}(x_i) \right)$$

$$P(\{x_1, \ldots, x_n\}) = E\left(\prod_{i=1}^{n} 1_{A}(x_i) \right)$$

By $1_{A^c} = 1 - 1_A$ and easy calculation, we obtain

$$Q(\{x_1, \ldots, x_n\}) = 1 - \sum_{i=1}^{n} P(\{x_i\}) + \sum_{i_1 < i_2} P(\{x_{i_1}, x_{i_2}\})$$

$$+ \cdots + (-1)^n P(\{x_1, \ldots, x_n\})$$

and an analogous relationship concerning the functional P.

If only one point $x \in E$ is involved, we simply find $Q(x)=1-P(x)$. In the case of a porous medium, we may consider the union A of the grains of the medium as a realization of a RACS, and thus $Q(x)$ is the *porosity* of the point $x \in E$. When two points x_1 and x_2 are involved, we obtain

$$Q(x_1,x_2)=1-P(x_1)-P(x_2)+P(x_1,x_2)$$

We shall say that the function $(x_1, x_2) \to P(x_1, x_2)$ is the (noncentered) *covariance* of the RACS A (or, more precisely, of its random indicator 1_A). If E is a euclidean space and the probability P is invariant under translation, the RACS A is said to be *stationary*. By the Choquet theorem, A is stationary if and only if its functional T is invariant on $\mathcal{K}$ under the translations. If A is stationary, the covariance $P(x_1, x_2)$ depends not on the points x_1 and x_2 separately, but only on their difference. Thus we put

$$C(h)=P(x, x+h)$$

and we say that the function $h \to C(h)$ is the (stationary) covariance of the RACS A.

Let us examine a few transformation formulae. Suppose that E is a euclidean space, and K a compact set in E. Then we may define the RACS $A \oplus \check{K}$ (dilatation of A by K), and $A \ominus \check{K}$ (erosion of A by K). We have seen that $x \in A \oplus \check{K}$ is equivalent to $K_x \cap A \neq \varnothing$, and $x \in A \ominus \check{K}$ to $K_x \subset A$; thus

$$P(K_x)=P(K_x \subset A)=P(x \in A \ominus \check{K})$$

$$Q(K_x)=P(K_x \cap A=\varnothing)=P(x \in A \oplus \check{K})$$

In the stationary case, these probabilities do not depend on the point $x \in E$. Unfortunately, it is not so easy to form explicit expressions for $P(x \in A^K)$ or $P(x \in A_k)$.

Now arises the question of the *specific area* of a RACS A. If $K \in \mathcal{K}$, denote $V(\rho)=V(K \oplus \rho B)$, the volume of the dilatation of K by a ball $B\rho$ of center o and radius $\rho \geqslant 0$. The simplest procedure used in integral geometry for defining the area $S(K)$ of the compact set K is to put $S(K)=\lim (V(\rho)-V(0))/\rho$ for $\rho \downarrow 0$ (if this limit exists). It would be pleasant to translate this definition into probabilistic terms by putting

$$\sigma(x)=\lim_{\rho \downarrow 0} \frac{P(x \in A \oplus \rho B)-P(x \in A)}{\rho}$$

(if this limit exists), and to say that $\sigma(x)$ is the value at $x \in E$ of the specific area of our RACS A. In particular, $\sigma(x)$ is constant if A is stationary. However, this definition is not really satisfactory, because it is not connected with precise geometrical properties of the realization of the RACS A. In the general case, it is not an easy task to prove that a.s. existence of a random variable associated (in any sense) with the area or the specific area of a random closed set. We shall examine only a few specific cases in Chapters 4 and 5.

Granulometries

Let us now examine the granulometries of a RACS A and its complementary set A^c in respect to a nonempty convex compact set $B \in C(\mathcal{K}')$. By Proposition 1-5-6, the function $x \to \Lambda_A(x) = \text{Sup}\ \{\lambda:\ x \in A_{\lambda B}\}$ is a random function, the realizations (or trajectories or sample paths) of which are u.s.c. and we have $P(\Lambda_A(x) \geqslant \lambda) = P(x \in A_{\lambda B})$ for any $\lambda > 0$. In the same way, $x \to \Lambda_{A^c}(x) = \text{Sup}\ \{\lambda:\ x \in A^c_{\lambda B}\}$ is a random function the realizations of which are l.s.c., and $P(\Lambda_{A^c}(x) > \lambda) = P(x \notin A^{\lambda B})$ for any $\lambda \geqslant 0$. Thus the *distribution functions* of the random variables $\Lambda_A(x)$ and $\Lambda_{A^c}(x)$ represent at each $x \in E$ the granulometries of the RACS A and its complementary set A^c, that is, the size distributions of these random sets as measured at $x \in E$ according to the standard B. In the stationary case, these granulometric distributions do not depend on the point $x \in E$, and represent intrinsic characteristics of the RACS A itself.

Except in very particular cases (see Chapter 6 for an example) it is extremely difficult to express these granulometric distributions in terms of the functionals P and Q (for instance). However, there is a case in which this is possible, namely, when the compact B is $B = \{\lambda u,\ 0 \leqslant \lambda \leqslant 1\}$ for a given unit vector u in the euclidean space E. The corresponding granulometries are often called *linear granulometries*. Let us examine only the case of a *stationary* RACS A in $E = \mathbf{R}^d$.

Let α be the direction of a unit vector u, and $P_\alpha(l)$ the probability (independent of $x \in E$ by stationarity) for the set $\{x + tu,\ 0 \leqslant t \leqslant l\}$ to be enclosed in A. In the same way, let $P_\alpha(l, l')$ be the probability for the set $\{x + tu,\ -l' \leqslant t \leqslant l\}$ to be enclosed in A. By the stationarity, we obtain

$$(a) \qquad P_\alpha(l, l') = P_\alpha(l + l')$$

This probability can be associated with the random variables L and L' equal, respectively, to the distance between x and the closed sets $\overline{A^c \cap D_\alpha}$

and $\overline{A^c \cap D_{-\alpha}}$ ($D_\alpha = \{tu, t \geqslant 0\}$, and $D_{-\alpha} = \{tu, t \leqslant 0\}$). Clearly, we have

$$P_\alpha(l, l') = P(\{L \geqslant l\} \cap \{L' \geqslant l'\})$$

$$P_\alpha(l) = P(\{L \geqslant l\}) = P(\{L' \geqslant l\})$$

Now, let $\varpi(\lambda) = E(\exp(-\lambda L))$ and $\psi(\lambda, \mu) = E(\exp(-\lambda L - \mu L'))$ (λ $\mu \geqslant 0$) be the Laplace transforms of these variables. By elementary calculation, (a) yields

$$\psi(\lambda, \mu) = \frac{\mu \varpi(\mu) - \lambda \varpi(\lambda)}{\mu - \lambda}$$

If μ converges toward λ, we obtain $\psi(\lambda, \lambda) = \varpi(\lambda) + \lambda \varpi'(\lambda)$. This function $\Phi(\lambda) = \psi(\lambda, \lambda) = E[\exp(-\lambda(L + M'))]$ is the Laplace transform of $L + L'$. Let us denote by F_α the probability law of $L + L'$, so that

$$(b) \qquad \Phi(\lambda) = \int_0^\infty \exp(-\lambda x) F_\alpha(dx) = \frac{d}{d\lambda}(\lambda \varpi(\lambda))$$

Note that the event $\{L + L' \geqslant l\}$ is equivalent to $\{x \in A_l\}$ (i.e., x belongs to the opening A_l of A by the set $\{tu, 0 \leqslant t \leqslant l\}$). Hence the linear granulometry $P(x \in A_l)$ is given by

$$(c) \qquad P(x \in A_l) = 1 - F_\alpha(l)$$

On the other hand, (b) implies

$$\varpi(\lambda) = \frac{1}{\lambda} \int_0^\lambda \Phi(\mu) d\mu = \frac{1}{\lambda} \int_0^\infty \frac{1 - \exp(-\lambda x)}{x} F_\alpha(dx)$$

and thus

$$\frac{1 - \varpi(\lambda)}{\lambda} = \frac{1}{\lambda^2} \int_0^\infty \frac{\exp(-\lambda x) - 1 + \lambda x}{x} F_\alpha(dx)$$
$$= \int_0^\infty \exp(-\lambda x) dx \int_x^\infty \frac{y - x}{y} F_\alpha(dy)$$

The left-hand side of this relationship is the Laplace transform of $P_\alpha(l)$, and we conclude that

$$(d) \qquad P_\alpha(l) = \int_l^\infty \frac{y - l}{y} F_\alpha(dy)$$

The intuitive meaning of (d) is as follows: if a given point $x \in E$ is known to belong to an intercept of $A \cap (D_\alpha \cup D_{-\alpha})$, the length of which is

y, one of the extremities of this intercept may fall anywhere on the segment $\{x+tu,\ 0 \leqslant t \leqslant y\}$ with a uniform probability. Hence the conditional probability of $\{L \geqslant l\}$ is $(y-l)/y$ if $l \leqslant y$ (and 0 if $l > y$).

By (*d*), the function $P_\alpha(l)$ is expressed in terms of the granulometry F_α. Conversely, it is also interesting to express the granulometry F_α in terms of $P_\alpha(l)$. From (*d*), we note that the function P_α admits a derivative $P'_\alpha(l)$ at each $l > 0$; and rewriting (*b*), we obtain $(1-\Phi(\lambda))/\lambda = (1-\bar{\omega}(\lambda))/\lambda - \bar{\omega}'(\lambda)$. Hence, we obtain the desired expression of the granulometry:

$$1 - F_\alpha(l) = P_\alpha(l) - lP'_\alpha(l)$$

Note that the linear granulometry F_α is *length weighted.* In other words, for estimating $1 - F_\alpha(l)$ from a finite number of measured intercepts, we must assign to each of these intercepts a weight proportional to its length l. The experimental histogram, in which each observed intercept is counted as an individual, is associated with the "number granulometry," that is, the probability law $(C/l)F_\alpha(dl)$, where the constant C is defined by $1/C = \int_0^\infty (1/l)F_\alpha(dl)$, provided that this integral is $< \infty$. But this integral is not necessarily convergent. If $\int_0^\infty (1/l)F_\alpha(dl) = \infty$, there exist an infinite number of infinitely small intercepts, and the "number granulometry" cannot exist (on the contrary, the length-weighted granulometry F_α always exists). Nevertheless, for any $l > 0$, the number $N(l)$ of intercept lengths $\geqslant l$ the origins of which fall into a given unit segment is $\leqslant 1/l$, and it is always possible to define a *truncated* number granulometry. In fact, the event $\{L' \in (l, l+dl)\}$ [the probability of which is $-P'_\alpha(l)dl$] is equivalent to {the interval $(x, x+dl)$ contains the origin of an intercept length $\geqslant l$ included in A}, and we obtain

$$E(N(l)) = -P'_\alpha(l)$$

On comparing with (*d*), we conclude that $E(N(l)) = \int_l^\infty (1/y)F_\alpha(dy)$. In particular, the average number of intercepts, that is, $E(N(0))$ is finite if and only if $1/C = \int_0^\infty (1/y)F_\alpha(dy) < \infty$. If so, the number granulometry is $(C/l)F_\alpha(dl)$, as follows from

$$\frac{E(N(l))}{E(N(0))} = C\int_l^\infty \frac{1}{y}F_\alpha(dy)$$

If $1/C = \infty$, the number granulometry truncated at l_0 may be defined by

$$\frac{E(N(l))}{E(N(l_0))} = \frac{P'_\alpha(l)}{P'_\alpha(l_0)}$$

CHAPTER 3

Infinitely Divisible Random Closed Sets (IDRACS)

In this chapter, we examine a particular class of RACS, namely, the RACS infinitely divisible for the union $\cup$—IDRACS, for brevity. The first section characterizes the class of the functional T or Q associated with the IDRACS, and in the second section it is shown that any IDRACS is the union of the closed sets belonging to a Poisson process in the space $\mathcal{F}' = \mathcal{F} \setminus \{\varnothing\}$, and conversely. As examples, we examine the case of a Poisson process in $\mathcal{K}'$ (boolean model with compact grains), the RACS stable for the union in a euclidean space, and the Poisson flats.

In this chapter, the space E is LCS but not compact, so that $\mathcal{F}' = \mathcal{F} \setminus \{\varnothing\}$ itself is LCS and not compact. Rather than T, we use the functional $Q = 1 - T$, defined by $Q(B) = P(\mathcal{F}^B) = P(A \cap B = \varnothing)$. In particular, Q is decreasing, l.s.c. on $\mathcal{K}$, and u.s.c. on $\mathcal{G}$. We also recall that two RACS A_1 and A_2, the functionals of which are Q_1 and Q_2, are independent if and only if, for any $K_1, K_2 \in \mathcal{K}$,

$$P(\{A_1 \in \mathcal{F}^{K_1}\} \cap \{A_2 \in \mathcal{F}^{K_2}\}) = Q_1(K_1) Q_2(K_2) \qquad (3\text{-}0)$$

3-1. CHARACTERIZATION OF THE FUNCTIONAL Q ASSOCIATED WITH AN IDRACS

Let A be a RACS and Q the functional on $\mathcal{K}$, defined by $Q(K) = P(\mathcal{F}^K) = P(\{A \cap K = \varnothing\})$, $K \in \mathcal{K}$. We say that A is infinitely divisible for $\cup$ or is an IDRACS if, for any integer $n > 0$, A is equivalent to the union $\cup A_i'$ of n independent RACS A_i', $i = 1, 2, \ldots, n$, equivalent to each other. This property depends only on the functional Q, for, when putting $Q_n(K) = P(\{A_i' \cap K = \varnothing\})$, we obtain by (3-0) $Q = (Q_n)^n$, so that A is an IDRACS if and only if the function $T_n = 1 - Q^{1/n}$ satisfies the conditions of the Choquet theorem.

In order to characterize these infinitely divisible functionals Q, we must first examine the case in which A admits *fixed points* [definition: x_0 is a fixed point for A if $P(\{x_0 \in A\})=1$, i.e., $Q(\{x_0\})=0$].

Lemma 3-1-1. *Let A be an IDRACS. Then the set F of its fixed points is closed in E, and the functional Q associated with A satisfies $Q(K)>0$ strictly for any $K \in \mathcal{K}$ disjoint of F.*

Proof. By the lower semicontinuity of Q on $\mathcal{K}' = \mathcal{K} \setminus \{\varnothing\}$, the set of the $K \in \mathcal{K}'$ such that $Q(K)=0$ is inductive with respect to the inclusion. Thus, by the Zorn theorem, if $Q(K)=0$, there exists a minimal $K_0 \in \mathcal{K}'$ such that $K_0 \subset K$ and $Q(K_0)=0$. If we prove that any minimal compact set K_0 is punctual (i.e., reduced to a single point which is obviously a fixed point), the lemma will follow, for the union F of these fixed points x_0 is a closed set F (because A is a RACS or Q is l.s.c.) and any $K \in \mathcal{K}$ such that $Q(K)=0$ will include a minimal compact, that is, a fixed point $x_0 \in F$, so that $K \cap F \neq \varnothing$.

Let K_0 be a minimal compact set such that $Q(K_0)=0$. We will show that K_0 is a point. Suppose that there exist two distinct points in K_0. Then there exist $K_1, K_2 \in \mathcal{K}'$ such that $K_0 = K_1 \cup K_2$, $K_1 \neq K_2$. But K_0 is minimal, and thus $Q(K_1)>0$ and $Q(K_2)>0$. By $\mathcal{F}^{K_0} = \mathcal{F}^{K_1 \cup K_2} = \varnothing$ a.s., we also have

$$P(\mathcal{F}^{K_1} \cup \mathcal{F}^{K_2}) = Q(K_1) + Q(K_2)$$

Now A is an IDRACS, and $Q_n = Q^{1/n}$ is the functional of a RACS A_n which obviously admits the same minimal compact sets as A itself. If P_n is the probability on σ_f associated with A_n, it follows that

$$P_n(\mathcal{F}^{K_1} \cup \mathcal{F}^{K_2}) = (Q(K_1))^{1/n} + (Q(K_2))^{1/n}$$

By $Q(K_1)>0$ and $Q(K_2)>0$, the right-hand side of this equation is >1 for n large enough, which is impossible. Thus K_0 is a point, and the lemma is proved.

By substituting $E \cap F^c$ (which is still a LCS space) for E and the RACS $A \cap F^c$ in $\mathcal{F}(E \cap F^c)$ for A, we may suppose that A is *without fixed points*, that is, $Q>0$ strictly on $\mathcal{K}$.

Lemma 3-1-2. *Let T be a capacity satisfying the conditions of the Choquet theorem, Theorem 2-2-1, and $G(s)=\Sigma_0^\infty p_n s^n$ be the generating function of a probability concentrated on the set of the integers $\geqslant 0$. Put $Q=1-T$. Then $1-G(Q)$ is a capacity and satisfies the conditions of the Choquet theorem.*

Proof. By Theorem 2-2-1, Q is the functional associated with a RACS, and, for any integer $N>0, Q^N$ is the functional associated with the union of N independent RACS equivalent to A. Let $\{A_n\}$ be a sequence of independent RACS equivalent to A, and N a random variable independent of the A_n such that $P(N=n)=P_n$. It is easy to verify that the mapping $(N,A_1,A_2,\ldots)\to A_1\cup\cdots\cup A_N$ from $\mathbf{N}\times\mathcal{F}^{\mathbf{N}}$ into $\mathcal{F}$ is measurable, and for $K\in\mathcal{K}$

$$P(\{A_1\cup\cdots\cup A_N\in\mathcal{F}^K\})=\sum p_n(Q(K))^n=G(Q(K))$$

By the Choquet theorem, the lemma follows.

Corollary. *If T is a capacity satisfying the conditions of the Choquet theorem, so also is $1-exp(-\lambda T)$ for any $\lambda>0$.*

Proof. Take $G(s)=\exp(-\lambda(1-s))$, that is, the generating function of the Poisson law.

Theorem 3-1-1. *Let Q be a functional on $\mathcal{K}$. There exists an IDRACS A without fixed points such that $Q(K)=P(\{A\cap K=\emptyset\}$, $K\in\mathcal{K}$, if and only if $Q=\exp(-\psi)$ for an alternating capacity of infinite order ψ with $\psi(\emptyset)=0$ and $\psi(K)<\infty$ for any $K\in\mathcal{K}$.*

Proof. Let A be an IDRACS without fixed points. By lemma 3-1-1, Q is >0 on $\mathcal{K}$, and $T_n=1-Q^{1/n}$ is a Choquet capacity. For any $K\in\mathcal{K},\psi(K)=-\log Q(K)=\lim(nT_n(K))$ for $n\uparrow\infty$, and $\psi(\emptyset)=0,\psi(K)<\infty$. The functional ψ is u.s.c., because Q is l.s.c. and >0 on $\mathcal{K}$. On the other hand, the functionals nT_n are Choquet capacities, so that the functions S associated with ψ by (2-1-3) are $\geqslant 0$, because they are the limits of the corresponding functions associated with nT_n. Thus ψ is a Choquet capacity.

Let us now prove the "if" part. Let ψ be a Choquet capacity satisfying $\psi(\emptyset)=0$ and $\psi(K)<\infty$ for any $K\in\mathcal{K}$. Put $Q=\exp(-\psi)$. Obviously, $Q(\emptyset)=1$ and $Q(K)\geqslant 0$, so that the first condition of Theorem 2-2-1 is satisfied by $1-Q$. Let $\{B_n\}\subset\mathcal{G}$ be a sequence such that $B_n\uparrow E$, $\overline{B}_n\in\mathcal{K}$, and $\overline{B}_n\subset B_{n+1}$ for any $n>0$. For each n, define a functional Q_n on $\mathcal{K}(\overline{B}_n)$ by putting $Q_n(K)=Q(K)=\exp(-\psi(K)),K\in\mathcal{K},K\subset\overline{B}_n$. For any $K\in\mathcal{K}(\overline{B}_n),\psi(K)\leqslant\psi(\overline{B}_n)<\infty$. If $\psi(\overline{B}_n)=0$ for all $n>0,Q=1$ is the functional associated with an IDRACS a.s. empty. Suppose that $\psi(\overline{B}_n)>0$ for n large enough. The capacity $\psi/\psi(\overline{B}_n)$ satisfies on $\mathcal{K}(\overline{B}_n)$ the conditions of Lemma 3-1-2, and by the corollary of this lemma, applied with $\lambda=\psi(\overline{B}_n),1-\exp(-\psi)$ is a Choquet capacity and satisfies on $\mathcal{K}(\overline{B}_n)$ the conditions of Theorem 2-2-1. Thus there exists a RACS A_n in $\mathcal{F}(\overline{B}_n)$ such that $Q_n(K)=P(\{A_n \cap K=\varnothing\})$, $K \in \mathcal{K}(\overline{B}_n)$. But condition (2-3-6) is satisfied by the sequence $\{Q_n\}$, so that Q is the functional on $\mathcal{K}(E)$ associated with the inductive limit A of the sequence $\{A_n\}$. A is an IDRACS, for, if n is an integer >0, ψ/n satisfies the same conditions as ψ itself, so that $Q^{1/n}=\exp(-\psi/n)$ is still the functional associated with a RACS. By Lemma 3-1-1, A admits no fixed points, because $\psi(K)<\infty$ for any $K \in \mathcal{K}$ implies $Q(K)>0$.

3-2. POISSON PROCESSES AND σ-FINITE MEASURES ON $\mathcal{F}'=\mathcal{F}\backslash\{\varnothing\}$

The space $\mathcal{F}'=\mathcal{F}\backslash\{\varnothing\}$ is LCS for the relative topology induced by $\mathcal{F}$, but it is not compact since E is assumed to be noncompact. In $\mathcal{F}$, the neighborhoods of $\varnothing$ are the $\mathcal{F}^K, K \in \mathcal{K}$, so that each compact subset of $\mathcal{F}'$ is included in a $\mathcal{F}_K, K \in \mathcal{K}$. Let $\{B_n\}$ be a sequence in $\mathcal{K}$ such that $B_n \subset \mathring{B}_{n+1}$ and $B_n \uparrow E$. Then a σ-finite measure θ on $\mathcal{F}'$ is determined if its restriction to $\mathcal{F}_{B_n}$ is known for each $n>0$. Conversely, if for any $n \geqslant 0$ a measure θ_n is given on $\mathcal{F}_{B_n}$ and $\theta_{n+m}(\mathcal{V}')=\theta_n(\mathcal{V}')$ for $n,m>0$ and $\mathcal{V}' \in \sigma_f$, $\mathcal{V}' \subset \mathcal{F}_{B_n}$, the formula

$$\theta(\mathcal{V})=\lim_n \theta_n(\mathcal{V} \cap \mathcal{F}_{B_n}) \tag{3-2-1}$$

defines a σ-finite measure θ on $\mathcal{F}'$.

With each positive σ-finite measure θ on $\mathcal{F}'$ is associated a Poisson process in $\mathcal{F}'$. The union A of the closed sets belonging to this poisson process is almost surely closed in E, because $\theta(\mathcal{F}_{B_n})<\infty$ implies that each B_n a.s. hits only a finite number of these closed sets. Moreover, A is a RACS, for $A \in \mathcal{F}^K, K \in \mathcal{K}$, is equivalent to {there is no element of the Poisson process in $\mathcal{F}_K$}, and the probability of this measurable event is

$$Q(K)=\exp[-\theta(\mathcal{F}_K)] \tag{3-2-2}$$

Obviously, A is an IDRACS without fixed points, because Q does not vanish on $\mathcal{K}$. Conversely, by Theorem 3-1-1, any IDRACS without fixed points is characterized by its functional $K \to Q(K)=\exp(-\psi(K)), K \in \mathcal{K}$, where ψ is a Choquet capacity such that $\psi(\varnothing)=0$ and $\psi<\infty$ on $\mathcal{K}$. By the Choquet theorem, Theorem 2-2-1, it is easy to show that there exists a unique σ-finite measure θ on $\mathcal{F}'$ satisfying $\theta(\mathcal{F}_K)=\psi(K), K \in \mathcal{K}$. Thus we may identify the class of the σ-finite measures $\geqslant 0$ on $\mathcal{F}'$ with the class of the IDRACS without fixed points and also with the class of the Poisson processes on $\mathcal{F}'$ such that, for any $K \in \mathcal{K}$, $\mathcal{F}_K$ a.s. contains only a finite number of elements of this process. Let us state these results.

Proposition 3-2-1.   *Let A be a RACS, and $K \to Q(K) = P(\{A \cap K = \varnothing\})$ be the associated functional Q on $\mathcal{K}$. Then the three following conditions are equivalent:*

a. A is an IDRACS without fixed points.

b. There exists a σ-finite measure $\theta \geqslant 0$ on $\mathcal{F}'$ (necessarily unique) such that $\theta(\mathcal{F}_K) = -\log Q(K)$ for any $K \in \mathcal{K}$.

c. A is equivalent to the union of a Poisson process locally finite in $\mathcal{F}'$ (i.e., for each $K \in \mathcal{K}, \mathcal{F}_K$ a.s. includes only a finite number of elements of this process).

Note. We may generalize the procedure used above for constructing an IDRACS. Let $\mathcal{V}$ be a closed subset of $\mathcal{F}'$. Then the union $V = \cup \{F \in \mathcal{V}\}$ is closed in E, and the mapping $\mathcal{V} \to V$ from $\mathcal{F}(\mathcal{F}')$ into $\mathcal{F}(E)$ is continuous, for $V \in \mathcal{F}^B$ is equivalent to $\mathcal{V} \cap \mathcal{F}_B = \varnothing$. Thus, to each RACS $\mathcal{A} = (\mathcal{F}(\mathcal{F}'), \sigma_f(\mathcal{F}'), \tilde{P})$ is associated the RACS $A = \cup \{F, F \in \mathcal{A}\}$, the functional Q of which is $K \to Q(K) = \tilde{P}(A \cap K = \varnothing) = \tilde{Q}(\mathcal{F}_K)$. In particular, if $\mathcal{A}$ is the Poisson process on $\mathcal{F}'$ associated with a σ-finite measure $\theta \geqslant 0$, we get $Q(K) = \tilde{Q}(\mathcal{F}_K) = \exp(-\theta(\mathcal{F}_K))$.

The results of Section 2-3 concerning conditional probabilities on σ_f hold partially for σ-finite measures and will be very useful in the sequel. Again let $\{B_n\} \subset \mathcal{K}$ be a sequence such that $B_n \subset \mathring{B}_{n+1}$ and $B_n \uparrow E$, and θ be a σ-finite positive measure on $\mathcal{F}'$. We suppose that $\theta \neq 0$, and denote by ψ the Choquet capacity $K \to \psi(K) = \theta(\mathcal{F}_K), K \in \mathcal{K}$. By $\theta \neq 0$, we have $\psi(B_n) > 0$ for n large enough, and we may suppose that $\psi(B_n) > 0$ for any $n > 0$. For each integer n, we define a probability P_n on σ_f by putting

$$P_n(\mathcal{V}) = \frac{\theta(\mathcal{V} \cap \mathcal{F}_{B_n})}{\psi(B_n)} \qquad (\mathcal{V} \in \sigma_f)$$

Let u be a measurable mapping from $(\mathcal{F}, \sigma_f)$ into a space $(\Omega, \mathcal{H})$, and F_n the probability on $\mathcal{H}$ defined by $F_n(H) = P_n(u^{-1}(H)), H \in \mathcal{H}$. By Section 2-3, there exists for F_n-almost every $\omega \in \Omega$ a probability $P_{u,n}(.; \omega)$ such that, for any $H \in \mathcal{H}$ and $\mathcal{V} \in \sigma_f$,

$$\theta(\mathcal{V} \cap \mathcal{F}_{B_n} \cap u^{-1}(H)) = \psi(B_n) \int_H P_{u,n}(\mathcal{V}; \omega) F_n(d\omega) \qquad (3\text{-}2\text{-}3)$$

Let us now define a probability G on $\mathcal{H}$ by putting

$$G = \sum_{n=1}^{\infty} \frac{1}{2^n} F_n$$

For each $n>0$, F_n is absolutely continuous with respect to G, so that there exists a sequence $\{\varphi_n\}$ of measurable functions on Ω such that $F_n=\varphi_n G$. If $\mathcal{V}\subset\mathcal{F}_{B_n}$ and $m>n$, (3-2-3) implies $\psi(B_n)P_{u,n}(\mathcal{V};.)F_n=\psi(B_m)$ $P_{u,m}(\mathcal{V};.)F_m$, so that the following relationship holds for G-almost every $\omega\in\Omega$ and any $\mathcal{V}\in\sigma_f$, $\mathcal{V}\subset\mathcal{F}_{B_n}$.

$$\varphi_n\psi(B_n)P_{u,n}(\mathcal{V};.)=\varphi_m\psi(B_m)P_{u,m}(\mathcal{V};.)$$

Thus, by putting for any $\mathcal{V}\in\sigma_f$

$$\theta_u(\mathcal{V};\omega)=\lim_{n\uparrow\infty}\varphi_n(\omega)P_{u,n}(\mathcal{V};\omega)\psi(B_n)$$

we define for G-almost every $\omega\in\Omega$ a σ-finite measure $\theta_u(.;\omega)\geqslant 0$ on $\mathcal{F}'$, and, by (3-2-3), for any $\mathcal{V}\in\sigma_f$ and $H\in\mathcal{H}$,

$$\theta(\mathcal{V}\cap u^{-1}(H))=\int_H\theta_u(\mathcal{V};\omega)G(d\omega) \tag{3-2-4}$$

By this relationship, we see that the notion of a conditional probability on σ_f may be extended to the case of the σ-finite measures $\geqslant 0$ on $\mathcal{F}'$. However, note that the probability G on $\mathcal{H}$ is no longer uniquely determined, for, if φ is a measurable function on Ω with $\varphi>0$ (G-a.s.) and $\int\varphi(\omega)G(d\omega)=1$, we may substitute φG and θ_u/φ for G and θ_u. Let us state these results.

Proposition 3-2-2. *Let θ be a σ-finite positive measure on $\mathcal{F}'$, and u a measurable mapping from $\mathcal{F}'$ into a measurable space $(\Omega,\mathcal{H})$. Then there exist a probability G on $\mathcal{H}$ and, for G-almost every $\omega\in\mathcal{H}$, a σ-finite measure $\theta_u(.;\omega)\geqslant 0$ on $\mathcal{F}'$ such that, for any $H\in\mathcal{H}$ and $\mathcal{V}\in\sigma_f$,*

$$\theta(\mathcal{V}\cap u^{-1}(H))=\int_H\theta_u(\mathcal{V};\omega)G(d\omega) \tag{3-2-4}$$

If $\theta(\mathcal{V})=0$, then $\theta_u(\mathcal{V};\omega)=0$ (G-a.s.). Moreover, for G-almost every $\omega\in\Omega$, we have $u(F)=\omega$ for $\theta_u(.;\omega)$-almost every $F\in\mathcal{F}'$.

Corollary. *Let φ be a measurable mapping from $\mathcal{F}'$ into itself. If θ is φ-invariant [i.e., $\theta(\mathcal{V})=\theta(\varphi^{-1}(\mathcal{V}))$ for any $\mathcal{V}\in\sigma_f$], and u is φ-invariant θ-a.e. [i.e., $\theta(\{u\circ\varphi\neq u\})=0$], then, for G-almost every $\omega\in\Omega$, the measure $\theta_u(.;\omega)$ is φ-invariant.*

Proof. Under the hypotheses stated in the corollary, for any $H\in\mathcal{H}$ and $\mathcal{V}\in\sigma_f$, we obtain $\theta(\mathcal{V}\cap u^{-1}(H)=\theta(\varphi^{-1}(\mathcal{V})\cap\varphi^{-1}u^{-1}(H))$ $=\theta(\varphi^{-1}(\mathcal{V})\cap u^{-1}(H))$, and thus, by (3-2-4), $\theta_u(\mathcal{V};\omega)=\theta_u(\varphi^{-1}(\mathcal{V});\omega)$ for G-almost every $\omega\in\Omega$.

In Proposition 3-2-2, θ is represented in terms of a probability G on $\mathcal{H}$ and a conditional σ-finite measure θ_u on $\mathcal{F}'$. In certain circumstances, however, it may be useful to proceed in the opposite manner, that is, to represent θ with the help of a σ-finite measure μ on Ω, and a conditional probability P_u on $\mathcal{F}'$. But this is not always possible.

Proposition 3-2-3. *With the same notations as in Proposition* 3-2-2, *suppose that there exists a sequence* $\{H_n\} \subset \mathcal{H}$ *such that* $H_n \uparrow \Omega$ *and* $0 < \theta(u^{-1}(H_n)) < \infty$ *for every* $n > 0$. *Then there exist a σ-finite measure* $\mu \geqslant 0$ *on* $\mathcal{H}$ *and, for μ-almost every* $\omega \in \Omega$, *a probability* $P_u(.;\omega)$ *on* σ_f *such that, for any* $H \in \mathcal{H}$ *and* $\mathcal{V} \in \sigma_f$,

$$\theta(\mathcal{V} \cap u^{-1}(H)) = \int_H P_u(\mathcal{V};\omega)\mu(d\omega) \tag{3-2-5}$$

Proof. Under the stated conditions, for each $n > 0$, we may define a probability P_n on σ_f by putting

$$P_n(\mathcal{V}) = \frac{\theta(\mathcal{V} \cap u^{-1}(H_n))}{\theta(u^{-1}(H_n))} \qquad (\mathcal{V} \in \sigma_f)$$

By Section 2-3, there exist a probability F_n on $\mathcal{H}$ and, for F_n-almost every $\omega \in \Omega$, a conditional probability $P_{u,n}(.;\omega)$ on σ_f such that

$$P_n(\mathcal{V} \cap u^{-1}(H)) = \int_H P_{u,n}(\mathcal{V};\omega) F_n(d\omega)$$

for any $\mathcal{V} \in \sigma_f$ and $H \in \mathcal{H}$. Now, we put $\mu_n = \theta(u^{-1}(H_n))F_n$. It follows for $H \subset H_n$ that

$$\theta(\mathcal{V} \cap u^{-1}(H)) = \int_H P_{u,n}(\mathcal{V},\omega)\mu_n(d\omega)$$

If $H \subset H_n$, we have $\mu_n(H) = \theta(u^{-1}(H))$, so that $\mu_n = 1_{H_n}\mu_{n+m}$ for any $m > 0$. Thus, by putting $\mu(H) = \lim \uparrow \mu_n(H \cap H_n)$, we define on $\mathcal{H}$ a σ-finite measure μ, the restriction of which to each H_n is μ_n, and

$$\theta(\mathcal{V} \cap u^{-1}(H)) = \int_H P_{u,n}(\mathcal{V};\omega)\mu(d\omega)$$

if $H \subset H_n$. It follows that there exists a probability P_u such that $P_{u,n} \uparrow P_u$, and (3-2-5) holds.

Corollary 1. *For any* $\mathcal{V} \in \sigma_f$, $\theta(\mathcal{V}) = 0$ *implied* $P_u(\mathcal{V};\omega) = 0$ *for μ-almost every* $\omega \in \Omega$.

Corollary 2. *Let $A_u(\omega)$ be the RACS defined by the probability $P_u(.;\omega)$. For μ-almost every $\omega \in \Omega, u(A_u(\omega)) = \omega(P_u(.;\omega)$-a.s.).*

Corollary 3. *Suppose that there exists a measurable mapping φ from $\mathcal{F}'$ into itself such that θ is φ-invariant, and a one-to-one mapping f from Ω into itself, f being measurable as well as the reciprocal mapping f^{-1}, such that $u \bigcirc \varphi = f \bigcirc u$ θ-a.e. in $\mathcal{F}'$. Then, for μ-almost every $\omega \in \Omega$, the following relationship holds for any $\mathcal{V} \in \sigma_f$:*

$$P_u(\mathcal{V};\omega) = P_u(\varphi^{-1}(\mathcal{V}); f^{-1}(\omega))$$

The proof is the same as for the analogous results stated in Section 2-3.

3-3. STATIONARY BOOLEAN MODELS IN A EUCLIDEAN SPACE

Now we suppose that E is a *euclidean space*, and the σ-finite measure $\theta \geqslant 0$ on $\mathcal{F}'$ is concentrated on $\mathcal{K}' = \mathcal{K} \setminus \{\varnothing\}$, that is, $\theta(\mathcal{F}' \setminus \mathcal{K}') = 0$. By Proposition 3-2-1, a Poisson process $\mathcal{Q}$ on $\mathcal{K}'$ is associated with θ, and the IDRACS $A = \cup \{K : K \in \mathcal{Q}\}$ is called a *boolean model*. If θ is translation-invariant, clearly the boolean model A is stationary.

Let $K \in \mathcal{K}'$ be a nonempty compact set, and $u(K)$ be the center of the ball circumscribing K. It is easy to verify that the mapping $K \to u(K)$ is continuous for the relative myope topology on $\mathcal{K}'$, and thus measurable for $\sigma_f \cap \mathcal{P}(\mathcal{K}')$. We get an extension of u on $\mathcal{F}'$ by putting, for instance, $u(F) = 0$ for any $F \in \mathcal{F}' \setminus \mathcal{K}'$, and then u is a measurable mapping from $\mathcal{F}'$ into E.

Now, we suppose that the IDRACS A is stationary, that is, the σ-finite measure θ is translation invariant. Then the condition of Proposition 3-2-3 is satisfied by the mapping $u: \mathcal{F}' \to E$, for let B be a bounded and semiopen cube in E, and $\{y_k\} \subset E$ a sequence such that, by putting $B_k = B \oplus \{y_k\}$ and $B'_k = B \oplus \{-y_k\}$, the sequences $\{B_k\}$ and $\{B'_k\}$ are two partitions of the euclidean space E. By the translation invariance, we may write

$$\infty > \theta(\mathcal{F}_B) = \sum_k \theta(u^{-1}(B_k) \cap \mathcal{F}_B) = \sum_k \theta\big(u^{-1}(B) \cap \mathcal{F}_{B'_k}\big)$$

$$\geqslant \theta\left((u^{-1}(B)) \cap \left(\bigcup_k \mathcal{F}_{B'_k}\right)\right) = \theta(u^{-1}(B))$$

It follows that $\theta(u^{-1}(K)) < \infty$ for any $K \in \mathcal{K}'$. Then, by Proposition 3-2-3, there exists a σ-finite measure $\mu \geqslant 0$ on E and, for μ-almost every

$x \in E$, a probability $P_u(.;x)$ on σ_f such that (3-2-5) holds. By Corollary 1 of Proposition 3-2-3, $P_u(.;x)$ is concentrated on $\mathcal{K}'$. For a given $h \in E$, let φ be the mapping $F \to F_h = F \oplus \{h\}$ from $\mathcal{F}'$ onto itself (i.e., the h-translation). By the hypothesis, θ is φ-invariant, and $u(K_h) = u(K) + h$ for any $K \in \mathcal{K}'$, so that $u \circ \varphi = f \circ u$ θ-a.e. on $\mathcal{F}'$ if f denotes the translation $x \to f(x) = x + h$. Thus, by Proposition 3-2-3, Corollary 3, μ is translation-invariant, that is, is *proportional to the Legesgue measure dx on E*, say $\mu(dx) = a\,dx$ for a constant $a > 0$, and the conditional probability P_u (defined a.e. in E) satisfies

$$P_u(\mathcal{V}; x) = P_u(\mathcal{V}_{-x}; 0)$$

($\mathcal{V} \in \sigma_f$, $\mathcal{V}_{-x} = \{F, F_x \in \mathcal{V} \cap \mathcal{F}'\}$). In other words, the RACS $A(x)$, the probability of which is $P_o(.;x)$, is equivalent to the translate $A_0 \oplus \{x\}$, where $A_0 = A(0)$ is the a.s. compact RACS associated with the probability $P_u(.,0)$.

The centers of the balls circumscribing the compact sets belonging to the Poisson process $\mathcal{Q}$ on $\mathcal{K}'$ constitute a Poisson point process in E, for the measure μ is proportional to the Lebesgue measure. The corresponding boolean model $A = \cup\{K : K \in \mathcal{Q}\}$ is associated with the functional $Q = \exp(-\psi)$, defined by

$$\psi(K) = \theta(\mathcal{F}_K) = a\int_E P_u(\mathcal{F}_K; x)\,dx = a\int_E P_u(\mathcal{F}_{K_{-x}}, 0)\,dx$$

Moreover, by Theorem 2-5-1, Corollary 1, we obtain

$$a\int P_u(\mathcal{F}_{K_{-x}}, 0)\,dx = a\int P(\{A_0 \cap K_{-x} \neq \varnothing\})\,dx = aE\big(V(A_0 \oplus \check{K})\big)$$

where V is the *volume*. Thus

$$Q(K) = \exp\Big[-aE\big(V(A_0 \oplus \check{K})\big)\Big] \qquad (K \in \mathcal{K}) \qquad \text{(3-3-1)}$$

In Chapter 5, we shall examine the boolean models from another point of view.

3-4. STABLE RACS IN A EUCLIDEAN SPACE

Let E again be a euclidean space. We say that a RACS A is $\cup$-*stable* (or simply stable) if, for any integer $n > 0$, there exists a real $\lambda_n > 0$ such that the union $A_1 \cup \cdots \cup A_n$ of n independent RACS A_i equivalent to A is equivalent to $\lambda_n A$. If Q: $K \to Q(\mathcal{F}^K)$ is the functional associated with A, then the functional associated with λA is $K \to Q_\lambda(K) = Q(K/\lambda)$. Thus A is stable if and only if there exists a sequence $\{\lambda_n\} \subset \mathbf{R}_+ \setminus \{o\}$ such that

$$(Q(K))^n = Q\left(\frac{K}{\lambda_n}\right) \qquad (n>0, K \in \mathcal{K})$$

If A is stable, it is equivalent to $\cup_{k=1}^n (A_i/\lambda_n)$, $n>0$, and thus A is an IDRACS. Suppose that A is without fixed points. Then, by Theorem 3-1-1, its functional Q is $Q = \exp(-\psi)$ for a Choquet capacity ψ such that $\psi(\varnothing)=0$ and $\psi(K)<\infty, K \in \mathcal{K}$. Then A is stable if and only if

$$n\psi(K) = \psi\left(\frac{K}{\lambda_n}\right) \qquad (n>0, K \in \mathcal{K}) \qquad (3\text{-}4\text{-}1)$$

The following theorem gives a more tractable criterion.

Theorem 3-4-1. *Let A be a RACS without fixed points and not a.s. empty in a euclidean space. Then A is $\cup$-stable if and only if the associated functional Q: $K \to P(\mathcal{F}^K)$ is $Q = \exp(-\psi)$ for a Choquet capacity ψ such that $\psi(\varnothing)=0$ and $\psi(\lambda K) = \lambda^\alpha \psi(K)(\lambda \geqslant 0, K \in \mathcal{K})$ for a real $\alpha > 0$.*

Proof. The "if" part is obvious by (3-4-1). Conversely, let A be a stable RACS not a.s. empty and without fixed points. Put $\psi = -\log Q$.

a. $\psi(\{o\}=0$, that is, $Q(\{o\})=1$, for $(Q(\{o\}))^n = Q(\{o\})$ implies $Q(\{o\}) = 0$ or 1. But $Q(\{o\})=0$ is impossible, because A is without fixed points.

b. Let B be the unit ball. If there exists a real $a \geqslant 0$ such that $\psi(aB) = \psi(B)$, then $a=1$, for $\psi(a^n B) = \psi(B) = \psi(a^{-n}B)$ and $a>1$, for instance, would imply $\psi(E) = \psi(\{o\})$ by $a^n B \uparrow E$ and $a^{-n}B \downarrow \{o\}$. Thus $\psi(E)=0$, by part *a*, and A would be a.s. empty.

c. By (3-4-1), it is easy to verify that for any rational $r>0$ there exists $\lambda_r > 0$ such that $r\psi(K) = \psi(K/\lambda_r), K \in \mathcal{K}$. This implies $\psi(K/\lambda_r \lambda_{r'}) = r\psi(K/\lambda_{r'}) = rr'\psi(K)$. Thus, by part *b*,

$$\lambda_{rr'} = \lambda_r \lambda_{r'} \qquad (r, r' \text{ rational} > 0) \qquad (3\text{-}4\text{-}2)$$

d. Choose $K \in C(\mathcal{K}_0)$, that is, K is convex and compact and $0 \in K$. For any $\lambda \geqslant 1$, the inclusion $K/\lambda \subset K$ holds, and, conversely, $K/\lambda \supset K$ if $0<\lambda \leqslant 1$. Let r be a rational >1. By $\psi(K/\lambda_r) = r\psi(K) > \psi(K)$, it follows that $K/\lambda_r \supset K$, because ψ is increasing, and thus $\lambda_r \leqslant 1$. By (3-4-1), the function $r \to \lambda_r$ is nonincreasing and thus continuous from the right by the upper semicontinuity of ψ. Then, by (3-4-2), we obtain $\lambda_r = r^{-1/\alpha}$ for a real $\alpha > 0$, and thus

$$\psi(r^{1/\alpha}K) = r\psi(K) \qquad (r \text{ rational} \geqslant 0) \tag{3-4-3}$$

e. It remains to show that (3-4-3) remains true for r real $\geqslant 0$. Let x be a postive real number, and $\{r_n\}$ a decreasing sequence of rational numbers such that $r_n \downarrow x$. By the continuity of the homothety, $x^{1/\alpha}K = \lim (r_n)^{1/\alpha}K$ for any $K \in \mathcal{K}$. But ψ is u.s.c., and thus, from (3-4-3),

$$x\psi(K) = \lim \psi(r_n{}^{1/\alpha}K) \leqslant \psi(x^{1/\alpha}K)$$

Conversely, let ϵ be a real number >0. By definition (1-4-1) of the Hausdorff metric, we obtain, for n large enough and the unit ball B,

$$x^{1/\alpha}K \subset (r_n^{1/\alpha}K) \oplus \epsilon B = (r_n)^{1/\alpha}(K \oplus \epsilon r_n{}^{-1/\alpha}B)$$

On the other hand, we have $r_n{}^{-1/\alpha} \leqslant x^{-1/\alpha}$, for the sequence $\{r_n\}$ is decreasing, and thus $x^{1/\alpha}K \subset r_n{}^{1/\alpha}(K \oplus \epsilon x^{-1/\alpha}B)$. But r_n is rational and it follows from (3-4-3) that $\psi(x^{1/\alpha}K) \leqslant r_n\psi(K \oplus \epsilon x^{-1/\alpha}B)$ and thus $\psi(x^{1/\alpha}K) \leqslant x\psi(K \oplus \epsilon x^{-1/\alpha}B)$. For $\epsilon \downarrow 0$, this implies $\psi(x^{1/\alpha}K) \leqslant x\psi(K)$, because ψ is u.s.c., and (3-4-3) holds for $r = x$.

Example. In the three-dimensional space $E = \mathbf{R}^3$, the conditions of Theorem 3-4-1 are satisfied by the *newtonian capacity*. Thus, if ψ is the newtonian capacity, there exists a RACS A in $\mathbf{R}^3$, called the harmonic RACS (see Choquet, 1953-54), the functional Q of which is $Q = \exp(-\psi)$. It is well known that ψ is translation invariant (and thus A is stationary), and also rotation invariant (and hence A is isotropic). If B_r and C_r are, respectively, a ball and a disk with radius r, we have $\psi(B_r) = r$ and $\psi(C_r) = 2r/\pi$. On the contrary, a straight line has a capacity $= 0$, and thus is a.s. not hit by A. Moreover, $\psi(\lambda K) = \lambda\psi(K)$ for any $K \in \mathcal{K}$, and thus A is stable.

Let us now give an explicit procedure for constructing this harmonic RACS, and investigate its relationship to the brownian motion. Let B be a ball with a radius as great as is wanted. By $0 < \psi(B) < \infty$, the functional T': $K \to T'(K) = \psi(B \cap K)/\psi(B), K \in \mathcal{K}$, satisfies the conditions of the Choquet theorem and is associated with a RACS $A' \subset B$. If $\{A'_n\}$ is a sequence of independent RACS equivalent to A', and N is a Poisson variable independent of $\{A'_n\}$ and such that $E(N) = \psi(B)$, then the RACS $A \cap B$ is equivalent to $A'_1 \cup \ldots \cup A'_N$, and A itself may be obtained by an inductive limit procedure if the radius of $B \uparrow \infty$.

But it is possible to construct the RACS A'. Let μ_B be the probability supported by B which realizes almost everywhere a constant potential $\cup_B$ on B (it is well known that μ_B is actually supported by the boundary ∂B). By the very definition of the Newtonian capacity, $\cup_B = 1/\psi(B)$. Then let $K \in \mathcal{K}$ be included in B, and μ_K be the measure supported by (the boundary of) K which realizes almost everywhere on K the same constant potential $\cup_B$, so that

$$\int \mu_K(dx) = \frac{\cup_B}{\cup_K} = \frac{\psi(K)}{\psi(B)}$$

In other words, this integral is $P(A' \cap K \neq \emptyset)$. But it is well known that μ_K is also the law of the first entering point in K of a brownian motion, the initial law of which is μ_B. In particular, $\int \mu_K(dx)$ is the probability for the trajectory of this brownian motion to hit K, so that A' is equivalent to this trajectory. Hence the following procedure: at time $t=0$, we let go a Poisson random number N of brownian particles with the same initial law μ_B, and take the union of their trajectories. This union is equivalent on B to the RACS $A \cap B$.

3-5. POISSON FLATS IN A EUCLIDEAN SPACE

As a third example, we shall now examine Poisson processes of linear varieties (or *Poisson flats*) in the euclidean space $E = \mathbf{R}^d$ (for a further investigation, see Chapters 4 and 6). Let $\mathcal{S}_k$ be the class of all *k-dimensional subspaces* in E. If $k=0$, $\mathcal{S}_k = \{o\}$; if $k=d$, $\mathcal{S}_d = \{\mathbf{R}^d\}$. For $0 \leqslant k \leqslant d$, $\mathcal{S}_k$ is closed in $\mathcal{F}$, and compact in $\mathcal{F}' = \mathcal{F} \setminus \{\emptyset\}$ by the inclusion $\mathcal{S}_k \subset \mathcal{F}_{\{o\}}$. In the same way, let $\mathcal{V}_k$ be the class of all *k-dimensional linear varieties* in $\mathbf{R}^d$, that is, each $V \in \mathcal{V}_k$ is a translate of an $S \in \mathcal{S}_k$. For $k=0$, $\mathcal{V}_0$ is the class of the punctual sets $\{x\}, x \in \mathbf{R}^d$, and may be identified with $\mathbf{R}^d$. For $k=d$, $\mathcal{V} = \{\mathbf{R}^d\}$. Note that $\mathcal{V}_k$ is closed in $\mathcal{F}'$, but not in $\mathcal{F}$, because there exist sequences $\{V_n\} \subset \mathcal{V}_k$ such that $\emptyset = \lim V_n$ in $\mathcal{F}$.

If S is an element in $\mathcal{S}_k$, we denote $S^\perp \in \mathcal{S}_{d-k}$ as its orthogonal subspace in E. Clearly, the mapping $S \to S^\perp$ is a homeomorphism from $\mathcal{S}_k$ onto $\mathcal{S}_{d-k}$ (for the relative $\mathcal{F}$-topologies). A variety $V \in \mathcal{V}_k$ is defined by its direction $S \in \mathcal{S}_k$ (i.e., the k-dimensional subspace parallel to V) and the unique hitting point $s \in S^\perp$ such that $\{s\} = V \cap S^\perp$. In particular, the linear varieties parallel to a fixed $S \in \mathcal{S}_k$ constitute a subspace of $\mathcal{V}_k$ homeomorphic to $\mathbf{R}^{d-k}$.

Then, with each σ-finite measure $\theta \geqslant 0$ concentrated on $\mathcal{V}_k$ [i.e., $\theta(\mathcal{F}'\setminus\mathcal{V}_k)=0$], is associated a *Poisson process* $\mathcal{Q}$ *in* $\mathcal{V}_k$, and the IDRACS $A=\cup\{F, F\in\mathcal{Q}\}$ is called a k-dimensional *Poisson flat network*. In the sequel we suppose that the measure θ is invariant under translations, and thus we consider only *stationary Poisson networks*. We denote by Q the usual functional $K\to Q(K)=P(\{A\cap K=\varnothing\})$ and put $\psi=-\log Q$, that is, $\psi(K)=\theta(\mathcal{F}_K), K\in\mathcal{K}$.

Our goal is to express the form of this functional ψ associated with a stationary Poisson network. Let θ be the corresponding σ-finite measure invariant under translation and concentrated on $\mathcal{V}_k$. If $k=0$, $\mathcal{V}_0$ is homeomorphic to E, and θ is the measure induced by this homeomorphism from a σ-finite measure on E invariant under the translations, that is, proportional to the Lebesgue measure on E. In other words, the corresponding IDRACS A is a point Poisson process in E. If $k=d$, the IDRACS A is a.s. $A=E$.

Suppose that $0<k<d$. We have seen that any $V\in\mathcal{V}_k$ is determined by the couple (S,s), where $S\in\mathbb{S}_k$ is the direction of V and $s\in S^\perp$ is the hitting point $s\in V\cap S^\perp$. Clearly, the mapping $V\to S$ is continuous on $\mathcal{V}_k$. By putting $u(V)=S, V\in\mathcal{V}_k$, we define θ-everywhere on $\mathcal{F}'$ a measurable mapping from $\mathcal{F}'$ into $\mathbb{S}_k$. By Proposition 3-2-2, there exist a probability G on $\mathbb{S}_k$ and, for G-almost every $S\in\mathbb{S}_k$, a σ-finite measure $\theta_u(.;S)\geqslant 0$ on σ_f which satisfies (3-2-4). Moreover, θ_u is concentrated on $\mathcal{V}_k$, and $u(V)=S, \theta_u(.;S)/$a.e. on $\mathcal{F}'$, so that $\theta_u(.;S)$ is concentrated on the subspace $\mathcal{V}_k(S)\subset\mathcal{V}_k$ defined by $\mathcal{V}_k(S)=\{V\colon V\in\mathcal{V}_k, u(V)=S\}$. If $s(V)$ is the hitting point $s(V)\in V\cap S^\perp$, the mapping $V\to s(V)$ is a homeomorphism from $\mathcal{V}_k(S)$ onto $S^\perp$, and $\theta_u(.;S)$ may be identified with a σ-finite measure μ_S on $S^\perp$ such that

$$\theta_u(\mathcal{F}_K;S)=\mu_S(\Pi_{S^\perp}K) \qquad (K\in\mathcal{K})$$

In this relationship, $\Pi_{S^\perp}K$ denotes the projection of K into $S^\perp$.

On the other hand, the mapping u is θ-a.e. translation invariant, and, by the corollary of Proposition 3-2-2, so are the measures θ_u and μ_S. Thus the measure μ_S is proportional to the Lebesgue measure on $S^\perp$. Let us denote by μ_{d-k} the *Lebesgue measure* on the euclidean space $\mathbf{R}^{d-k}$ (identified with $S^\perp$). Then, for G-almost every $S\in\mathbb{S}_k$, we have

$$\theta_u(\mathcal{F}_K;S)=\varphi(S)\mu_{d-k}(\Pi_{S^\perp}K) \qquad (K\in\mathcal{K})$$

where $\varphi(S)$ is the proportionality constant, and thus, by (3-2-4),

$$\theta(\mathcal{F}_K)=\int_{\mathbb{S}_k}\varphi(S)\mu_{d-k}(\Pi_{S^\perp}K)G(dS)$$

In particular, the function $S \rightarrow \varphi(S)$ is G-integrable on $\mathbb{S}_k$ [because $\mu_{d-k}(\Pi_{S^\perp}K) = C^{\text{ste}}$ if K is the unit ball in $\mathbf{R}^d$]. By substituting $\varphi G/a$ for G $[a = \int \varphi(S)\, G(dS)]$, we finally obtain the following representation of ψ:

$$\psi(K) = a \int_{\mathbb{S}_k} \mu_{d-k}(\Pi_{S^\perp}K) G(dS) \qquad (K \in \mathcal{K}) \tag{3-5-1}$$

Conversely, let G be a probability on $\mathbb{S}_k$, and a a real number >0, and define a function $K \rightarrow \psi(K)$ by (3-5-1). Then $K_n \downarrow K$ in $\mathcal{K}$ implies $\Pi_{S^\perp}K_n \downarrow \Pi_{S^\perp}K$; thus $\mu_{d-k}(\Pi_{S^\perp}K_n) \downarrow \mu_{d-k}(\Pi_{S^\perp}K)$, and finally $\psi(K_n) \downarrow \psi(K)$ by the continuity of the integral. It follows that ψ is u.s.c. on $\mathcal{K}$. By the obvious formula

$$\Pi_{S^\perp}(K \cup K') = (\Pi_{S^\perp}K) \cup (\Pi_{S^\perp}K')$$

the functional $K \rightarrow \mu_{d-k}(\Pi_{S^\perp}K)$ satisfies for each $S \in \mathbb{S}_k$ the positivity conditions required in order to be a Choquet capacity, and it follows that ψ itself a Choquet capacity. Then, by Theorem 3-1-1, there exists an IDRACS A such that $P(\{A \cap K = \varnothing\}) = \exp(-\psi(K))$, $K \in \mathcal{K}$.

It remains to verify that A is a k-dimensional Poisson flat network. For any $S \in \mathbb{S}_k$, let θ_S be the σ-finite positive measure on $\mathcal{F}'$ such that $\theta_S(\mathcal{F}_K) = a\mu_{d-k}(\Pi_{S^\perp}K)$. Obviously, θ_S is concentrated on $\mathcal{V}_k(S)$. Thus $\theta(.) = \int_{\mathbb{S}_k} \theta_S(.) G(dS)$ itself is concentrated on $\mathcal{V}_k$, and A is a stationary Poisson network. With a little change of notation, we may state:

Proposition 3-5-1. *A RACS A is a stationary k-dimensional Poisson flat network if and only if its functional $\psi = -\log Q$ admits the representation*

$$\psi(K) = \int_{\mathbb{S}_k} \mu_{d-k}(\Pi_{S^\perp}K) G(dS) \qquad (K \in \mathcal{K}) \tag{3-5-1'}$$

for a measure $G \geqslant 0$ on $\mathbb{S}_k$ ($\Pi_{S^\perp}K$ is the projection of K into the orthogonal subspace $S^\perp \in \mathbb{S}_{d-k}$, and μ_{d-k} is the Lebesgue measure on the space $S^\perp$ identified with $\mathbf{R}^{d-k}$).

Results Concerning the Space $C(\mathcal{K}')$

The space $C(\mathcal{K}')$ of the nonempty convex compact subsets in $E = \mathbf{R}^d$ will be thoroughly investigated in Chapter 4 simultaneously from the points of view of integral geometry and RACS theory. In order to prepare for this more general study, we shall now give a few results concerning the restriction to $C(\mathcal{K}')$ of the functional ψ defined in (3-5-1').

Definition 3-5-1. *A functional ψ on $C(\mathcal{K}')$ is said to be C-additive if K, K', and $K \cup K' \in C(\mathcal{K}')$ implies*

$$\psi(K \cup K') + \psi(K \cap K') = \psi(K) + \psi(K') \tag{3-5-2}$$

Proposition 3-5-2. *The functional ψ defined by* (3-5-1′) *is continuous and C-additive on $C(\mathcal{K}')$.*

Proof. For any $K, K' \in C(\mathcal{K}')$ and $S \in \mathbb{S}_k$, we have $\Pi_{S^\perp}(K \cup K') = (\Pi_{S^\perp}K) \cup (\Pi_{S^\perp}K')$. Moreover, if $K \cup K'$ is convex, it is easy to verify that the relationship $\Pi_{S^\perp}(K \cap K') = (\Pi_{S^\perp}K) \cap (\Pi_{S^\perp}K')$ also holds. Then, the Lebesgue measure μ_{d-k} being additive, we obtain

$$\mu_{d-k}(\Pi_{S^\perp}(K \cup K')) + \mu_{d-k}(\Pi_{S^\perp}(K \cap K')) = \mu_{d-k}(\Pi_{S^\perp}K) + \mu_{d-k}(\Pi_{S^\perp}K')$$

By (3-5-1′), it follows that ψ is C-additive on $C(\mathcal{K}')$.

To prove the continuity of ψ, we shall use three lemmas.

Lemma 3-5-1. *The mapping $F \to \mathring{F}$ from $C(\mathcal{F})$ into $C(\mathcal{G})$ is continuous.*

Proof of Lemma 3-5-1. By Proposition 1-2-4, Corollary 2, the mapping $F \to \complement\mathring{F}$ from $\mathcal{F}$ into itself is l.s.c., and thus we must prove that the restriction to $C(\mathcal{F})$ of this mapping is u.s.c. Let $\{F_n\} \subset C(\mathcal{F})$ be a sequence and $F = \lim F_n$ in $\mathcal{F}$. By Proposition 1-5-4, $F \in C(\mathcal{F})$. Let $\{F_{n_k}\}$ be a subsequence and, for each k let there be a point $x_{n_k} \notin \mathring{F}_{n_k}$ such that $\lim x_{n_k} = x$ in E. For each $k > 0$, there exists a closed half space H_{n_k} such that $x_{n_k} \in H_{n_k}$ and $F_{n_k} \cap \mathring{H}_{n_k} = \emptyset$. The sequence $\{\mathring{H}_{n_k}, H_{n_k}\}$ admits in the compact space $\mathcal{K}$ an adherent point $\{\mathring{H}, H\}$, where H is still a closed half space (or eventually $H = \mathbf{R}^d$ if $F = \emptyset$). By $x_{n_k} \in H_{n_k}$ and $F_{n_k} \subset \complement\mathring{H}_{n_k}$, we obtain $x \in H$ and $F \cap \mathring{H} = \emptyset$, that is $\mathring{F} \cap H = \emptyset$.

Thus $x \notin \mathring{F}$, and $\overline{\lim}\ \complement\mathring{F}_n \subset \complement\mathring{F}$. By Proposition 1-2-4, it follows that $F \to \complement\mathring{F}$ is u.s.c. on $C(\mathcal{F})$.

Lemma 3-5-2. *The Lebesgue measure is continuous on $C(\mathcal{K}')$.*

Proof of Lemma 3-5-2. Let μ be the Lebesgue measure on the euclidean space $E = \mathbf{R}^d$. For any $K \in C(\mathcal{K}')$, we have $\mu(K) = \mu(\mathring{K})$. If $\mathring{K} = \emptyset$, K is underdimensioned and $\mu(K) = \mu(\mathring{K}) = 0$. If $\mathring{K} \neq \emptyset$, we may suppose that $0 \in \mathring{K}$. For any $\lambda > 0$, we have $\mu(\lambda K) = \lambda^d \mu(K)$, and $\mathring{K} = \lim \lambda K, \lambda \uparrow 1$, implies $\mu(\mathring{K}) = \lim \mu(\lambda K) = \mu(K)$.

On the other hand, μ is u.s.c. on $\mathcal{K}$ and l.s.c. on $\mathcal{G}$. Let $\{K_n\}$ be a sequence in $C(\mathcal{K}')$ and $K=\lim K_n$. By Lemma 3-5-1, it follows that $\lim \mathring{K}_n = \mathring{K}$ in $\mathcal{G}$, and thus $\mu(K)=\mu(\mathring{K}) \leqslant \underline{\lim}\,\mu(\mathring{K}_n) \leqslant \overline{\lim}\,\mu(K_n) \leqslant \mu(K)$. Hence $\mu(K)=\lim \mu(K_n)$.

Lemma 3-5-3. *The mapping $(S,K)\to\Pi_{S^\perp}K$ from $\mathbb{S}_k \times \mathcal{K}'$ into $\mathcal{K}'$ is continuous.*

Proof of Lemma 3-5-3. Let $\{S_n\}\subset \mathbb{S}_k$ and $\{K_n\}\subset\mathcal{K}'$ be two sequences, and $S=\lim S_n, K=\lim K_n$, respectively, in $\mathbb{S}_k$ and $\mathcal{K}'$. Let $k\to x_{nk}\in\Pi_{S^\perp_{Nk}}K_{n_k}$ be a sequence in E such that $\lim x_{n_k}=x$. For each $k>0$, $x_{n_k}\in S^\perp_{n_k}$ and there exists a point $y_{n_k}\in S_{n_k}$ such that $z_{n_k}=x_{n_k}+y_{n_k}\in K_{n_k}$. But the compact sets K_{n_k} remain included in a fixed $K_0\in\mathcal{K}'$, so that the sequences $\{z_{n_k}\}$, $\{x_{n_k}\}$, and $\{y_{n_k}\}$ admit adherent points z, x, and y, respectively, such that $z=x+y$. It follows that $y\in S$, $x\in S^\perp$, and $z\in K$; thus $x\in\Pi_{S^\perp}K$ and the mapping $(S,K)\to\Pi_{S^\perp}K$ is u.s.c.

Now let x be a point in $\Pi_{S^\perp}K$, so that $x\in S^\perp$ and $x+y\in K$ for a point $y\in S$. Thus, for each $n>0$, there exists a point $z_n\in K_n$ such that $x+y=\lim z_n$ in E, and $z_n=x_n+y_n, x_n=\Pi_{S_n}z_n\in S_n, y_n=\Pi_{S_n^\perp}z_n\in S_n^\perp$. It is easy to verify that $z=\lim z_n$ implies $\lim x_n=x_0\in S_n^\perp$, $\lim y_n=y_0\in S_n$, and $z=x_0+y_0$. By the uniqueness of the orthogonal decomposition of a vector $z\in\mathbf{R}^d$, we conclude that $x_0=x$ and $y_0=y$. Thus $x=\lim x_n$ for a sequence $n\to x_n\in\Pi_{S_n^\perp}K_n$, and the mapping $(S,K)\to\Pi_{S^\perp}K$ is l.s.c.

Proof of Proposition 3-5-2 (end). Let b_k be the volume of the unit ball in $\mathbf{R}^k$. For any $K\in\mathcal{K}'$ and $S\in\mathbb{S}_k$ we have

$$\mu_{d-k}(\Pi_{S^\perp K})=\frac{1}{b_k}\mu_d((\Pi_{S^\perp}K)\oplus(B\cap S))$$

(B is the unit ball in $\mathbf{R}^d$). By Proposition 1-5-1 (i.e., the continuity of the Minkowski sum $\oplus$) and Lemmas 3-5-2 and 3-5-3, it follows that the mapping $(S,K)\to\mu_{d-k}(\Pi_{S^\perp}K)$ is continuous on $\mathbb{S}_k\times C(\mathcal{K}')$. But the space $\mathbb{S}_k$ is compact, and thus the function $K\to\mu_{d-k}(\Pi_{S^\perp}K)$ is continuous on $C(\mathcal{K}')$ uniformly with respect to $S\in\mathbb{S}_k$. Then, by (3-5-1′), ψ is continuous on $C(\mathcal{K}')$.

Remark. The Lebesgue measure μ_{d-k} on $\mathbf{R}^{d-k}$ is homogeneous of degree $d-k$ by (3-5-1′). Thus the corresponding Poisson network is a stable RACS (Theorem 3-4-1). We shall see in Chapter 5 that the C-additivity of ψ is equivalent to a semi-markovian property of the corresponding IDRACS, so that any stationary Poisson network is stable and semi-markovian. Conversely, we shall prove (Theorem 5-4-2) that any stable, stationary, and semi-markovian RACS is equivalent to a Poisson network, so that these conditions characterize the Poisson network.

Random Cross Sections of a Compact Set B

Let $A = \cup \mathcal{Q}$ be a k-dimensional Poisson flat, and B a compact set such that $\psi(B) = \theta(\mathcal{F}_B) > 0$ [and thus, by (3-5-1′), B is not included in a linear variety with dimension $< d-k$]. The probability for B to hit one and only one of the linear varieties belonging to the Poisson process $\mathcal{Q}$ on $\mathcal{V}_k$ is $\psi(B)\exp(-\psi(B))$. Conditionally, if so, denote by A' the unique (random) linear variety hitting B. Thus $A' \cap B$ is a RACS on $\mathcal{F}(B)$, called the *random cross section* (of B by the Poisson network). In integral geometry, the random cross sections associated with an isotropic Poisson network have been thoroughly investigated. As a matter of fact, the probability of the random cross section $A' \cap B$ depends only on the measure $G \geqslant 0$ of relationship (3-5-1)—more precisely, only on the probability $G/\int G(dS)$ on $\mathcal{S}_k$—and in the particular case considered in integral geometry, this probability is the unique probability on $\mathcal{S}_k$ invariant under the translations.

The probability of the event "exactly one linear variety falls in $\mathcal{F}_B$ and none in $\mathcal{F}_B^K$" is $\psi(K)\exp(-\psi(B))$ for any $K \in \mathcal{K}(B)$ (i.e., $K \subset B$), so that the conditional law T_B: $K \to T_B(K) = P(A' \cap B \in \mathcal{F}_K)$ is defined by

$$T_B(K) = \frac{\psi(K)}{\psi(B)} \qquad [K \in \mathcal{K}(B)]$$

Instead of $A' \cap B$, we may consider the linear variety A' itself. Its law, also denoted by T_B, must be defined on the entire space $\mathcal{K}$, and not only on $\mathcal{K}(B)$. If $K \not\subset B$, note that the event "exactly one variety falls in $\mathcal{F}_{B,K}$ and none in $\mathcal{F}_B^K$" admits the probability

$$\theta(\mathcal{F}_B \cap \mathcal{F}_K)\exp[-\theta(\mathcal{F}_B \cap \mathcal{F}_K) - \theta(\mathcal{F}_B \cap \mathcal{F}^K)]$$
$$= \theta(\mathcal{F}_B \cap \mathcal{F}_K)\exp[-\theta(\mathcal{F}_B)]$$

Thus, by $\theta(\mathcal{F}_B \cap \mathcal{F}_K) = \psi(B) + \psi(K) - \psi(B \cup K)$, the law T_B is defined by

$$T_B(K) = \frac{\psi(B) + \psi(K) - \psi(B \cup K)}{\psi(B)} \qquad (K \in \mathcal{K}) \qquad (3\text{-}5\text{-}3)$$

On the other hand, the variety $A' \in \mathcal{V}_k$ may be identified with a (random) pair (S,s), where $S \in \mathbb{S}_k$ is the direction of A' and $x \in S^{\perp}$ is the hitting point $\{s\} = A' \cap S^{\perp}$. Obviously, $s \in \Pi_{S^\perp} B$ a.s., because A' hits B. By Section 2-3, there exist a unique probability F_B on $\mathbb{S}_k$ and, for F_B-almost every $S \in \mathbb{S}_k$, a conditional RACS A'_S [which is a.s. in $\mathcal{V}_k(S)$], and its law T_S satisfies

$$T_B(K) = \int_{\mathbb{S}_k} T_S(K) F_B(dS) \qquad (K \in \mathcal{K})$$

By comparing (3-5-3) and (3-5-1), and taking into account the uniqueness of the laws F_B and (for F_B-almost every $S \in \mathbb{S}_k$)T_S, we obtain the following explicit expressions:

$$F_B(dS) = \frac{\mu_{d-k}(\Pi_{S^\perp} B) G(dS)}{\int \mu_{d-k}(\Pi_{S^\perp} B) G(dS)}$$

$$T_S(K) = \frac{\mu_{d-k}((\Pi_{S^\perp} B) \cap (\Pi_{S^\perp} K))}{\mu_{d-k}(\Pi_{S^\perp} B)} \qquad (K \in \mathcal{K}) \tag{3-5-4}$$

In other words, *the direction S of the random cross section A' admits the law F_B defined by* (3-5-4), *that is, the measure G weighted by the measure of the projection $\Pi_{S^\perp} B$, and, for a fixed S, the hitting point $s \in A' \cap S^\perp$ is uniformly distributed on the projection $\Pi_{S^\perp} B$*, that is, admits the probability law $(1_{\Pi_{S^\perp} B}(s) / \mu_{d-k}(\Pi_{S^\perp} B)) \mu_{d-k}(ds)$.

Poisson Network Induced on a Linear Variety

Let p be an integer such that $d-k \leqslant p < d$, $V_p \in \mathcal{V}_p$ be a p-dimensional linear variety in $E = \mathbf{R}^d$, and $S_p \in \mathbb{S}_p$ be its direction. We may identify V_p with the euclidean space $\mathbf{R}^p$. Then consider the mapping α from $\mathcal{F}(\mathbf{R}^d)$ into $\mathcal{F}(\mathbf{R}^p)$ defined by $\alpha(F) = F \cap V_p, F \in \mathcal{F}$. This mapping is u.s.c. and thus measurable. If θ is a σ-finite measure on $\mathcal{F}'(\mathbf{R}^d)$, its image under the mapping α is the σ-finite measure θ_p on $\mathcal{F}'(V_r)$, defined by $\theta_r(\mathcal{V}) = \theta(\alpha^{-1}(\mathcal{V}))$ for any $\mathcal{V} \in \sigma_f(V_p)$. Clearly, if θ is invariant for the translations in $\mathbf{R}^d$, so also is θ_p for the translations in $V_p = \mathbf{R}^p$. Then, if A is an IDRACS on $\mathbf{R}^d$ associated with the measure $\theta, A_p = A \cap V_p$ is an IDRACS on $\mathbf{R}^p$ associated with θ_p, and A_p is stationary if A is already stationary in $\mathbf{R}^d$. Moreover, the probability of the stationary IDRACS A_p depends only on the direction S_p of $V_p \in \mathcal{V}_p$ (and not on the hitting point $s \in V_p \cap S_p^{\perp}$).

In particular, if A is a stationary Poisson network in $\mathbf{R}^d$, A_p will be a stationary Poisson network in $V_p = \mathbf{R}^p$, called the *Poisson network induced* on V_p by A, for θ is translation invariant and concentrated on $\mathcal{V}_k(\mathbf{R}^d)$, and, for θ-almost every $F \in \mathcal{V}_p(\mathbf{R}^d), \alpha(F) = F \cap V_p \in \mathcal{V}_{k-d+p}(\mathbf{R}^p)(0 \leqslant k-d+p < k)$, so that θ_p itself is translation invariant and concentrated on $\mathcal{V}_{k-d+p}(\mathrm{R}^p)$. Moreover, the Poisson network induced on V_p is $(k+p-d)$-dimensional. In particular, for $p = d-k$, the induced network is an ordinary stationary point Poisson process in $\mathbf{R}^p$ and is characterized by a constant "density." We shall first investigate this particular case.

The following notations will be very useful in the sequel. Let S, S' be two subspaces in $E = \mathbf{R}^d$ (with eventually different dimensions). Then *we denote* by $|S, S'|$ *the absolute value of the determinant of the linear mapping* $s \to \Pi_{S'} s$ from S into its range in S', if the dimension of its range is Dim S, and $|S, S'| = 0$ otherwise. In other words, if $S \in \mathbb{S}_p, |S, S'|$ is the constant $\geqslant 0$ such that

$$\mu_p(\Pi_{S'} K) = |S, S'| \, \mu_p(K) \qquad (3\text{-}5\text{-}5)$$

for any compact set $K \subset S$. Clearly, $|S, S'| = |S, \Pi_{S'} S|$ and $|S, S'| = 0$ if and only if Dim $\Pi_{S'} S < \operatorname{Dim} S$. In particular, $|S, S'| = 0$ if $\operatorname{Dim} S' < \operatorname{Dim} S$. If S and S' have the same dimension, obviously $|S, S'| = |S', S|$.

Point Processes Induced on the $(d-k)$-Dimensional Varieties

Let $V \in \mathcal{V}_{d-k}$ be a $(d-k)$-dimensional variety, and $\sigma \in \mathbb{S}_{d-k}$ its direction. Then the point Poisson process induced on V by A is stationary and characterized by its density $\varphi(\sigma)$, that is, the real number $\varphi(\sigma) \geqslant 0$ so that $\psi(K) = \varphi(\sigma)\mu_{d-k}(K)$ for any compact set $K \subset V$ (or, which is the same, $K \subset \sigma$). We must find the expression of the function $\sigma \to \varphi(\sigma)$ representing the density of the point Poisson process induced on each $\sigma \in \mathbb{S}_{d-k}$.

Suppose first that the measure $G \geqslant 0$ in (3-5-1′) is the Dirac measure δ_S for a given $S \in \mathbb{S}_k$. In this case, the Poisson network in $\mathbf{R}^d$ is constituted by parallel linear varieties with the same direction S, and (3-5-1′) simply gives $\psi(K) = \mu_{d-k}(\Pi_{S^\perp} K)$. For any $\sigma \in \mathbb{S}_{d-k}$ and $K \in \mathcal{K}, K \subset \sigma$, it follows from (3-5-5) that $\mu_{d-k}(\Pi_{S^\perp} K) = |\sigma, S^\perp| \, \mu_{d-k}(K)$. Hence the density $\varphi(\sigma)$ of the point process induced on σ is $\varphi(\sigma) = |\sigma, S^\perp|$. In the general case, G is a positive measure on $\mathbb{S}_{d-k}$, and a conditioning argument yields

$$\varphi(\sigma) = \int_{\mathbb{S}_k} |\sigma, S^\perp| G(dS) \qquad (\sigma \in \mathbb{S}_{d-k}) \qquad (3\text{-}5\text{-}6)$$

Conversely, let φ be a function on $\mathbb{S}_{d-k}$ admitting such a representation. Then, for each $\sigma \in \mathbb{S}_{d-k}, \varphi(\sigma)$ is the density of the point process induced on σ by a Poisson network associated with the positive measure G according to (3-5-1′). We shall denote by $\mathcal{R}_{d-k}$ the class of these functions. It is easy to verify that $\mathcal{R}_{d-k}$ is a convex cone in the space $\mathbf{C}(\mathbb{S}_{d-k})$ of the continuous functions on $\mathbb{S}_{d-k}$ (topologized by the uniform convergence) and admits a compact base. Then two questions arise.

1. First, what kind of conditions must be satisfied by a given function $\varphi \in \mathbf{C}(\mathbb{S}_{d-k})$ in order to represent the densities of the point processes induced on each $\sigma \in \mathbb{S}_{d-k}$ by a Poisson network in $\mathbf{R}^d$?

2. The second question concerns the uniqueness of the integral representation (3-5-6). In other words, is a Poisson network in $\mathbf{R}^d$ uniquely determined if the induced density $\varphi(\sigma)$ is known for each $\sigma \in \mathbb{S}_{d-k}$? In other words, is the convex cone $\mathcal{R}_{d-k}$ a *simplex* in $\mathbf{C}(\mathbb{S}_{d-k})$? Obviously, $\mathcal{R}_{d-k}$ will be a simplex if and only if the vector space spanned in $\mathbf{C}(\mathbb{S}_{d-k})$ by the functions $\sigma \to |\sigma, S^{\perp}|, S \in \mathbb{S}_k$, is *dense* in $\mathbf{C}(\mathbb{S}_{d-k})$.

In the next chapter, we shall be able to give a (positive) answer to the second question, but only in the cases $k=1$ and $k=d-1$ (i.e., for the line networks and the hyperplane networks), and a complete characterization of the simplexes $\mathcal{R}_1$ and $\mathcal{R}_{d-1}$. We may conjecture that this uniqueness theorem remains true if $1<k<d-1 (d \geqslant 4)$.

Networks Induced on the Varieties with Dimension $>d-k$

Let us now examine the case of a p-dimensional variety $V \in \mathcal{V}_p, d-k<p<d$, and let $S_p \in \mathbb{S}_p$ be its direction. The Poisson network induced on V is constituted by varieties of dimension $k+p-d>0$. By (3-5-1), this induced network is characterized by a measure $G_p(.\,; S_p) \geqslant 0$ on the space $\mathbb{S}_{k+p-d}(S_p)$ of the $k+p-d$ subspaces in S_p. In order to determine G_p, let $K \subset V$ be a compact set, so that the events $\{A \cap K = \varnothing\}$ and $\{A_p \cap K = \varnothing\}$ are identical. In other words, the following relationship holds:

$$\psi(K) = \int_{\mathbb{S}_{k+p-d}(S_p)} \mu_{d-k}(\Pi'_{\sigma^{\perp}} K) G_p(d\sigma; S_p) \qquad (K \in \mathcal{K}, K \subset S_p) \quad (3\text{-}5\text{-}7)$$

for any compact set $K \subset S_p$. The notation $\Pi'_{\sigma^{\perp}}$ denotes the projection on the subspace $\sigma^{\perp} \subset S_p$ orthogonal to σ in S_p, the dimension of which is $d-k$. By comparing (3-5-1) and (3-5-7), we shall obtain an explicit expression of G_p.

If the measure G in (3-5-1′) is δ_{S_0} for a given $S_0 \in \mathbb{S}_k$, then the Poisson varieties are parallel to the direction S_0, and in this case $\psi(K) = \mu_{d-k}(\Pi_{S_0^\perp} K)$ for any $K \in \mathcal{K}$. Obviously, the network induced on $S_p \in \mathbb{S}_p$ is constituted by varieties parallel to $\sigma_0 = S_0 \cap S_p$. If $\operatorname{Dim} \sigma_0 > k+p-d$, $\mu_{d-k}(\Pi_{S_0^\perp} K) = 0$ for any compact set $K \subset S_p$), and $G_p = 0$, so that the induced network vanishes. Denote by $\sigma_0^\perp = \Pi_{S_p} S_0^\perp$ the subspace of S_p orthogonal to σ_0 in S_p. For any $K \in \mathcal{K}(S_p)$, it follows from (3-5-5) that

$$\mu_{d-k}(\Pi_{S_0^\perp} K) = \mu_{d-k}(\Pi_{S_0^\perp} \Pi'_{\sigma_0^\perp} K) = |S_0^\perp, \sigma_0^\perp| \, \mu_{d-k}(\Pi'_{\sigma_0^\perp} K).$$

But $\sigma_0^\perp = \Pi_{S_p} S_0^\perp$ implies $|S_0^\perp, \sigma_0^\perp| = |S_0^\perp S_p|$, and $|S_0^\perp, S_p| = 0$ if $\operatorname{Dim} \sigma_0 > k + p - d$, so that

$$\mu_{d-k}(\Pi_{S_0^\perp} K) = |S_0^\perp, S_p| \, \mu_{d-k}(\Pi'_{\sigma_0^\perp} K)$$

holds in all cases. Thus $G_p = |S_0^\perp, S_p| \delta_{\sigma_0}$ (with the convention $G_p = 0$ if $|S_0^\perp, S_p| = 0$), and G_p is characterized by the relationship $\int f(\sigma) G_p(d\sigma) = |S_0^\perp, S_p| f(S_0 \cap S_p)$ for any continuous function f on $\mathbb{S}_{k+p-d}(S_p)$. By (3-5-1), we obtain in the general case (i.e., for any measure $G \geqslant 0$ on $\mathbb{S}_k$)

$$\int_{\mathbb{S}_{k+p-d}} f(\sigma) G_p(d\sigma; S_p) = \int_{\mathbb{S}_k} |S^\perp, S_p| f(S \cap S_p) G(dS)$$

for any continuous function f on $\mathbb{S}_{k+p-d}(S_p)$, with the convention $|S^\perp, S_p| f(S \cap S_p) = 0$ if $|S^\perp, S_p| = 0$. Note that the function $S \to |S^\perp, S_p| f(S \cap S_p)$ defined in this way is continuous on $\mathbb{S}_k$. If $d-k < p < d$, the network induced on S_p itself induces on each $S_{d-k} \in \mathbb{S}_{d-k}(S_p)$ a point process, the density $\varphi(S_{d-k})$ of which is given by (3-5-6), and also by the analogous relationship written with G_p instead of G and S_p instead of $\mathbf{R}^d$, that is,

$$\varphi(S_{d-k}) = \int_{\mathbb{S}_{k+p-d}(S_p)} |\sigma^\perp, S_{d-k}| G_p(d\sigma; S_p) \tag{3-5-8}$$

for each $S_{d-k} \in \mathbb{S}_{d-k}$ and $S_p \in \mathbb{S}_p$ such that $S_p \supset S_{d-k}$. In this relationship, $\sigma^\perp$ is the subspace of S_p orthogonal to S_{d-k}. Hence the following condition must be satisfied by the family $G_p(.; S_p)$, $S_p \in \mathbb{S}_p$, of the positive measures associated with the networks induced by the same Poisson network on $\mathbf{R}^d$: for each $S_{d-k} \in \mathbb{S}_{d-k}$, the integral on the right-hand side of (3-5-8) does not depend on the choice of the subspace $S_p \supset S_{d-k}$.

CHAPTER 4

Convexity

In this chapter, the space E will be the d-dimensional euclidean space $E=\mathbf{R}^d$. In the first section, we define the Minkowski functionals in connection with the results concerning the Poisson networks. Our reference book for the Integral Geometry will be the volume by Hadwiger (1957). Then we show that the RACS a.s. convex are characterized by a property of their functionals T, that is, the C-additivity. The third section is devoted to the linear granulometries. Then it will be shown that the space $C_0(\mathcal{K})$ of the convex compact sets including the origin is homeomorphic to a convex cone $\mathfrak{R}$ with a compact basis in the space $\mathbf{C}(S_0)$ of the continuous function on the unit sphere. Although $\mathfrak{R}$ is not a simplex, it encloses a simplex $\mathfrak{R}_1$, associated with an interesting class of symmetrical convex compact sets, that is, the Steiner class, the elements of which satisfy a generalized version of the famous Steiner formula. The results concerning the simplex $\mathfrak{R}_1$ will lead to a uniqueness theorem, by which a Poisson line (or hyperplane) network is determined if the densities of the point processes induced on the hyperplanes (or the lines) are given. The same theorem also yields a characterization of the tangential cones at O of the covariance functions in $\mathbf{R}^d$. In the case of the euclidean space $\mathbf{R}^2$, the Steiner class is identical to the class of the convex compact sets symmetrical about the origin O, but this is no longer true if $d>2$. Finally, we give results concerning the interpretation of the Minkowski functionals in terms of measures on $\mathbf{R}^d$, and two possible extensions to the convex ring $\mathfrak{S}$: the first one is based on the celebrated Euler-Poincaré characteristic, and the second one on the convexity number. The second extension may also be interpreted in terms of measures on $\mathbf{R}^d$, and applied to certain RACS, so that it will be possible to define the densities in $\mathbf{R}^d$ of the corresponding Minkowski functionals.

4-1. THE MINKOWSKI FUNCTIONALS

In integral geometry, special attention is paid to the functionals defined on the space $C(\mathcal{K}')$ of the nonempty convex compacts sets in E, which are invariant for the euclidean transformations, that is, under rotation and translation in the euclidean space $E=\mathbf{R}^d$. On the other hand, we know that a RACS is *stationary* if and only if the functional T associated with it by the Choquet theorem is invariant under translation. In the same way, we shall say that a RACS is isotropic if its probability is invariant by the orthogonal group $\Omega=\Omega(E)$, that is, the group of the rotations in E with center at O, and obviously a RACS is isotropic if and only if its functional T itself is invariant by the orthogonal group. Thus there will exist interesting relationships between the stationary and isotropic RACS and the classical functionals of the integral geometry.

Let us first examine a simple procedure which enables us to construct isotropic functionals on $\mathcal{K}'$. The orthogonal group Ω is compact, and it is well known (see Nachbin, 1965) that there exists on Ω a bounded Haar measure unique up to a multiplicative constant, that is, a bounded positive measure invariant under rotation. In other words, there exists *on the orthogonal group* Ω *a unique probability invariant under rotation*, which will be denoted by ϖ.

Then, if K is a compact (or closed) set, obviously the mapping $\omega \to \omega K$ from Ω into $\mathcal{K}$ (or $\mathcal{F}$) is continuous (for the natural topology on Ω). Thus, for any measurable function φ on $\mathcal{K}$, the function $\bar{\varphi}$ defined by putting

$$\bar{\varphi}(K)=\int_{\Omega}\varphi(\omega K)\varpi(d\omega) \qquad (K\in\mathcal{K}')$$

is an isotropic function on $\mathcal{K}'$, which will be called the *rotation average* of φ. It is easy to see that $\bar{\varphi}$ is continuous if φ itself is continuous (because the Hausdorff metric on $\mathcal{K}'$ is invariant by the orthogonal group).

In the same way, if $S_0\in\mathbb{S}_k$ is a given k-dimensional subspace in $E=\mathbf{R}^d(0<k<d)$, the mapping $\omega\to\omega S_0$ from Ω onto $\mathbb{S}_k$ is continuous, and the probability induced from ϖ by this mapping is invariant under rotation and does not depend on the choice of the particular subspace $S_0\in\mathbb{S}_k$. This invariant probability on $\mathbb{S}_k$ will be denoted by $\varpi_k{}^d$, or simply ϖ_k if there is no ambiguity concerning the dimension of E. It may be shown that ϖ_k is the unique probability on $\mathbb{S}_k$ invariant under the rotations.

Now let $\psi:\mathcal{K}(\mathbf{R}^k)\to\mathbf{R}$ be an isotropic functional on the space $\mathcal{K}(\mathbf{R}^k)$, where $\mathbf{R}^k$ is the k-dimensional euclidean space, and $0<k<d$. For each $S\in\mathbb{S}_k$, $\psi(\Pi_S K)$ is well defined (Π_S denotes the projection into S), for it is possible to identify a given $S_0\in\mathbb{S}_k$ with $\mathbf{R}^k$, and for any $S\in\mathbb{S}_k$ there exists a rotation $\omega_S\in\Omega$ such that $\omega_S S=S_0$. By the invariance of ψ under rotation, the expression $\psi(\omega_S\Pi_S K)$ does not depend on the choice of the rotation ω_S, so that we may put $\psi(\Pi_S K)=\psi(\omega_S\Pi_S K)$. If we suppose that ψ is measurable, the mapping $(S,K)\to\psi(\Pi_S K)$ itself is measurable on $\mathbb{S}_k\times$

$\mathcal{K}(\mathbf{R}^d)$ by Lemma 3-5-3, and is continuous if ψ itself is continuous on $\mathbf{R}^k$. In particular, we may define the rotation average ψ^d of ψ in $\mathbf{R}^d$ by putting, for any $K \in \mathcal{K}'$,

$$\psi^d(K) = \int_{\mathbb{S}_k} \psi(\Pi_S K)\varpi_k(dS) = \int_\Omega \psi(\Pi_{\omega S_0} K)\varpi(d\omega) \qquad \text{(4-1-1)}$$

Then ψ^d is an isotropic measurable function on $\mathbf{R}^d$ and is continuous if ψ itself is continuous on $\mathbf{R}^k$.

Now let k' be an integer such that $k < k' < d$. In the same way, we may define the rotation average $\psi^{k'}$ of ψ under the rotations in $\mathbf{R}^{k'}: \psi^{k'}$ is an isotropic and measurable functional on $\mathcal{K}(\mathbf{R}^{k'})$. With $\psi^{k'}$, we can again associate its rotation average $\bar{\psi}^{k'}$ under the rotation in $\mathbf{R}^d$, which is an isotropic functional on $\mathcal{K}(\mathbf{R}^d)$. It turns out that this new functional is identical with ψ^d. In other words, if $0 < k < k' < d$, then, for any $K \in \mathcal{K}(E)$,

$$\psi^d(K) = \int_{\mathbb{S}_{k'}} \psi^{k'}(\Pi_{S'} K)\varpi_{k'}(dS') = \int_\Omega \psi^{k'}(\omega S_0')\varpi(d\omega) \qquad \text{(4-1-2)}$$

(S_0' is an arbitrary element in $\mathbb{S}_k$).

Proof of Relationship (4-1-2). Choose subspaces $S_0 \in \mathbb{S}_k$ and $S_0' \in \mathbb{S}_{k'}$ such that $S_0 \subset S_0'$, and identify S_0 with $\mathbf{R}^k$ and S_0' with $\mathbf{R}^{k'}$. By the continuous mapping from Ω into $\mathbb{S}_k \times \mathbb{S}_{k'}$ defined by $\omega \to (\omega S_0, \omega S_0')$, a probability $G(dS, dS')$ is induced from ϖ on $\mathbb{S}_k \times \mathbb{S}_{k'}$ and is invariant under the rotations. We have $G(\mathbb{S}_k; dS') = \varpi_{k'}(dS')$, and $G(dS, dS') = \varpi_{k'}(dS')G_{S'}(dS)$ for a conditional probability $G_{S'}(dS)$ on $\mathbb{S}_k$, defined for $\varpi_{k'}$-almost every $S' \in \mathbb{S}_{k'}$ and concentrated on S' (i.e., $S \subset S'$ a.s. for $G_{S'}$). Moreover, $G_{S'}$ is invariant by the subgroup ($\Omega(S')$, that is, the rotations in S'. In other words, if $\omega' \in \Omega$ is such that $\omega' S' = S_0'$, $G_{S'}$ is induced by ω' from the invariant probability $\varpi_k^{k'}$ on $\mathbb{S}_k(\mathbf{R}^{k'})$. On the other hand, by the definition of $\psi^{k'}$,

$$\int \psi(\Pi_S K) G_{S'}(dS) = \int \psi(\Pi_S \Pi_{S'} K) G_{S'}(dS)$$
$$= \int \psi(\Pi_S \Pi_{S'} K)\varpi_k^{k'}(dS) = \psi^{k'}(\Pi_{S'} K)$$

It follows that

$$\int_{\mathbb{S}_{k'}} \psi^{k'}(\Pi_{S'} K)\varpi_{k'}(dS') = \int_{\mathbb{S}_k \times \mathbb{S}_{k'}} \psi(\Pi_S K) G(dS, dS') = \psi^d(K)$$

that is, (4-1-2) holds. This relationship is very useful in integral geometry.

It is now possible to define the famous *Minkowski functionals*. A stationary k-dimensional Poisson flat network is *isotropic* if and only if the measure G on $\mathbb{S}_k$ in (3-5-1′) is proportional to the unique probability ϖ_k invariant under rotation. Thus we define $\psi_k{}^d$ (or simply ψ_k if there is no ambiguity) by taking the rotation average of $\mu_{d-k}(\Pi_{S^\perp}K)$, that is,

$$\psi_k(K)=\int_{\mathbb{S}_k}\mu_{d-k}(\Pi_{S^\perp}K)\varpi_k(dS) \qquad (K\in\mathcal{K}) \tag{4-1-3}$$

if $0<k<d$, and $\psi_0(K)=\mu_d(K)$ [i.e., $\psi_0(K)$ is the volume of K]. Up to a multiplicative constant, the Minkowski functionals W_k are these rotation averages restricted to the domain $C(\mathcal{K}')$. More precisely, let us denote by b_k *the volume of the unit ball in* $\mathbf{R}^k$, that is,

$$b_k=\frac{\pi^{k/2}}{\Gamma(1+k/2)} \tag{4-1-4}$$

Then the kth Minkowski functional $W_k{}^d$ (or simply W_k if there is no ambiguity), $0\leqslant k\leqslant d$, is defined by

$$\left.\begin{aligned} W_k(K)&=\frac{b_d}{b_{d-k}}\psi_k(K)\\ W_d(K)&=b_d\end{aligned}\right\} \qquad [K\in C(\mathcal{K}')] \tag{4-1-5}$$

By the choice of the normalizing constants, we have $W_k(B)=b_d$, $k=0,1,\ldots,d$, if B is the unit ball in $\mathbf{R}^d$.

By the results of Chapter 3, and specifically by Proposition 3-5-2, we may state:

Proposition 4-1-1. *For* $k=0,1,\ldots,d$, *the Minkowski functional* W_k *on* $C(\mathcal{K}')$ *is increasing, continuous, C-additive, homogeneous with degree* $d-k$, *and invariant under the euclidean transformations.*

By applying to $\psi_k{}^d$ relationship (4-1-2), we have

$$\psi_k{}^d(K)=\int_{\mathbb{S}_{d'}}\psi_{k'}{}^{d'}(\Pi_S K)\varpi_{d'}{}^d(dS)$$

$(d>d'>d-k,\ k'=d'-d+k)$. Thus:

Proposition 4-1-2. *If* $d>d'>d-k$ *and* $k'=d'-d+k$, *the functional* W_k^d *on* $\mathcal{K}(\mathbf{R}^d)$ *is proportional to the rotation average of the functional* $W_{k'}^{d'}$ *on*

$\mathcal{K}(\mathbf{R}^{d'})$, *according to the following formula*:

$$W_k{}^d(K) = \frac{b_d}{b_{d'}} \int_{\mathbb{S}_{d'}} W_{k'}{}^{d'}(\Pi_S K)\varpi_{d'}(dS) \tag{4-1-6}$$

Among these Minkowski functionals, the most important are W_0, W_1, W_2, and W_{d-1} (W_d is simply the constant b_d). We have already seen that W_0 is the *volume*. The functional W_1 is proportional to the *surface area*. More precisely, if we denote by $F(K)$ the surface area of a $K \in C(\mathcal{K}')$, we have

$$W_1{}^d(K) = \frac{1}{d} F(K)$$

Proof. From Definition (4-1-3), it follows that

$$\psi_1(K) = \int_{\mathbb{S}_1} \mu_{d-1}(\Pi_{S^\perp} K)\varpi_1(dS)$$

so that $\psi_1(K)$ is the average of the $(d-1)$-volume $\mu_{d-1}(\Pi_{S^\perp}K)$ of the projection of K into the hyperplanes in $\mathbf{R}^d$. If K is a convex polyhedron, the surface area $F(K)$ of which is defined in an elementary way, it is easy to see that $\psi_1(K) = (b_{d-1}/db_d)F(K)$. This relationship (due to Cauchy) may be extended by continuity to the whole space $\mathcal{K}(\mathbf{R}^d)$, because the convex polyhedra constitute a dense subspace in $\mathcal{K}$. The above formula follows.

If the *boundary* ∂K of $K \in C(\mathcal{K}')$ is regular enough so that it is possible to define at each point $x \in \partial K$ the $d-1$ main curvatures, the Minkowski functionals admit simple integral representations in which symmetric functions of these curvatures intervene. For instance, W_2 is proportional to the integral M of the mean curvature, that is, $W_2 = M/d$.

Finally, the quantity $N(K) = dW_{d-1}(K)$ is called the *norm* of $K \in C(\mathcal{K}')$. In the definition formula (4-1-3), $\mu_1(\Pi_{S^\perp}K)$ is the *breadth* of the convex compact set K in the direction $S^\perp \in \mathbb{S}_1$, that is, the distance between the two supporting hyperplanes parallel to S. Thus the norm $N(K)$ is *proportional to the average breadth* $\bar{b}(K)$, that is,

$$N(K) = \frac{db_d}{2}\bar{b}(K)$$

Moreover, the *norm* and the *functional* W_{d-1} are linear with respect to the Minkowski sum, that is,

$$N(\alpha K \oplus \beta K') = \alpha N(K) + \beta N(K') \qquad [\alpha, \beta \geqslant 0;\ K, K' \in C(\mathcal{K}')] \tag{4-1-7}$$

It is easy to prove (4-1-7) by observing that the breadth $\mu_1(\Pi_{S^\perp}K)$ in a given direction $S^\perp \in \mathbb{S}_1$ is already linear. This property, in fact, characterizes the norm, for it can be shown that any functional φ, translation and rotation invariant, continuous, and linear on $C(\mathcal{K})$ is necessarily of the form $\varphi = a + bN$ for two constants a and b (see Hadwiger, 1957).

If the dimension $d \leqslant 3$, the Minkowski functionals admit very simple geometrical interpretations.

For $d=1$, $W_0 = N$ is the *length*, and $W_1 = F = b_1 = 2$ is constant.

For $d=2$, W_0 is the two-dimensional volume (or area), $N = 2W_1$ is the *perimeter*, and $M = 2W_2 = 2\pi$.

Finally, for $d=3$, W_0 is the usual *volume, $3W_1 = F$ is the surface area,* and the norm $N = M = 3W_2$ is the integral of the mean curvature.

The Minkowski functionals are characterized by the conditions stated in Proposition 4-1-1. More precisely, we have the following theorem, for the proof of which the reader is referred to Hadwiger (1957, pp. 221–225).

Theorem 4-1-1. *a. If a function φ on $C(\mathcal{K})$ is continuous, C-additive, and invariant under euclidean transformation, then $\varphi = \Sigma_0^d a_k W_k$ for suitable constants a_k.*

b. If a function φ on $C(\mathcal{K})$ is increasing, C-additive, and invariant under euclidean transformations, then $\varphi = \Sigma_0^d a_k W_k$ for suitable constants $a_k \geqslant 0$.

Remark. By analogy with Theorem 4-1-1, we may ask whether any functional φ that is C-additive, continuous (or increasing), and invariant under translations (but no longer under rotations) on $C(\mathcal{K})$ is necessarily $\varphi = \Sigma_0^d a_k \psi_k$, where each ψ_k is a functional on $C(\mathcal{K})$ defined by (3-5-1′) for a measure $G_k \geqslant 0$ on $\mathbb{S}_k$. We shall see in Section 4-5 that this is *not* true, so that Theorem 4-1-1 cannot be generalized.

In the same way, the Steiner formula (4-1-8) may be generalized by taking an arbitrary convex compact set A instead of the unit ball B. Then we have

$$V(K \oplus \rho A) = \sum_0^d \binom{d}{k} W_k(A,K)\rho^k$$

and these "mixed functionals" $(A,K) \to W_k(A,K)$ are continuous on the product space $C(\mathcal{K}') \times C(\mathcal{K}')$, invariant under translations, and C-additive with respect to each of their arguments A and K. But $K \to W_k(A,K)$ does not admit a representation in the form (3-5-1), except when A belongs to the Steiner class (Section 4-5).

Theorem 4-1-1 implies, as corollaries, the useful and celebrated Steiner and Crofton formulae.

The Steiner Formula

The volume $V(K \oplus \rho B)$ of the dilatation of $K \in C(\mathcal{K}')$ by the ball ρB with radius $\rho \geqslant 0$ is

$$V(K \oplus \rho B) = \sum_{k=0}^{d} \binom{d}{k} W_k(K) \rho^k \tag{4-1-8}$$

More generally, the Minkowski functional $W_i (0 \leqslant i \leqslant k)$ satisfies

$$W_i(K \oplus \rho B) = \sum_{k=0}^{d-i} \binom{d-i}{k} W_{k+i}(K) \rho^k \tag{4-1-9}$$

Proof. The mapping $K \to W_i(K \oplus \rho B)$ is continuous on $C(\mathcal{K}')$ and invariant under euclidean transformations. It is also C-additive, by the relationship $(K \cup K') \oplus A = (K \oplus A) \cup (K' \oplus A)$, which holds for any K, K', and A in $\mathcal{P}(E)$, and

$$(K \cap K') \oplus A = (K \oplus A) \cap (K' \oplus A)$$

$$[K, K', A, \text{ and } K \cup K' \in C(\mathcal{K})] \tag{4-1-10}$$

which holds if K, K', A, and $K \cup K'$ are convex. Then it follows from Theorem 4-1-1 that $W_i(K \oplus \rho B) = \Sigma a_k(\rho) W_k(K)$. By taking $K = \lambda B$, we get $(\rho + \lambda)^{d-i} = \Sigma a_k(\rho) \lambda^{d-k}$, and an easy calculation yields (4-1-8) and (4-1-9).

The Steiner formula may be generalized in the following way: by substituting an arbitrary $K' \in C(\mathcal{K})$ for the ball ρB and taking the rotation average of $V(K \oplus \omega K')$, $\omega \in \Omega(E)$, we define a functional on $C(\mathcal{K}') \times C(\mathcal{K}')$ which satisfies the conditions of Theorem 4-1-1 with respect to each of its arguments K and K'. By identifying the coefficients, we get, after calculations,

$$\int_{\Omega} W_i(K \oplus \omega K') \varpi(d\omega) = \frac{1}{b_d} \sum_{k=0}^{d-i} W_{k+i}(K) W_{d-k}(K') \tag{4-1-11}$$

The Crofton Formula

Taking $K \in C(\mathcal{K})$ and $S \in \mathbb{S}_k$, we may consider the cross sections $K \cap (S \oplus \{s\}) = K \cap S_s$, where $s \in S^{\perp}$, and compute the translation integrals $\int_{S^{\perp}} W_{k'}{}^{k}(K \cap S_s)\mu_{d-k}(dS)$ (μ_{d-k} is the Lebesgue measure in $S^{\perp}$ identified with $\mathbf{R}^{d-k}$, and $W_{k'}{}^{k}$ is the Minkowski functional in $\mathbf{R}^{k}$). Thus we obtain a functional invariant under translations. If we now take the rotation average, we get an invariant functional on $C(\mathcal{K})$, and we can verify that it is increasing and C-additive, so that statement b of Theorem 4-1-1 may be applied. It is easy to compute the numerical coefficients by taking $K = B$, that is, the unit ball. Finally, we obtain

$$\int_{\mathbb{S}_k} \varpi_k(dS) \int_{S^{\perp}} W_{k'}{}^{k}(K \cap S_s)\mu_{d-k}(ds) = \frac{b_k b_{d-k'}}{b_d b_{k-k'}} W_{k'}{}^{d}(K) \quad (4\text{-}1\text{-}12)$$

4-2. THE RACS ALMOST SURELY CONVEX

A closed set F in the euclidean space $E = \mathbf{R}^d$ is convex if $x, y \in F$ implies $\lambda x + (1-\lambda)y \in F$ for $0 \leqslant \lambda \leqslant 1$. More generally, use the following definition.

4-2-1. *Let K, K', and C be three compact sets in E. Then we say that K and K' are separated by C if, for any $x \in K$ and $x' \in K'$ there exists a real λ such that $0 \leqslant \lambda \leqslant 1$ and $\lambda x + (1-\lambda)x' \in C$.*

Note also the following result.

Proposition 4-2-1. *Let K and K' be two compact sets, the union $K \cup K'$ of which is convex. Then K and K' are separated by $K \cap K'$.*

Proof. Take $x \in K$ and $x' \in K'$, and put $x(\alpha) = \alpha x + (1-\alpha)x'$, $0 \leqslant \alpha \leqslant 1$. We have $x(\alpha) \in K \cup K'$, for $K \cup K'$ is convex. Put $\alpha_0 = \text{Sup}\{\alpha, x(\alpha) \in K'\}$. Then $x(\alpha_0) \in K'$, because K' is compact. If $\alpha_0 = 1$, $x(\alpha_0) \in K \cap K'$ and the proposition holds. If $\alpha_0 < 1$, we have $x(\alpha_0 + \epsilon) \in K$ for any $\epsilon > 0$ small enough, and thus $x(\alpha_0) \in K$ because K is compact. Then $x(\alpha_0) \in K \cap K'$.

By Definition 4-2-1, if K and K' are separated by C, a convex set cannot hit K and K' without hitting C. In other words, $\mathcal{F}^{C}_{K,K'}$ and $C(\mathcal{F})$ are disjoint in $\mathcal{F}$. In particular, a RACS A a.s. convex will satisfy $P(\mathcal{F}^{C}_{K,K'}) = 0$, that is,

$$T(K \cup K' \cup C) + T(C) = T(K \cup C) + T(K' \cup C)$$

and by Proposition 4-2-1 its functional T is C-additive on $C(\mathcal{K})$.

The converse is also true:

Theorem 4-2-1. *Let A be a RACS in a euclidean space E, and T the functional $K \to T(K) = P(\{A \cap K \neq \phi\})$, $K \in \mathcal{K}$. Then the three following conditions are equivalent:*

a. The RACS A is a.s. convex.

b. For any K, K', and $C \in \mathcal{K}$ such that K and K' are separated by C, the following relationship holds:

$$T(K \cup K' \cup C) + T(C) = T(K \cup C) + T(K' \cup C) \qquad (4\text{-}2\text{-}1)$$

c. The functional T is C-additive on $C(\mathcal{K}')$, that is $T(K \cup K') + T(K \cap K') = T(K) + T(K')$ for K, K', and $K \cup K' \in C(\mathcal{K}')$.

Proof. Condition a implies b by Definition 4-2-1, and condition b implies c by Proposition 4-2-1. It remains to show that condition c implies a.

If $F \in \mathcal{F}$ is not convex, there exist $x \in F$, $x' \in F$, and λ such that $0 < \lambda < 1$ and $y = \lambda x + (1-\lambda)x' \notin F$. Because F is closed, there exists a ball $B_\epsilon(y_0)$ with center y_0 rational and radius $\epsilon > 0$ *rational* such that $y \in B_\epsilon(y_0)$ and $F \cap B_\epsilon(y_0) = \varnothing$ Then we can choose two points x_0 and x_0' with rational coordinates in $E = \mathbf{R}^d$ such that x_0, x_0', and y_0 lie on the same straight line, and $x \in B_\epsilon(x_0)$, $x' \in B_\epsilon(x_0')$. Let C be the convex hull of $B_\epsilon(x_0) \cup B_\epsilon(y_0)$, and C' the convex hull of $B_\epsilon(x_0') \cup B_\epsilon(y_0)$. Obviously, $C \cup C'$ is then the convex hull of $B_\epsilon(x_0) \cup B_\epsilon(x_0')$. Thus C, C', $C \cup C' \in C(\mathcal{K})$, and $F \in \mathcal{F}^{C \cap C'}_{C,C'}$. But the class $\mathbf{C}$ of the possible pairs (C, C') is *countable*. Thus a RACS A is a.s. convex if and only if $P(\mathcal{F}^{C \cap C'}_{C,C'}) = 0$ for any $(C, C') \in \mathbf{C}$. By $P(\mathcal{F}^{C \cap C'}_{C,C'}) = T(C) + T(C') - T(C \cup C') - T(C \cap C')$, it follows that condition c implies a.

If a RACS A is a.s. convex and compact, $W_k(A)$ is a random variable, because the Minkowski functional is continuous, and thus measurable, on $C(\mathcal{K})$. We impose the condition of a.s. compactness, because we have not yet defined the extension of W_k on $C(\mathcal{F})$. However, the volume $V = W_0$ is defined without ambiguity by $V(F) = \mu_d(F)$, $F \in \mathcal{F}$ (μ_d is the Lebesgue measure). It is easy to verify that a closed set $F \in \mathcal{F}$ is compact if and only if the dilatation $F \oplus \epsilon B$ of F by a ball with radius $\epsilon > 0$ admits a finite volume. It follows that a RACS A is a.s. compact if and only if the random variable $V(A \oplus \epsilon B)$ is a.s. finite for an $\epsilon > 0$. By the Steiner formula, if A is a RACS a.s. convex and compact, we have

$$V(A \oplus \epsilon B) = \sum_{k=0}^{d} \binom{d}{k} W_k(A) \epsilon^k \quad \text{(a.s.)}$$

In particular, if $E(V(A\oplus\epsilon B))<\infty$ for a given $\epsilon>0$, then A is a.s. compact and all the $W_k(A)$ admit finite expectations $(k=0,1,\ldots,d)$. Moreover, $E(V(A\oplus\rho B))<\infty$ for any $\rho>0$. The converse is obvious. Furthermore, by Theorem 2-5-1, Corollary 1, we have

$$E(V(A\oplus B))=\int P(x\in A\oplus B)\,dx=\int T(B_x)\,dx$$

and we may state:

Proposition 4-2-2. *Let A be an a.s. convex RACS, and T the functional $K\to T(K)=P(\{A\cap K\neq\varnothing\})$. Then A is a.s. compact, and the random variables $W_k:A\to W_k(A)$ admit finite expectations $E(W_k)<\infty$ $(k=0,1,\ldots,d)$ if and only if $\int T((\epsilon B)_x)\,dx<\infty$ for a ball ϵB with radius $\epsilon>0$. If so, for any $\rho\geqslant 0$,*

$$E[V(A\oplus\rho B)]=\int T((\rho B)_x)\,dx=\sum_{k=0}^{d}\binom{d}{k}E(W_k)\rho^k$$

The procedure used above for constructing a rotation-invariant functional from a given arbitrary functional may be used also for *isotropizing* a given RACS A. In fact, if Ω denotes the orthogonal group, the mapping $(\omega,F)\to\omega F$ is continuous on $\Omega\times\mathfrak{F}$ and thus measurable. Let ϖ be the unique probability on Ω invariant under rotation, P the probability on σ_f associated with A, and $\varpi\otimes P$ the product probability on $\Omega\times\mathfrak{F}$. Then the probability P' on σ_f associated with the RACS A': $(\omega,F)\to\omega(F)$ is invariant under the rotations, and A' is isotropic. In particular, the functional T': $K\to T'(K)=P(\{A'\cap K\neq\varnothing\})$ is given by

$$T'(K)=\int_{\Omega}T(\omega^{-1}K)\varpi(d\omega)=\int_{\Omega}T(\omega K)\varpi(d\omega)\qquad(K\in\mathcal{K})$$

In other words, *T' is the rotation average of T.* We shall say that A' is the transform of A by *isotropizing*.

Proposition 4-2-3. *Any stationary a.s. convex RACS A is a.s. $\varnothing$ or E [i.e., $P(\{A=\varnothing\}\cup\{A=E\})=1$].*

Proof. Let A be a stationary RACS a.s. convex. Then the isotropized set A' is stationary, isotropic, and a.s. convex, so that its functional T' satisfies the conditions of Theorem 4-1-1 (statement *b*). But W_d is the only bounded Minkowski functional, so that T' is constant on $C(\mathcal{K})$. It follows that A' is a.s. $\varnothing$ or E, and so also is A.

By (4-1-11), we may also state:

Proposition 4-2-4. *If a RACS A is isotropic and a.s. convex, then for any* $K \in C(\mathcal{K})$,

$$E(V(A \oplus K)) = \frac{1}{b_d} \sum_{k=0}^{d} \binom{d}{k} E(W_k(A)) W_{d-k}(K)$$

In particular, if $d=2$ (i.e., E is the euclidean plane), the norm N and the perimeter are identical, and

$$E(V(A \oplus K)) = E(V(A)) + \frac{1}{2\pi} N(K) E(N(A)) + V(K)$$

In the same way, if $d=3$, we denote by F and N the surface area and the norm, respectively, and we have

$$E(V(A \oplus K)) = E(V(A)) + \frac{1}{4\pi} E(F(A)) N(K) + \frac{1}{4\pi} E(N(A)) F(K) + V(K)$$

4-3. THE LINEAR GRANULOMETRIES

Let K be a convex compact set in $E = \mathbf{R}^d$. For any vector $h \in \mathbf{R}^d$, we denote by $g_K(h)$ the volume of the intersection $K \cap K_h$, where $K_h = K \oplus \{h\}$ is the translate of K by h. Thus, if 1_K is the indicator of K, we have

$$g_K(h) = \mu_d(K \cap K_h) = \int_E 1_K(x) 1_K(x+h)\, dx \tag{4-3-1}$$

The following relationships are obvious:

$$g_K(o) = V(K) = \mu_d(K)$$

$$\left.\begin{array}{l} 0 \leqslant g_K(h) \leqslant g_K(o) \\ \int_E g_K(h)\, dh = (V(K))^2 \end{array}\right\} \quad (h \in E) \tag{4-3-2}$$

The mapping $(K,h) \to g_K(h)$ is *continuous* on $C(\mathcal{K}') \times E$.

Proof. This mapping is u.s.c., for the translation $h \to K_h$ is continuous, the intersection $(K,h) \to K \cap K_h$ is u.s.c., and the Lebesgue measure is u.s.c. on $\mathcal{K}$. On the other hand, $g_K(h) = \mu_d(\mathring{K} \cap \mathring{K}_h)$, for $K \cap K_h$ is convex and its interior is $\mathring{K}\cap\mathring{K}_h$. By Lemma 3-5-1, the mapping $K\to\mathring{K}$ is continuous on $C(\mathcal{K})$. The intersection is continuous on $\mathcal{G}$, and the Lebesgue measure is l.s.c. on $\mathcal{G}$. It follows that $(K,h)\to g_K(h)$ is l.s.c. and thus continuous on $C(\mathcal{K}')\times E$.

Now we consider $K\in C(\mathcal{K}')$ as a fixed element, and write g instead of g_K. It will be convenient to use *polar coordinates*: let S_0 be the unit sphere in $\mathbf{R}^d$. Then any $h\in\mathbf{R}^d\setminus\{o\}$ is determined by the couple (r,u), where $r=|h|$ is a real number >0 and $u\in S_0$ is the unit vector $u=h/r$. We shall write $g(r,u)$ instead of $g(ru)$. By the above considerations, for a fixed $u\in S_0$, $r\to g(r,u)$ is a continuous function on $\mathbf{R}^+$. Moreover, this function possesses differentiability properties that we will now examine.

Let us denote by

$$\gamma(r,u)=\mu_{d-1}(\Pi_{u^\perp}(K\cap K_{ru}))$$

the $(d-1)$-volume of the projection of the set $K\cap K_{ru}$ into the hyperplane orthogonal to the direction u of the translation vector ru.

For a given $u\in S_0$, $r\to\gamma(r,u)$ is continuous from the left at any $r>0$, for $r_n\uparrow r$ implies $K\cap K_{r_n u}\downarrow K\cap K_{ru}$ (because the intersection is u.s.c.) and $\Pi_{u^\perp}(K\cap K_{r_n u})\downarrow\Pi_{u^\perp}(\mathring{K}\cap K_{ru})$ by Lemma 3-5-3. Thus $\gamma(r_n,u)\downarrow\gamma(r,u)$ by sequential monotone continuity.

In the same way, $r\to\gamma(r,u)$ is continuous from the right at any $r\geqslant 0$ such that $g(r;u)>0$, for $\mathring{K}\cap\mathring{K}_{ru}$ is not empty and its projection on $u^\perp$ is identical with the interior of $\Pi_{u^\perp}(K\cap K_{ru})$ (considered as a subset in the euclidean space $u^\perp=\mathbf{R}^{d-1}$). It follows that $r\to\gamma(r,u)$ is l.s.c. and thus continuous from the right.

For any $r>0$ and ϵ such that $r>\infty>0$, the obvious inequalities

$$\epsilon\gamma(r-\epsilon,u)\geqslant g(r-\epsilon,u)-g(r,u)\geqslant\epsilon\gamma(r,u)$$

and the continuity from the left of $r\to\gamma(r,u)$ imply that g itself admits a derivative from the left which is $-\gamma(r;u)$. In the same way, the inequalities

$$\epsilon\gamma(r,u)\geqslant g(r,u)-g(r+\epsilon,u)\geqslant\epsilon\gamma(r+\epsilon,u)$$

imply that $g(r,u)$ admits a derivative from the right which is again $-\gamma(r,u)$, at any point r such that $g(r,u)>0$. Let us state these results.

Proposition 4-3-1. *Let K be a convex compact set and $u\in S_0$ a unit vector. For any $r\geqslant 0$, put $g(r,u)=\mu_d(K\cap K_{ru})$ and $\gamma(r,u)=\mu_{d-1}(\Pi_{u^\perp}(K\cap K_{ru}))$. Then:*

a. The function $r\to g(r,u)$ is nonincreasing, continuous at any $r\geqslant 0$, and admits a continuous derivative, equal to $-\gamma(r,u)$ on the open interval defined by $r>0$, $g(r,u)>0$.

b. The function $r \to \gamma(r,u)$ is continuous from the left at any $r > 0$, and from the right at any $r \geqslant 0$ such that $g(r,u) > 0$. Moreover, if $\mathring{K} \neq \emptyset$, $r \to g(r,u)$ admits at $r = 0$ a derivative from the right equal to $-\gamma(o,u)$, and at any $r > 0$ a derivative from the left equal to $-\gamma(r,u)$.

The Random Intercepts

If $K \in C(\mathcal{K}')$ has a nonempty interior $\mathring{K}$, the function $r \to \gamma(r,u)$ admits a simple probabilistic interpretation. Let x be a random point in the hyperplane $u^\perp$, uniformly distributed on $\Pi_{u^\perp} K$ (i.e., its probability is $(1_{\Pi_{u^\perp} K} / \gamma(o,u))\mu_{d-1}(dx))$ and denote by $L(x)$ the length of the intersection $K \cap \{x + \lambda u, \lambda \in \mathbf{R}\}$. Then $x \to L(x)$ is a random variable, the law of which is defined by

$$P(L \geqslant l) = \frac{\gamma(l;u)}{\gamma(o,u)}$$

We say that L is the *random intercept* of K for the given direction $u \in S_0$.

We may also define a random intercept with a random direction. The most natural way is to use a stationary Poisson network isotropic in E and the definition of the random secant given in Section 3-5. Then the random intercept is the length L of the intersection $K \cap D$, where D denotes the random secant. By (3-5-4), the probability of the direction $S \in \mathbb{S}_1$ of the random secant D is not the invariant law ϖ_1, but rather

$$\frac{\gamma(o,u)\varpi_1(du)}{\int_{\mathbb{S}_1} \gamma(o,u)\varpi_1(du)} = \frac{db_d}{b_{d-1}} \frac{\gamma(o,u)\varpi_1(du)}{F(K)}$$

[In this formula, we implicitly identify a vector $u \in S_0$ with the element $S \in \mathbb{S}_1$ parallel to u; this is legitimate, because the function $u \to \gamma(o,u)$ is symmetrical, i.e., $\gamma(o,u) = \gamma(o,-u)$.]

In the same way, conditionally for a fixed direction $u \in S_0$, the hitting point $s \in u^\perp \cap D$ of D into the orthogonal $u^\perp$ is uniformly distributed in $\Pi_{u^\perp} K$ as above, so that $P_u(L \geqslant l) = \gamma(l,u)/\gamma(o,u)$. It follows that the corresponding unconditional probability is

$$P(L \geqslant l) = \left(\frac{db_d}{b_{d-1} F(K)} \right) \int_{\mathbb{S}_1} \gamma(l,u)\varpi_1(du) \tag{4-3-3}$$

This relationship leads to an easy calculation of the expectation $E(L) = \int_0^\infty P(L \geqslant l)\,dl$. By Proposition 4-3-1, we have $\gamma(l,u) = -g_l'(l,u)$ and thus $\int_0^\infty \gamma(l,u)\,du = g(o,u) = V(K)$. In other words, the expectation of the random intercept is proportional to the ratio $V(K)/F(K)$ = volume/surface area. Explicitly,

$$E(L) = \frac{db_d}{b_{d-1}} \frac{V(K)}{F(K)} \tag{4-3-4}$$

On the other hand, let us consider the third relationship (4-3-2), that is,

$$(V(K))^2 = db_d \int_{\mathbb{S}_1} \varpi_1(du) \int_0^\infty r^{d-1} g(r,u)\,dr$$

Taking Proposition 4-3-1 into account, we obtain by an easy calculation

$$\int_0^\infty r^{d-1} g(r,u)\,dr = \frac{1}{d} \int_0^\infty r^d \gamma(r,u)\,dr$$

and by substituting this result in $E(L^{d+1}) = (d+1)\int_0^\infty l^d P(L \geqslant l)\,dl$, we get

$$E(L^{d+1}) = \frac{d(d+1)}{b_{d-1}} \frac{(V(K))^2}{F(K)}$$

4-4. THE CONVEX CONE $\mathfrak{R} = C_0(\mathcal{K})$

We will now characterize the space $C_0(\mathcal{K})$ (i.e., the space of the compact convex sets in E containing the origin O) as a convex cone with a compact basis in a functional space (for a general theory of convex cones, see, e.g., Alfsen, 1971). Let S_0 be the unit sphere in $E = \mathbf{R}^d$, and $\mathbf{C}(S_0)$ the space of the functions continuous on S_0, topologized by uniform convergence. For a given $K \in C_0(\mathcal{K})$, we may associate with each $u \in S_0$ the distance $r_K(u)$ between the origin O and the supporting hyperplane, the positive normal of which is u. Conversely, it is well known that the convex compact K itself may be defined as the intersection $K = \bigcap_{u \in S_0} \{x: \langle x,u \rangle \leqslant r_K(u)\}$ of the closed half spaces associated with its supporting hyperplanes ($\langle x,u \rangle$ denotes the scalar product in $\mathbf{R}^d$).

We shall say that the function $u \to r_K(u)$ is the *supporting function* of $K \in C_0(\mathcal{K})$. For each $K \in C_0(\mathcal{K})$, the supporting function r_K is continuous on S_0, that is, $r_K \in \mathbf{C}(S_0)$.

We denote by α the one-to-one mapping $K \to r_K$ from $C_0(\mathcal{K})$ into $\mathbf{C}(S_0)$, and by $\mathcal{R}$ the range of α in $\mathbf{C}(S_0)$, that is, the class of the supporting functions.

The mapping α is a *homeomorphism* from $C_0(\mathcal{K})$ onto its range $\mathcal{R} \subset \mathbf{C}(S_0)$, for if B is the unit ball in $\mathbf{R}^d$,

$$\operatorname{Sup}_{u \in S_0} |r_K(u) - r_{K'}(u)| \leqslant \epsilon$$

is equivalent to $K \subset K' \oplus \epsilon B$ and $K' \subset K \oplus \epsilon B$, that is, $\rho(K, K') \leqslant \epsilon$ if ρ is the Hausdorff metric.

On the other hand, by the obvious relationships

$$r_{K \oplus K'} = r_K + r_{K'}; \qquad r_{\alpha K} = \alpha r_K$$

$$[\alpha \geqslant 0, K, K' \in C_0(\mathcal{K}),\ K \subset K' \text{ if and only if } r_K \leqslant r_{K'}]$$

the mapping α is also an *isomorphism* from $C_0(\mathcal{K})$ into $\mathcal{R}$ with respect to the following operations: Minkowski's addition in $C_0(\mathcal{K})$ and ordinary addition in $\mathbf{C}(S_0)$; positive homethetics in $C_0(\mathcal{K})$ and multiplication by positive constants in $\mathbf{C}(S_0)$; orderings $\subset$ in $C_0(\mathcal{K})$ and $\leqslant$ in $\mathbf{C}(S_0)$. Hence $C_0(\mathcal{K})$ may be *identified with the convex cone* $\mathcal{R}$, closed in $\mathbf{C}(S_0)$ and admitting a compact base.

Proposition 4-4-1. *The cone $\mathcal{R}$ is max-stable, that is, f, $g \in \mathcal{R}$ implies* $\operatorname{Sup}(f,g) \in \mathcal{R}$. *Moreover, if K_i, $i \in I$, is a class of compact convex sets enclosed in a fixed compact set, the supporting functions r_{K_i} are uniformly bounded, and* $\operatorname{Sup}\{r_{K_i}, i \in I\}$ *is the supporting function of the closed convex hull $C(\overline{\cup K_i}) \in C_0(\mathcal{K})$.*

Proof. Let r_K, $r_{K'}$ be the supporting functions of K, $K' \in C_0(\mathcal{K})$. Obviously, $r_{C(K \cup K')} = \operatorname{Sup}(r_K, r_{K'}) \in \mathcal{R}$. If a family K_i, $i \in I$, is bounded in $C_0(\mathcal{K})$, put $K_J = C(\cup_{i \in J} K_i)$ for any finite subset $J \subset I$. Thus the filtering family K_J converges in $\mathcal{K}$ toward $K_0 = C(\overline{\cup_{i \in I} K_i})$, by Theorem 1-2-2, Corollary 3, and the continuity of C on $\mathcal{K}$ (Proposition 1-5-4). Thus the family r_J of the corresponding supporting functions converges in $\mathbf{C}(S_0)$ toward the supporting function r_{K_0} of $K_0 \in C_0(\mathcal{K})$, by the homeomorphism $K \to r_K$. Then $r_J = \operatorname{Sup}\{r_{K_i}, i \in J\}$ implies $r_{K_0} = \operatorname{Sup}\{r_{K_i}, i \in I\} \in \mathcal{R}$.

It is possible to improve Proposition 4-4-1. Let u_1 and u_2 be two unit vectors, $\langle u_1, u_2 \rangle$ be their scalar product in $\mathbf{R}^d$, and $\langle u_1, u_2 \rangle_+ = \langle u_1, u_2 \rangle$ if $\langle u_1, u_2 \rangle \geqslant 0$ and 0 otherwise. For a given $u_0 \in S_0$, the function $u \to \langle u, u_0 \rangle_+$ is the supporting function of the segment $\{\lambda u_0, 0 \leqslant \lambda \leqslant 1\} \in C_0(\mathcal{K})$ and thus belongs to $\mathcal{R}$. Let (λ_i, u_i), $i \in I$, be a subset in $\mathbf{R}_+ \times S_0$ such that $\mathrm{Sup}_i \lambda_i < \infty$. Then, by the proposition, the function r: $u \to r(u) = \mathrm{Sup}\{\lambda_i \langle u, u_i \rangle_+, i \in I\}$ belongs to $\mathcal{R}$. Conversely, each supporting function admits such a representation, because any $K \in C_0(\mathcal{K})$ is the closed convex hull of its extremal points. It is even possible to take a countable set I, by choosing the family (λ_i, u_i), $i \in I$, so that $\{\lambda_i u_i, i \in I\}$ is a countable subset dense in ∂K. Hence:

Corollary 1. *A function r on S_0 belongs to $\mathcal{R}$ if and only if it admits the representation $r(u) = \mathrm{Sup}\{\lambda_i \langle u, u_i \rangle_+, i \in I\}$, $u \in S_0$, for a countable family (λ_i, u_i), $i \in I$, in $\mathbf{R}_+ \times S_0$.*

Corollary 2. *A function r on S_0 belongs to $\mathcal{R}$ if and only if the set $F = \{x: |x| r(x/|x|) \leqslant 1\}$ is a closed and convex neighborhood of O in $E = \mathbf{R}^d$.*

Proof of Corollary 2. By Corollary 1, $r \in \mathcal{R}$ if and only if $r(u) = \mathrm{Sup}\{\lambda_i \langle u, u_i \rangle_+, i \in I\}$ $(u \in S_0)$ for a family (λ_i, u_i), $i \in I$, in $\mathbf{R}_+ \times S_0$ such that $\gamma = \mathrm{Sup}\,\lambda_i < \infty$. But this is equivalent to $F = \cap_{i \in I}\{x: \langle x, u_i \rangle_+ \leqslant 1/\lambda_i\}$ and $\mathrm{Inf}(1/\lambda_i) > 0$, and this condition is satisfied if and only if F is a closed and convex neighborhood of O.

Note that the closed set F is compact if and only if $r > 0$ strictly on S_0, that is, if and only if o is an interior point of the set $K \in C_0(\mathcal{K})$, the supporting function of which is r.

The cone $\mathcal{R}$ is max-stable by Proposition 4-1-1, but it is not closed under the operation Inf. In order to characterize this operation, we shall use the following lemma.

Lemma 4-4-1. *If K, $K' \in C(\mathcal{K})$, then $K \cup K' \in C(\mathcal{K})$ if and only if $K \oplus K' = (K \cup K') \oplus (K \cap K')$.*

Proof. The inclusion $(K \cup K') \oplus K \cap K' \subset K \oplus K'$ holds for arbitrary sets K, $K' \in \mathcal{P}(E)$, and we must prove that the converse inclusion is equivalent to $K \cup K' \in C(\mathcal{K})$.

Suppose that $K \cup K' \in C(\mathcal{K})$, and let $y \in K \oplus K'$, that is, $y = x + x'$ with $x \in K, x' \in K'$. By Proposition 4-2-1, there exists $\lambda \in [0,1]$ such that $\lambda x + (1-\lambda)x' \in K \cap K'$. Thus $y = z_1 + z_2$, with $z_1 = \lambda x + (1-\lambda)x' \in K \cap K'$ and $z_2 = (1-\lambda)x + \lambda x' \in K \cup K'$ (because $K \cup K'$ is convex). It follows that $y \in (K \cup K') \oplus (K \cap K')$, and the desired inclusion holds.

Conversely, suppose that $(K \cup K') \oplus (K \cap K') = K \oplus K'$, and take $x \in K, x' \in K'$. We have $x + x' = y + z$ for a point $y \in K \cup K'$ and a point

$z \in K \cap K'$. It follows that $(x+x')/2 = (y+z)/2 \in K \cup K'$. By repeating this procedure, we see that $\{\lambda x + (1-\lambda)x', 0 \leqslant \lambda \leqslant 1\}$ is included in the closed set $K \cup K'$. Hence $K \cup K' \in C(\mathcal{K})$.

Proposition 4-4-2. *Let K, K' be in $C_0(\mathcal{K})$, and r_K, $r_{K'}$ be their supporting functions. Then the supporting function of $K \cap K'$ is $r_{K \cap K'} = \mathrm{Sup}\{r\colon r \in \mathcal{R}, r \leqslant \mathrm{Inf}(r_K, r_{K'})\}$. We have* Inf $(r_K, r_{K'}) \in \mathcal{R}$ *if and only if $K \cup K'$ is convex, and in this case $r_{K \cap K'} =$* Inf $((r_K, r_{K'})$.

Proof. The first statement follows from Proposition 4-4-1, because $K \cap K'$ is the largest compact convex set included in K and K'. By $r_K + r_{K'} = \mathrm{Sup}(r_K, r_{K'}) + \mathrm{Inf}(r_K, r_{K'})$, we have $K \oplus K' = (K \cup K') \oplus (K \cap K')$ if and only if $r_{K \cap K'} = \mathrm{Inf}(r_K, r_{K'})$, that is, by the first statement if and only if $\mathrm{Inf}(r_K, r_{K'}) \in \mathcal{R}$. By Lemma 4-4-1, the proposition follows.

Note. If a functional L on $C_0(\mathcal{K})$ satisfies $L(K \oplus K') = L(K) + L(K')$, $K, K' \in C_0(\mathcal{K})$, then L is C-additive.

It follows from Definition 3-5-1 and Lemma 4-4-1.

The Ordering $\geqslant$ in $\mathcal{R} = C_0(\mathcal{K})$

The space $\mathbf{C}(S_0)$ may be provided with the (partial) ordering $\geqslant$ defined by $f \geqslant f'$ if $f - f' \in \mathcal{R}$, and, in the same way, $C_0(\mathcal{K})$ itself may be provided with the ordering induced by $\geqslant$. By the definition, $A \geqslant B$ in $C_0(\mathcal{K})$ is equivalent to $r_A - r_B \in \mathcal{R}$, and this relationship is verified if and only if there exists a $C \in C_0(\mathcal{K})$ such that $r_A = r_B + r_C$, that is, $A = B \oplus C$. In other words, when using the terminology defined in Section 1-5, *$A \geqslant B$ if and only if A is open with respect to B*, that is, $A_B = A$.

A convex cone is a simplex if every point in it admits a unique representation by a boundary integral (see, e.g., Choquet, 1960, or Alfsen, 1971). But in $\mathbf{R}^2$ a regular hexagon is the Minkowski sum of two equilateral triangles, as well as the sum of three line segments, and triangles and line segments are boundary elements in $C_0(\mathcal{K})$. We conclude that *the convex cone $C_0(\mathcal{K})$ is not a simplex*.

In the following section, we shall define another convex cone $\mathcal{R}_1 \subset \mathcal{R}$ identified with a subset in $C_0(\mathcal{K})$ called the *Steiner class*, and $\mathcal{R}_1$ will be a simplex. In the euclidean plane $\mathbf{R}^2$, the Steiner class will be identical with the class $\mathcal{R}_s$ of the symmetrical compact convex sets, but the inclusion $\mathcal{R}_1 \subset \mathcal{R}_s$ is strict if $d > 2$.

The Minkowski Space $\mathfrak{M}(S_0)$

The vector space spanned in $\mathbf{C}(S_0)$ by the cone $\mathfrak{R}$ is called the *Minkowski space* and denoted by $\mathfrak{M}(S_0)$. In other words, $\varphi \in \mathfrak{M}(S_0)$ if and only if there exists $f_1, f_2 \in \mathfrak{R}$ such that $\varphi = f_1 - f_2$. The Minkowski space is dense in (but not identical with) $\mathbf{C}(S_0)$. This follows from Lemma 4-4-2, in which $\mathbf{C}_2(S_0)$ denotes the space of the functions twice continuously differentiable on S_0.

Lemma 4-4-2. *For any $f \in \mathbf{C}_2(S_0)$, there exists a constant $C \geqslant 0$ such that $f + C \in \mathfrak{R}$, and $C_2(S_0) \subset \mathfrak{M}(S_0)$.*

Proof. The positive constants C are in $\mathfrak{R}$, as the supporting functions of the balls with center O, so that the second statement follows from the first one. Let f be a function in $\mathbf{C}_2(S_0)$, ∇f its second differential (continuous on S_0), and γ the metric tensor on the unit sphere S_0. Then the tensor $\gamma f + \nabla f$ is continuous on the compact space S_0, and thus there exists a constant $C \geqslant 0$ such that $\gamma(f + C) + \nabla f$ is positive definite on S_0. But it is well known that a function $\varphi \in \mathbf{C}_2(S_0)$ is a supporting function (i.e., $\in \mathfrak{R}$) if and only if $\varphi\gamma + \nabla f$ is positive definite. Hence $f + C \in \mathfrak{R}$.

Extension of a Positively Linear Functional on $\mathfrak{R}$

Let L be a functional defined on $\mathfrak{R} = C_0(\mathcal{K})$. We say that L is *increasing* if $r \leqslant r'$ in $\mathfrak{R}$ implies $L(r) \leqslant L(r')$, and is *positively linear* if $L(\alpha r + \beta r') = \alpha L(r) + \beta L(r')$, for any α, $\beta \geqslant 0$ and $r, r' \in \mathfrak{R}$. If L is increasing and positively linear, $L(o) = 0$ and $L(r) \geqslant 0$ on $\mathfrak{R}$. By the following proposition, such a functional may be identified with a positive Radon measure on S_0.

Proposition 4-4-3. *Let L be an increasing and positively linear functional on $\mathfrak{R}$. Then there exists a unique linear and continuous extension of L on $\mathbf{C}(S_0)$. In other words, there exists a unique measure $\lambda \geqslant 0$ on S_0 such that*

$$L(r) = \int_{S_0} r(u)\lambda(du) \qquad (r \in \mathfrak{R})$$

Proof. Let f be a function in $\mathbf{C}_2(S_0)$, and $C \geqslant 0$ so that $f + C \in \mathfrak{R}$ (Lemma 4-4-2). $L(f + C) - L(C)$ does not depend on the choice of this constant C, for L is positively linear. Thus we may define a functional L_2 on $\mathbf{C}_2(S_0)$ by putting $L_2(f) = L(f + C) - L(C)$, and L_2 is identical with L

on $\mathcal{R} \cap \mathbf{C}_2(S_0)$. Clearly, L_2 is additive, $L_2(\alpha f) = \alpha L_2(f)$, $\alpha \geqslant 0$, and $L_2(-f) = -L_2(f)$, so that L_2 is linear on $\mathbf{C}_2(S_0)$. Moreover, L_2 is increasing; for, if $f \leqslant f'$ in $\mathbf{C}_2(S_0)$, let C be a constant $< \infty$ such that $f + C$ and $f' + C$ belong to $\mathcal{R}$ (Lemma 4-4-2). Then $L_2(f+C) \leqslant L_2(f'+C)$, and thus $L_2(f) \leqslant L_2(f')$. From a classical theorem (see Bourbaki, 1965b, Chapter III, Section 1, Proposition 9), it follows that L_2 admits an extension by a measure $\lambda \geqslant 0$ on $\mathbf{C}(S_0)$, and λ is unique, because $\mathbf{C}_2(S_0)$ is dense in $\mathbf{C}(S_0)$. Clearly, this extension is identical with L on $\mathcal{R}$.

4-5. THE CONVEX CONE $\mathcal{R}_1$ AND THE STEINER CLASS

In this section, we consider only the class $\mathcal{R}_s \subset \mathcal{R}$ of the nonempty compact convex sets *symmetric with respect to the origin O*. $\mathcal{R}_s$ is still a convex cone and admits a compact base. The functions $r \in \mathcal{R}_s$ are the symmetric supporting functions, that is, $r(u) = r(-u)$ for any $u \in S_0$. The quotient space $S_0/\equiv$, where $\equiv$ is the equivalence relationship defined by $u \equiv u'$ if $u = u'$ or $u = -u'$, may be identified with the space $\mathbb{S}_1 \subset \mathcal{F}$, that is, the space of the one-dimensional subspaces in $\mathbf{R}^d$, and in the same way the symmetric functions and measures on S_0 may be identified with the functions and measures on $\mathbb{S}_1$. Thus, in what follows, $\mathcal{R}_s$ will be considered as a convex cone admitting a compact basis in the space $\mathbf{C}(\mathbb{S}_1)$ of the continuous functions on $\mathbb{S}_1$ (topologized by uniform convergence).

For a given $u_0 \in S_0$, the segment $\{\lambda u_0, |\lambda| \leqslant 1\}$ is a symmetric compact convex set, and its supporting function $u \to |\langle u, u_0 \rangle|$ is symmetric in $\mathcal{R}$. If $L_0 \in \mathbb{S}_1$ is the one-dimensional subspace $\{\lambda u_0, \lambda \in R\} \in \mathbb{S}_1$, the element in $\mathcal{R}_s$ associated with this segment is the function $L \to |L, L_0|$ on $\mathbb{S}_1$ [with the notation $|L, L_0|$ defined in (3-5-5)]. We shall denote by $\mathcal{R}_1$ *the class* of functions φ on $\mathbb{S}_1$ admitting the *integral representation*

$$\varphi(L) = \int_{\mathbb{S}_1} |L, L'| G(dL') \qquad (L \in \mathbb{S}_1) \tag{4-5-1}$$

for a measure $G \geqslant 0$ *on* $\mathbb{S}_1$.

If G is concentrated on a finite subset in $\mathbb{S}_1$, φ is the supporting function of the *Minkowski sum* of a *finite number* of *symmetric line segments*. More generally, we shall say that a $K \in C_0(\mathcal{K})$ is a *Steiner compact*, or belongs to the Steiner class, if there exists a sequence $\{K_n\} \subset \mathcal{K}$ such that $K = \lim K_n$ in $\mathcal{K}$, and, for each $n > 0$, K_n is a finite Minkowski sum of symmetric line segments. Then *a function φ on $\mathbb{S}_1$ belongs to $\mathcal{R}_1$ if and only if φ is the supporting function of a Steiner compact*, so that the Steiner class in $C_0(\mathcal{K})$ may be identified with the convex cone $\mathcal{R}_1$.

Proof. If a sequence $\{G_n\}$ of positive measures on $\mathbb{S}_1$ weakly converges toward a measure G, the sequence $\{\varphi_n\} \subset \mathcal{R}_1$ defined by $\varphi_n(L) = \int |L, L'| G_n(dL')$ converges uniformly toward the function φ defined by (4-5-1). On the other hand, any positive measure on $\mathbb{S}_1$ is the weak limit of a sequence of finite measures. Thus $\varphi \subset \mathcal{R}_1$ if and only if φ is the supporting function of a Steiner compact.

The convex cone $\mathcal{R}_1$ is *closed* in $\mathbf{C}(\mathbb{S}_1)$ and admits a compact basis.

Proof. Let $\{\varphi_n\} \subset \mathcal{R}_1$ be a sequence uniformly converging toward $\varphi \subset \mathbf{C}(\mathbb{S}_1)$, and, for any $n > 0$, G_n be the positive measure such that $\varphi_n(L) = \int |L, L'| G_n(dL')$, $L \in \mathbb{S}_1$. If ϖ denotes the invariant probability on $\mathbb{S}_1$, the relationship $\int \varphi_n(L) \varpi(dL) = C \int G_n(dL')$ $[C = \int |L, L'| \varpi(dL)$ does not depend on $L']$ implies that the sequence $\{G_n\}$ is *bounded*. Hence there exists a subsequence $\{G_{n_k}\}$ weakly converging toward a measure $G \geqslant 0$, and the sequence $\{\varphi_{n_k}\}$ uniformly converges toward φ': $L \to \varphi'(L) = \int |L, L'| G(dL')$. Thus $\varphi = \varphi' \in \mathcal{R}_1$, and $\mathcal{R}_1$ is closed. Clearly, the subset in $\mathcal{R}_1$ constituted by the functions φ associated with the probabilities on $\mathbb{S}_1$ [i.e., $\int G(dL) = 1$ in (4-5-1)] is a compact basis for the closed convex cone $\mathcal{R}_1$.

In what follows, we shall show that $\mathcal{R}_1$ is a *simplex* [i.e., that the integral representation (4-5-1) is unique], or that the mapping $G \to \varphi$ defined by (4-5-1) is a homeomorphism from the space $\mathfrak{M}^+(S_0)$ of the measures $\geqslant 0$ on S_0 onto the Steiner cone $\mathcal{R}_1$.

First, note that, if $d=2$, *the Steiner class is identical to the class of the symmetric compact convex sets*, that is, $\mathcal{R}_1 = \mathcal{R}_s$, for any symmetric polygon in $\mathbf{R}^2$ is a finite Minkowski sum of symmetric segments, and the class of the symmetric polygons is dense in $\mathcal{R}_s$. But *this is not true for* $d > 2$, and the inclusion $\mathcal{R}_1 \subset \mathcal{R}_s$ is strict.

Proof. Let T be an equilateral triangle enclosed into a two-dimensional linear variety not including O, and $T' = T$ its symmetric set with respect to the origin. Then the convex hull $C(T \cup T') \in \mathcal{R}_s$ does not belong to $\mathcal{R}_1$, for any two-dimensional face of a polygon in $\mathcal{R}_1$ is a finite Minkowski sum of line segments and cannot be a triangle.

The Uniqueness Theorem

In order to prove the uniqueness of the integral representation (4-5-1), we shall use a procedure based on harmonic analysis. First, let us recall some classical results. A real function $h \to K(h)$ on $\mathbf{R}^d$ is said to be *conditionally positive definite order* 0 (or, for brevity, conditionally positive definite) if $\int\int \lambda(dx)K(x-y)\lambda(dy) \geqslant 0$ for any finite measure λ such that $\int \lambda(dx)=0$. (A finite measure is a measure the support of which is a finite subset in $\mathbf{R}^d$.)

Proposition 4-5-1. *a. A real function K on* $\mathbf{R}^d$ *is conditionally positive definite if and only if the function* $C=\exp(tK)$ *is positive definite for any* $t>0$. *If so, K is a symmetric function.*

b. A real continuous function K on $\mathbf{R}^d$ *is conditionally positive definite if and only if K admits the representation*

$$K(h)=\int \frac{\cos(2\pi\langle uh\rangle)-1}{4\pi^2|u|^2}\chi(du)+a+Q(h)$$

for a real number a, a positive quadratic form Q, and a symmetric measure $\chi \geqslant 0$ *on* $\mathbf{R}^d$, *necessarily unique, without an atom at the origin and such that* $\int(1/(1+4\pi^2|u|^2))\chi(du)<\infty$ ($\langle\rangle$ *is the scalar product in* $\mathbf{R}^d$).

Proof. a. If $\exp(tK)$ is positive definite for any $t>0$, the function $K_t=(\exp(tK)-1)/t$ is conditionally positive definite, and thus also $K=\lim K_t$, $t\downarrow 0$. Conversely, suppose that K is conditionally positive definite on $\mathbf{R}^d$. Then the function σ defined on $\mathbf{R}^d\times\mathbf{R}^d$ by $\sigma(x,y)=K(x-y)-K(x)-K(y)+K(o)$ is such that $\int\mu(dx)\sigma(x,y)\mu(dy)\geqslant 0$ for any finite measure μ, as can be verified by elementary calculations. Thus there exists an order-2 random function Y [i.e., a mapping from $\mathbf{R}^d$ into a Hilbert space $L^2(\Omega,\mathcal{Q},P)$] the covariance of which is σ, that is, $\sigma(x,y)=\langle Y(x),Y(y)\rangle$. Let $\{Y_n\}$ be a sequence of independent random functions equivalent to Y, and N a Poisson random variable independent of $\{Y_n\}$ and such that $E(N)=t$. Then the mapping Z: $\mathbf{R}^d\to L^2(\Omega,\mathcal{Q},P)$ defined by $Z=1$ if $N=0$ and by $Z=\Sigma_1^n Y_i$ if $N=n$ is an order-2 random function, the covariance of which is $(x,y)\to\langle Z(x),Z(y)\rangle=\exp(t\sigma(x,y)-1)$. Hence the random function $x\to Z(x)/\|Z(x)\|$ admits the covariance $(x,y)\to\exp(tK(x-y)-1)$, and thus the function $C=\exp(tK)$ is positive definite on $\mathbf{R}^d$.

b. The second statement follows immediately from part *a* by taking into account the classical results concerning the infinitely divisible probability laws.

These results can be used for investigating the behavior near the origin O of a real continuous *covariance* C on $\mathbf{R}^d$ (i.e., C is a positive definite continuous real function on $\mathbf{R}^d$).

More precisely, suppose that there exists a number α (and then, necessarily, $0\leqslant\alpha\leqslant 2$, as is well known) such that the limit $C_0(h)\equiv\lim[(C(ah)-C(o))/a^\alpha]$ exists for any $h\in\mathbf{R}^d$ if $a\downarrow 0$. For $\lambda>0$, we obviously have $C_0(\lambda h)=\lambda^\alpha C_0(h)$, so that there exists a symmetric function $\varphi \geqslant 0$ on the unit sphere S_0 such that $C_0(h)=-|h|^\alpha \varphi(h/|h|)$, $h \neq 0$. We examine only the case in which φ is continuous on S_0, that is, $\varphi \in \mathbf{C}(S_0)$. If $\alpha=2$, it is well known that $C_0(h)$ is a quadratic form $\geqslant 0$ on $\mathbf{R}^d$; and if $\alpha=0$, then $C_0(h)=0$. Thus the interesting case is $0<\alpha<1$. In the sequel, we shall use only the results concerning $\alpha=1$, and the following uniqueness theorem is more general than needed.

Theorem 4-5-1. *Let α be a real number such that $0<\alpha<2$, and $\varphi \in \mathbf{C}(S_0)$ be a continuous function on the unit sphere S_0 in $\mathbf{R}^d$. Then the three following conditions are equivalent (and imply that φ is a symmetric function):*

a. There exists a real covariance C on $\mathbf{R}^d$ such that, for any $h \in \mathbf{R}^d$, $\lim[(C(o)-C(ah))/a^\alpha]=|h|^\alpha \varphi(h/|\mathring{h}|)$ if $a \downarrow 0$.

b. The function K: $h \to K(h)=-|h|^\alpha \varphi(h/|h|)$ is conditionally positive definite on $\mathbf{R}^d$.

c. There exists a necessarily unique symmetric measure $\mu \geqslant 0$ on S_0 such that

$$\varphi(s)=\int_{S_0} |\langle s,s' \rangle|^\alpha \mu(ds') \qquad (s \in S_0)$$

Proof. Suppose that C is a covariance. Then, for any $a>0$, the function K_a: $h \to (C(ah)-C(o))/a^\alpha$ is conditionally positive definite, and so also is its limit K if $a \downarrow 0$. Thus condition *a* implies *b*.

The proof of the implication $b \Rightarrow c$ is more intricate. Suppose that K: $h \to K(h)=-|h|^\alpha \varphi(h/|h|)$ is conditionally positive definite. Then K admits the representation

$$K(h)=\int \frac{\cos(2\pi\langle uh \rangle)-1}{4\pi^2|u|^2}\chi(du) \tag{4-5-2}$$

for a *necessarily unique* symmetric measure $\chi \geqslant 0$ on $\mathbf{R}^d$, without an atom at the origin O and such that $\int(1/(1+4\pi^2|u|^2))\chi(du)<\infty$. [This follows from Proposition 4-5-1 and the relationships $K(o)=0$, $\lim(K(h)/|h|^2)=0$ if $|h|\uparrow\infty$.]

By $K(ah)=a^\alpha K(h)$, $a \geqslant 0$, the measure χ in (4-5-2) satisfies $\chi(a\,du) = a^{2-\alpha}\chi(du)$, because the spectral measure associated with a conditionally definite function is unique. For any borelian set $B \subset S_0$, put $\tilde{B}=\cup\{aB, 0 \leqslant a \leqslant 1\}$, and $\nu(B)=\chi(\tilde{B})/\chi(\tilde{S}_0)$, so that ν is a symmetrical probability on S_0. By $\chi(a\tilde{B})/\chi(S_0)=a^{2-\alpha}\nu(B)$, we have

$$\int f(u)\chi(du)=A\int_{S_0}\nu(d\sigma)\int_0^\infty f(\rho\sigma)\rho^{1-\alpha}\,d\rho$$

for any continuous function f with compact support in $\mathbf{R}^d$, $A=(2-\alpha)\chi(\tilde{S}_0)$, and ν is the *unique probability* on S_0 such that this relationship holds. Moreover, by the uniqueness of the spectral representation (4-5-2), this relationship is equivalent to

$$r^\alpha\varphi(s)=A\int_{S_0}\nu(d\sigma)\int_0^\infty \frac{1-\cos(2\pi\rho r\langle\sigma,s\rangle)}{4\pi^2\rho^2}\rho^{1-\alpha}\,d\rho \tag{4-5-3}$$

On the other hand, the function $h\to-|h|^\alpha$ is conditionally positive definite on $\mathbf{R}^1$ and admits the representation

$$|h|^\alpha=B\int_0^\infty \frac{1-\cos 2\pi\rho h}{4\pi^2\rho^2}\rho^{1-\alpha}\,d\rho$$

for a suitable constant $B>0$. By putting $h=r\langle\sigma,s\rangle$ and substituting in (4-5-3), we obtain the following relationship, which is *equivalent* to (4-5-3):

$$\varphi(s)=A'\int_{S_0}|\langle\sigma,s\rangle|^\alpha\nu(d\sigma) \tag{4-5-4}$$

for a constant $A'>0$ obviously uniquely determined.

Hence the function φ admits the representation stated in condition c (with $\mu=A'\nu$), and the uniqueness of the probability ν implies the uniqueness of μ. Then condition c is true.

Conversely, condition c implies b, because (4-5-4) and (4-5-3) are equivalent. Finally, condition b implies a by Proposition 4-5-1, statement a when taking $C(h)=\exp(-|h|^\alpha\varphi(h/|h|))$.

Taking into account the homeomorphism between the continuous functions (respectively the measures) on $\mathbb{S}_1$ and the continuous symmetric functions (respectively the symmetric measures) on S_0, we can restate the uniqueness theorem as follows.

Corollary 1. *Let α be a real number such that $0<\alpha<2$, and $\mathfrak{R}_\alpha\subset\mathbf{C}(\mathbb{S}_1)$ be the convex cone defined by $\varphi\in\mathfrak{R}_\alpha$ if $\varphi(L)=\int_{\mathbb{S}_1}|L,L'|^\alpha G(dL')$, $L\in\mathbb{S}_1$, for a measure $G\geqslant 0$ on $\mathbb{S}_1$. Then $\mathfrak{R}_\alpha$ is a simplex, and the mapping $G\to\varphi$ is a homeomorphism from $\mathfrak{M}^+(\mathbb{S}_1)$ onto $\mathfrak{R}_\alpha$. In particular, the linear space spanned by the functions $L\to|L,L'|^\alpha$, $L'\in\mathbb{S}_1$, is dense in $\mathbf{C}(\mathbb{S}_1)$.*

Proof. The first statements immediately follow from the uniqueness theorem (Theorem 4-5-1). Suppose that a measure μ is orthogonal to the functions $L\to|L,L'|^\alpha$, $L'\in\mathbb{S}_1$, and let $\mu=\mu_+-\mu_-$ be its Jordan decomposition into two positive measures, μ_+ and μ_-. Then $\int\mu_+(dL')|L,L'|^\alpha=\int\mu_-(dL')|L,L'|^\alpha$ implies $\mu=\mu_+-\mu_-=0$, by the uniqueness theorem, and the corollary holds.

Geometrical Interpretation

Let C be a covariance in $\mathbf{R}^d$ by which condition c in Theorem 4-5-1 is satisfied for $\alpha=1$. Then the graph of $h \to C(h)$ in $\mathbf{R}^d \times \mathbf{R}$ admits at $h=o$ a *tangential cone*, the basis of which in $\mathbf{R}^d$ is the boundary ∂F of the set $F=\{x: |x|\varphi(x/|x|) \leqslant C(o)\}$. By Proposition 4-4-1, Corollary 2, F is a symmetric closed convex neighborhood of O in $\mathbf{R}^d$. If $d=2$, the converse is true, because the Steiner class in $\mathbf{R}^2$ is identical to the class $\mathcal{R}_s$ of the symmetric compact convex sets. We may state:

Corollary 2. *Let F be a subset in the euclidean plane $\mathbf{R}^2$. There exists a continuous covariance in $\mathbf{R}^2$ such that its tangential cone at $h=0$ admits the basis ∂F if and only if F is a symmetric closed and convex neighborhood of* 0 *in* $\mathbf{R}^2$.

In $\mathbf{R}^d$, $d>2$, this condition becomes "there exists a Steiner compact set the supporting function φ of which is such that $F=\{x: |x|\varphi(x/|x|) \leqslant 1\}$." In other words, a function φ on the unit sphere S_0 is the supporting function of a Steiner compact set if and only if the function $h \to \exp(-|h|\varphi(h/|h|))$ is a covariance on $\mathbf{R}^d$.

Characterization of a Line or Hyperplane Poisson Network

Let A be the hyperplane (resp. line) Poisson network associated by (3-5-1′) with a measure $G \geqslant 0$ on $\mathbb{S}_{d-1}$ (resp. $\mathbb{S}_1$). We have seen that the Poisson point process induced by A on a line with direction $S \in \mathbb{S}_1$ (resp. on a hyperplane with direction $S \in \mathbb{S}_{d-1}$) admits the constant *density*

$$\varphi(S)=\int |S,\sigma^{\perp}|G(d\sigma)$$

Taking into account the homeomorphism $S \to S^{\perp}$ from $\mathbb{S}_1$ onto $\mathbb{S}_{d-1}$ and the relationship $|S_1,S_2|=|S_1^{\perp},S_2^{\perp}|$, it follows from Theorem 4-5-1, Corollary 2, that these functions φ are exactly those which belong to $\mathcal{R}_1$, and the Poisson network is determined if the corresponding function φ is given. Let us give the precise statement only in the case of Poisson hyperplanes.

Corollary 3. *Let φ be a function on $\mathbb{S}_1$, identified with the symmetrical function $u \to \varphi(S)$ on S_0, where $S \in \mathbb{S}_1$ is the direction of the unit vector $u \in S_0$. Then the four following conditions are equivalent*:

a. There exists a stationary Poisson hyperplane network A in $\mathbf{R}^d$ such that, *for any $S \in \mathbb{S}_1$, $\varphi(S)$ is the density of the Poisson point process induced by A on the S-directed lines.*

b. There exists a necessarily unique measure $G \geqslant 0$ on $\mathbb{S}_1$ such that $\varphi(S) = \int |S, S'| G(dS')$ for any $S \in \mathbb{S}_1$.

c. The function K: $h \to -|h| \varphi(h/|h|)$ is continuous and conditionally positive definite on $\mathbf{R}^d$.

d. The function $C = \exp(K)$ is a continuous covariance on $\mathbf{R}^d$.

Moreover, if one of these conditions is satisfied, G is the positive measure associated by (3-5-1′) with the Poisson network A by which the densities $\varphi(S)$ are induced on each $S \in \mathbb{S}_1$, and, for any $x, y \in \mathbf{R}^d$, $C(x-y)$ is the probability for the points x and y not to be separated by A.

The Surface Measure G_{d-1}^K

Before investigating the Steiner class further, let us show that with any $K \in C_0(\mathcal{K})$ is associated a necessarily unique measure $G_{d-1}^K \geqslant 0$ on $\mathbb{S}_{d-1}$ such that

$$\mu_{d-1}(\Pi_S K) = \int_{\mathbb{S}_{d-1}} |S, S'| G_{d-1}^K(dS') \qquad (S \in \mathbb{S}_{d-1}) \qquad \text{(4-5-5)}$$

This unique measure G_{d-1}^K is called the *surface measure* associated with K.

Proof. The uniqueness follows from Theorem 4-5-1. In order to prove the existence of G_{d-1}^K, we suppose first that K is a polyhedron. Let $u_i \in S_0$, $i = 1, \ldots, p$, be the positive normals associated with the $d-1$ faces of K, and F_i, $i = 1, 2, \ldots, p$ be the $(d-1)$-volume of these faces. Put $F = \frac{1}{2} \sum F_i \delta_{u_i}$. Then F is a positive measure on S_0, and $\int F(du) = dW_1(K)$, that is, is the surface area of K. Moreover, $\mu_{d-1}(\Pi_{u^\perp} K) = \int |\langle u, u' \rangle| F(du')$ for any $u \in S_0$. If K is now an arbitrary compact convex set in $C_0(\mathcal{K})$, there exists a sequence $\{K_n\}$ converging toward K in $C_0(\mathcal{K})$, such that, for each $n > 0$, K_n is a polyhedron, with which is associated a measure F_n satisfying the above condition. By the continuity of the Minkowski functionals W_1, $\operatorname{Sup}_n \int F_n < \infty$, and the sequence $\{F_n\}$ is bounded in $\mathfrak{M}^+(S_0)$. Thus there exists a subsequence $\{F_{n_k}\}$ weakly converging toward a measure $F \geqslant 0$ on S_0. By Lemmas 3-5-2 and 3-5-3, the mapping $K \to \mu_{d-1(\Pi_{u^\perp} K)}$ is continuous, so that the relationship $\mu_{d-1}(\Pi_{u^\perp} K) = \int |\langle u, u' \rangle| F(du')$ is still satisfied by F.

Then let us denote by $G = G_{d-1}^K$ the measure on $\mathbb{S}_{d-1}$ defined by $\int \varphi(S) G(dS) = \int f(u) F(du)$, $\varphi \in \mathbf{C}(\mathbb{S}_{d-1})$, and f being the symmetric function $u \to f(u) = \varphi(u^\perp)$ on $\mathbb{S}_1$. This measure G satisfies condition (4-5-5).

This surface measure is related to the mixed Minkowski functionals W_k defined on $C(\mathcal{K}')\times C(\mathcal{K}')$ by

$$\mu_d(K\oplus\rho K')=\mu_d(K)+\sum_{k=1}^{d-1}\binom{d}{k}\rho^k W_k(K,K')+\rho^d\mu_d(K') \qquad (4\text{-}5\text{-}6)$$

(see the remark following Theorem 4-1-1). These functionals are C-additive and invariant under translations with respect to each of their arguments K and K'. In particular, the mixed functional W_1 is such that, for any K, $K'\in C(\mathcal{K}')$,

$$dW_1(K,K')=\lim_{\rho\downarrow 0}\frac{\mu_d(K\oplus\rho K')-\mu_d(K)}{\rho}$$

For a given $K\in C_0(\mathcal{K})$, the mapping $K'\to W_1(K,K')$ is continuous, C-additive, and increasing on $C_0(\mathcal{K})$, and is also additive with respect to the Minkowski sum $\oplus$ and homogeneous with degree 1, that is, positively linear. Thus, by Proposition 4-4-4, there exists a unique measure $\lambda_K\geqslant 0$ on the unit sphere S_0 such that

$$dW_1(K,K')=\int_{S_0} r_{K'}(u)\lambda_K(du)$$

But the measure λ_K is symmetric on S_0, and $\mu_1(\Pi_{S^\perp}K)=r_K(u)+r_K(-u)$, if $S^\perp$ is the direction of the unit vectors u and $-u$. Thus there exists a measure $H_K\geqslant 0$ on S_{d-1} such that

$$dW_1(K,K')=\int_{\mathbb{S}_{d-1}}\mu_1(\Pi_{S^\perp}K')H_K(dS)$$

If K′ is a segment with unit length and direction $S_1^\perp$, $S_1\in\mathbb{S}_{d-1}$, we have

$$dW_1(K,K')=\mu_{d-1}(\Pi_{S_1}K)=\int_{\mathbb{S}_{d-1}}|S,S_1|H_K(dS)$$

By the uniqueness theorem, it follows that $H_K=G_{d-1}^K$, and we may state:

Proposition 4-5-2. *The surface measure associated with a* $K\in C_0(\mathcal{K})$ *by* (4-5-5) *is such that, for any* $K'\in C_0(K)$,

$$dW_1(K,K')=\lim_{\rho\downarrow 0}\frac{\mu_d(K\oplus\rho K')-\mu_d(K)}{\rho}=\int_{\mathbb{S}_{d-1}}\mu_1(\Pi_{S^\perp}K')G_{d-1}^K(dS)$$

An Extension of the Steiner Formula

We may ask to what extent there exist measures $G_{d-k}^K \geqslant 0$ on $\mathcal{S}_{d-k}$, $k=2,\ldots,d$, such that the functionals $K' \to W_k(K, K')$ admit integral representations analogous to that of Proposition 4-5-2. However, this is not true in general, even if K is a symmetric compact convex set (except in the case $d=2$).

Theorem 4-5-2. *Let K be a symmetric compact convex set in $\mathbf{R}^d$. Then there exists for each $k=1, 2,\ldots,d-1$ a positive measure G_{d-k}^K on $\mathcal{S}_{d-k}$ such that, for any $\rho \geqslant 0$ and $K' \in C_0(\mathcal{K})$,*

$$\mu_d(K \oplus \rho K') = \mu_d(K) + \sum_{k=1}^{d-1} \rho^k \int_{\mathcal{S}_{d-k}} \mu_k(\Pi_S \perp K') G_{d-k}^K(dS) + \rho^d \mu_d(K') \tag{4-5-7}$$

if and only if K is a Steiner compact (i.e., $K \in \mathfrak{R}_1$). If so, for $k=2,\ldots,d-1$, $G_k{}^K$ is induced by the mapping $(S_1, S_2,\ldots,S_k) \to S_1 \oplus S_2 \oplus \ldots \oplus S_k$ from the measure $(1/k!)V(u_1,\ldots,u_k) \times G_1{}^K(dS_1) G_1{}^K(dS_2) \ldots G_1{}^K(dS_k)$ on $(\mathcal{S}_1)^k$, $V(u_1,\ldots,u_k)$ denoting the volume of the parallelotope defined by unit vectors $u_i \in S_0$, $i=1, 2,\ldots,k$, such that S_i is parallel to u_i.

Proof. First consider the "only if" part. By (4-5-6), the mixed functionals W_k on $C(\mathcal{K}') \times C(\mathcal{K}')$ are such that $W_k(K,K') = W_{d-k}(K',K)$. Thus it follows from Proposition 4-5-2 that

$$dW_{d-1}(K,K') = dW_1(K',K) = \int_{\mathcal{S}_{d-1}} \mu_1(\Pi_S \perp K) G_{d-1}^{K'}(dS)$$

$$= \int_{\mathcal{S}_1} \mu_{d-1}(\Pi_S \perp K') G_1{}^K(dS)$$

if (4-5-7) holds. Moreover, if $K' \subset S_1 \in \mathcal{S}_{d-1}$ and $\mu_{d-1}(K')=1$, we get $G_{d-1}^{K'} = \delta_{S_1}$ and $\mu_{d-1}(\Pi_S \perp K') = |S_1, S^{\perp}|$. Thus the supporting function r_K: $u \to \frac{1}{2}\mu_1(\Pi_u K)$, $u \in S_0$, of the symmetrical compact convex set K verifies that $r_k(u_0) = \frac{1}{2}\int |S_1, S^{\perp}| G_1{}^K(dS)$ if $S_1^{\perp} \in \mathcal{S}_1$ is the direction of a unit vector $u_0 \in S_0$. In other words, $r_K \in \mathfrak{R}_1$ and K is a Steiner compact.

The "if" part is a little more intricate, and we shall use an induction argument. Let

$$A = \bigoplus_{i=1}^{n} l_i L_i \quad (l_i \geqslant 0,\ L_i = \{\lambda u_i,\ |\lambda| \leqslant \tfrac{1}{2}\} \text{ for unit vectors } u_i \in S_0,\ i=1, 2,\ldots,n)$$

that is, $A \in \mathcal{R}_1$ is a finite Minkowski sum of n line segments, and prove that (4-5-7) is satisfied by A. We denote by $V(u_1,\ldots,u_n)$ the n-volume of the parallelotope $u_1 \oplus \ldots \oplus u_n$. The relationship

$$\mu_k(\Pi_\sigma(K \oplus lL)) = \mu_k(\Pi_\sigma K) + l|u,\sigma|\, \mu_{k-1}(\Pi_{S\perp\cap\sigma}K) \qquad \text{(4-5-8)}$$

holds for any $K \in C(\mathcal{K})$, any $\sigma \in \mathbb{S}_k$, and any unit vector u identified with its direction S in $\mathbb{S}_1$(L is the segment $\{\lambda u,\ |\lambda| \leqslant \frac{1}{2}\}$). If $\sigma = E$, we simply have

$$\mu_d(K \oplus lL) = \mu_d(K) + l\mu_{d-1}(\Pi_u \perp K)$$

It follows that (4-5-7) is satisfied by $lL \in \mathcal{R}_1$ for $G_1^{lL} = l\delta_{S_u}$ (S_u is the direction of $u \in S_0$) and $G_2 = G_3 = \ldots = 0$.

Suppose now that (4-5-7) is true for any $A = \oplus_{i=1}^n l_i L_i \in \mathcal{R}_1$, which is the Minkowski sum of at most n line segments, for the measures G_k defined as follows:

a. $G_1 = \sum l_i \delta_{S_i}$ (S_i is the direction of $u_i \in S_0$).

b. For $k = 1, 2,\ldots,d-1$, G_k is induced from the measure $\nu_k = (1/k!)\, V(u_1,\ldots,u_k) G_1(dS_1) G_1(dS_2) \ldots G_1(dS_k)$ on the product space $(\mathbb{S}_1)^k$ by the mapping $(S_1,\ldots,S_k) \to S_1 \oplus \ldots \oplus S_k$ from $(\mathbb{S}_1)^k$ into $\mathbb{S}_k$ [actually, this mapping is defined ν_k-a.e. in $(\mathbb{S}_1)^k$, because the dimension of $S_1 \oplus \ldots \oplus S_k$ is $<k$ if and only if $V(u_1,\ldots,u_k) = 0$].

Let us show that (4-5-7) is still satisfied by $A \oplus lL$ ($L = \{\lambda u,\ |\lambda| \leqslant \frac{1}{2}\}$, $u \in S_0$) for measures G'_k constructed according to rules *a* and *b*, so that the theorem will be proved for any Minkowski sum of finite line segments. We start from the recurrence hypothesis:

$$\begin{aligned}\mu_d(A \oplus K) = {} & V(K) + \sum_i l_i \mu_{d-1}(\Pi_{u_i} \perp K) + \cdots \\ & + \sum_{i_1 < \cdots < i_k} l_{i_1} l_{i_2} \ldots l_{i_k} V(u_{i_1} \ldots u_{i_k}) \mu_{d-k}(\Pi_{u_{i1}\perp\cap\cdots\cap u_{ik}\perp} K) \\ & + \sum_{i_1 < \cdots < i_d} l_{i_1} l_{i_2} \ldots l_{i_d} V(u_{i_1} \ldots u_{i_d}) \end{aligned} \qquad \text{(4-5-9)}$$

and substitute $K \oplus lL$ for K. By (4-5-8), we have

$$\mu_{d-k}(\Pi_{\sigma_{i1,\ldots,ik}}(K \oplus lL)) = \mu_{d-k}(\Pi_{\sigma_{i1,\ldots,ik}} K)$$

$$+\, l|u, \sigma_{i_1,\ldots,i_k}|\, \mu_{d-k-1}(\Pi_{u\perp\cap\sigma_{i1,\ldots,ik}} K)$$

with $\sigma_{i_1,\ldots,i_k} = u_{i_1}^{\perp} \cap \cdots \cap u_{i_k}^{\perp}$. But we have $|u, \sigma_{i_1,\ldots,i_k}| = V(u_{i_1},\ldots,u_{i_k}, u)$ and $u^{\perp} \cap \sigma_{i_1,\ldots,i_k} = u_{i_1}^{\perp} \cap \cdots \cap u_{i_k}^{\perp} \cap u^{\perp}$. Hence, by substituting these expressions in (4-5-9), we obtain the same formula (4-5-9) written with $n+1$ instead of n, and $u_{n+1} = u$, $l_{n+1} = l$.

Thus the theorem is proved if $A \in \mathcal{R}_1$ is a finite Minkowski sum of line segments. If K is an arbitrary Steiner compact, there exists a sequence $\{A_n\} \subset \mathcal{R}_1$ converging toward K in $C_0(\mathcal{K})$ and such that, for each $n>0$, A_n is a finite Minkowski sum of line segments. Let also $G_k^{A_n}$ be the measures associated with A_n according to (4-5-7). The sequences $\{G_k^{A_n}\}$ are bounded by $\int G_{d-k}^{A_n}(dS) = C_k W_k(A_n)$, W_k denoting the Minkowski functional and C_k a suitable constant $\geqslant 0$. Thus there exists a subsequence $\{A_{n_p}\}$ such that each sequence $\{G_k^{A_{n_p}}\}$ is weakly converging toward a measure $G_k^K \geqslant 0$ on $\mathbb{S}_k$. Hence (4-5-7) is still satisfied by these measures G_k^K, and the theorem is proved.

The following corollaries will be very useful for the applications.

Corollary 1. *Let A be a Steiner compact, p an integer such that $0<p<d$, $S \in \mathbb{S}_p$, and $K \in C_0(\mathcal{K})$. Then:*

$$\mu_{d-p}(\Pi_S \perp A \oplus K)$$

$$= \mu_{d-p}(\Pi_S \perp K) + \sum_{k=1}^{d-p} \int |\sigma, S^{\perp}| \mu_{d-k-p}(\Pi_{S^{\perp} \cap \sigma^{\perp}} K) G_k^A(d\sigma)$$

and, in particular,

$$\mu_{d-p}(\Pi_S \perp A) = \int_{\mathbb{S}_{d-p}} |\sigma, S^{\perp}| G_{d-p}^A(d\sigma) \tag{4-5-10}$$

The proof is exactly the same as for the theorem itself. By the symmetry $W_k(K,K') = W_{d-k}(K',K)$ of the mixed functionals in (4-5-6), we obtain the following:

Corollary 2. *For K, $K' \in \mathcal{R}_1$ and $0<k<d$, we have*

$$\binom{d}{k} W_k(K,K') = \int_{\mathbb{S}_{d-k}} \mu_k(\Pi_S \perp K') G_{d-k}^K(dS) = \int_{\mathbb{S}_k} \mu_{d-k}(\Pi_\sigma \perp K) G_k^{K'}(d\sigma)$$

If B is the *unit ball* in $\mathbf{R}^d$, $B \in \mathcal{R}_1$, and the measures G_k^B are proportional to the invariant probabilities ϖ_k, the corresponding coefficients are easily computed by comparing (4-5-7) and (4-1-8). Hence:

Corollary 3. *The unit ball B is a Steiner compact, and*

$$G_k{}^B = \binom{d}{k}\frac{b_d}{b_{d-k}}\varpi_k \qquad (k=1,2,\ldots,d-1)$$

By Corollary 2 written for $K = A \in \mathfrak{R}_1$ and $K' = B$, it follows that

$$b_k \int_{\mathbb{S}_{d-k}} G_{d-k}^A(dS) = \binom{d}{k} W_k(A)$$

Corollary 4. *For any $A \in \mathfrak{R}_1$ and $k = 1, 2, \ldots, d-1$, the Minkowski functional W_k satisfies*

$$W_k(A) = \frac{b_k}{\binom{d}{k}} \int_{\mathbb{S}_{d-k}} G_{d-k}^A(dS)$$

By rewriting the relationship of Corollary 1 with $A = B$ and $K = \rho B$, and identifying with the Steiner formula, we obtain the following:

Corollary 5. *For a given $S \in \mathbb{S}_p$ and $0 < k \leqslant p$, the rotation average of the function $\sigma \to |\sigma, S|$ on $\mathbb{S}_k$ is*

$$\int_{\mathbb{S}_k} |\sigma, S| \varpi_k(d\sigma) = \frac{\binom{p}{k} b_p b_{d-k}}{\binom{d}{k} b_d b_{p-k}}$$

and does not depend on $S \in \mathbb{S}_p$. If $k = p$, this constant is $b_p b_{d-p}(\binom{d}{p} b_d)$.

Corollary 6. *Let d, d', and k be integers such that $0 < k < d' < d$, and $K \in C(\mathcal{K}')$ be a compact convex set in $\mathbf{R}^d$ such that $K \subset S_{d'}$ for a given subspace $S_{d'} \in \mathbb{S}_{d'}$. Then, for $k = 0, 1, \ldots, d'$, the Minkowski functionals W_{d-k}^d and $W_{d'-k}^{d'}$ on $\mathbf{R}^d$ and $S_{d'} = R^{d'}$, respectively, satisfy the following relationship:*

$$W_{d-k}^d(K) = \frac{\binom{d'}{k} b_{d-k}}{\binom{d}{k} b_{d'-k}} W_{d'-k}^{d'}(K)$$

Proof. The conditions of Theorem 4-1-1 are satisfied by the mapping $K \to W_{d-k}^d(K)$; $K \subset S_{d'}$ with respect to the space $S_{d'}$ identified with $\mathbf{R}^{d'}$, and thus $W_{d-k}^d(K) = C W_{d'-k}^d(K)$. In order to calculate the constant C, we put $K = \Pi_{S_{d'}} B$, where B is the unit ball in $\mathbf{R}^d$. Then, by Corollary 5,

$$W_{d-k}^d(\Pi_{S_{d'}} B) = b_d \int_{\mathbb{S}_k} |\sigma, S_{d'}| \varpi_k(d\sigma) = \frac{\binom{d'}{k} b_{d'} b_{d-k}}{\binom{d}{k} b_{d'-k}}$$

On the other hand, $W_{d'-k}^{d'}(\Pi_{S_{d'}} B) = b_{d'}$, so that $C = \binom{d'}{k} b_{d-k} / (\binom{d}{k} b_{d'-k})$.

4-6. THE RANDOM VARIABLES $|S,S_p|$

We recall that, for any $S\in\mathbb{S}_k$ and $S_p\in\mathbb{S}_p$, $|S,S_p|$ is the absolute value of the determinant of the linear mapping Π_{S_p} from S into S_p. Clearly, $k>p$ implies $|S,S_p|=0$, so that we suppose $0<k\leqslant p<d$. We shall give an explicit expression for the probability law of the random variable $S\to|S,S_p|$ on $(\mathbb{S}_k, \varpi_k)$, ϖ_k denoting as usual the unique invariant probability on $\mathbb{S}_k$.

For brevity, we write $X\overset{\mathcal{L}}{\equiv}Y$ if the two random variables X and Y are equivalent, that is, admit the same probability law, and we denote by $Z_{\alpha,\beta}$ a random variable the law of which is beta with the parameters α, $\beta>0$, that is, admits on $[0,1]$ the density

$$f_{\sigma,\beta}(z)=\frac{\Gamma(\alpha+\beta)}{\Gamma(\alpha)\Gamma(\beta)}z^{\alpha-1}(1-z)^{\beta-1}\qquad(z\in[0,1])$$

and the moments of which are:

$$E\left((Z_{\sigma,\beta})^{\lambda}\right)=\frac{\Gamma(\alpha+\beta)}{\Gamma(\alpha)}\,\frac{\Gamma(\alpha+\lambda)}{\Gamma(\alpha+\beta+\lambda)}\qquad(\lambda\geqslant 0)\qquad\text{(4-6-1)}$$

Let us first examine the case $k=1$, and denote by Y_p a random variable equivalent to $|S,\ S_p|$. From the invariance of the law ϖ_1 under rotations, the law of Y_p does not depend on the particular choice of the subspace $S_p\in\mathbb{S}_p$. In other words, if S and S_p are random and independent on $\mathbb{S}_1\times\mathbb{S}_p$ with the probability $\varpi_1\otimes P_p$ (where P_p is an arbitrary probability on $\mathbb{S}_p$), the *random variable* $|S,S_p|$ *is independent of* S_p.

In order to determine the law of Y_p, we denote by $X=(X_1,\dots,X_d)$ a random vector in $\mathbf{R}^d$, with the gaussian law the density of which is

$$g_d(x)=\frac{1}{(2\pi)^{d/2}}\exp\left(\frac{-|x|^2}{2}\right)$$

and by $X'=(X_1,\dots,X_p)$, $X''=(X_{p+1},\dots,X_d)$ its projections into the subspaces spanned, respectively, by the p first and the $d-p$ last coordinate axes. Obviously, X' and X'' are independent, and their laws admit the densities g_p and g_{d-p}, respectively. On the other hand, the random variable $|X'|/|X|$ is equivalent to Y_p and independent of $|X|^2=|X'|^2+|X''|^2$. In the same way, the random variables $|X'|^2$ and $|X''|^2$ are independent, and admit the gamma laws with parameters $p/2$ and $(d-p)/2$, respectively. By elementary calculation, it follows that $|X'|^2/|X|^2$ is beta with $\alpha=p/2$ and

$\beta=(d-p)/2$, that is,

$$Y_p \overset{\mathfrak{L}}{\equiv} \left(Z_{p/2,\,(d-p)/2}\right)^{1/2}$$

or, which is the same,

$$E\left((Y_p)^{\lambda}\right)=\frac{\Gamma(d/2)}{\Gamma(p/2)}\,\frac{\Gamma((\lambda+p)/2)}{\Gamma((\lambda+d)/2)} \qquad (\lambda\geqslant 0) \qquad (4\text{-}6\text{-}2)$$

Suppose now that $1<k\leqslant p<d$, and consider the random variable $S\to|S,S_p|$ on $(\mathbb{S}_k,\varpi_k)$, for a given $S_p\in\mathbb{S}_p$. The law of $|S,S_p|$ still does not depend on $S_p\in\mathbb{S}_p$. On the other hand, the law ϖ_k on $\mathbb{S}_k$ is induced from the probability $\varpi_1(dL_1)\varpi_1(dL_2)\dots\varpi_1(dL_k)$ on $(\mathbb{S}_1)^k$ by the (a.e. defined) mapping $(L_1,L_2,\dots,L_k)\to L_1\oplus L_2\oplus\cdots\oplus L_k$. Thus, if $u_1,\dots,u_k$ are k independent random unit vectors with the same invariant law on S_0, the random variable $|S,S_p|$ is equivalent to

$$\frac{V(\Pi_{S_p}u_1,\dots,\Pi_{S_p}u_k)}{V(u_1,\dots,u_k)}$$

where $\mathrm{V}(\Pi_{S_p}u_1,\dots,\Pi_{S_p}u_k)$ and $V(u_1,\dots,u_k)$, respectively, denote the volumes of the corresponding parallelotopes. On the other hand, we may write

$$V(u_1,\dots,u_k)=|u_1|\times|\Pi_{u_1^\perp}u_2|\times\cdots\times|\Pi_{u_1^\perp\cap\cdots\cap u_{k-1}^\perp}u_k|$$

and

$$V(\Pi_{S_p}u_1,\dots,\Pi_{S_p}u_k)=|\Pi_{S_p}u_1|\times|\Pi_{S_p\cap u_1^\perp}u_2|\times\cdots\times|\Pi_{S_p\cap u_1^\perp\cap\cdots\cap u_{k-1}^\perp}u_k|$$

But the independence and isotropy of the random unit vector u_i again imply that the factors occuring in the preceding formulae are independent random variables. Then it follows from (4-6-2) that

$$V(u_1,\dots,u_k)\overset{\mathfrak{L}}{\equiv}\prod_{j=d-k+1}^{d}Y_j$$

$$(4\text{-}6\text{-}3)$$

$$V(\Pi_{S_p}u_1,\dots,\Pi_{S_p}u_k)\overset{\mathfrak{L}}{\equiv}\prod_{p-k+1}^{p}Y_j$$

where the Y_j are independent random variables, the laws of which are (4-6-2) (and $Y_d=1$ a.s.).

On the other hand, the random variable $V(u_1,u_2,\dots,u_k)$ is invariant under rotations and thus independent of $S=L_1\oplus\cdots\oplus L_k$. On the contrary, the ratio $V(\Pi_{S_p}u_1,\dots,\Pi_{S_p}u_k)/V(u_1,\dots,u_k)=|S,S_p|$ depends only on S

(for a fixed $S_p \in \mathcal{S}_p$). Thus we have the following equivalence, where the Y_j, Y_j' are independent random variables the laws of which are (4-6-2):

$$|S,S_p| \prod_{d-k+1}^{d} Y_j \overset{\mathcal{L}}{\equiv} \prod_{p-k+1}^{p} Y_j'$$

In other words, by (4-6-2) we have, for any real $\lambda \geqslant 0$,

$$E(|S,S_p|^{\lambda}) = \prod_{p-k+1}^{p} \frac{E((Y_j)^{\lambda})}{E((Y_{j+d-p})^{\lambda})}$$

But it follows from (4-6-1) and (4-6-2) that

$$\frac{E((Y_j)^{\lambda})}{E((Y_{j+d-p})^{\lambda})} = \frac{\Gamma((j+d-p)/2)}{\Gamma(j/2)} \frac{\Gamma((j+\lambda)/2)}{\Gamma((j+d-p+\lambda)/2)}$$
$$= E((Z_{j/2,(d-p)/2})^{\lambda})$$

Thus we finally get the equivalence

$$|S,S_p| \equiv \prod_{p-k+1}^{p} (Z_{j/2,(d-p)/2})^{1/2}$$

In other words, $|S,S_p|^2$ *is equivalent to the product of* k *independent beta random variables* $Z_{j/2,(d-p)/2}$, that is, explicitly, for any $\lambda \geqslant 0$,

$$E(|S,S_p|^{\lambda}) = \prod_{j=p-k+1}^{p} \frac{\Gamma((j+d-p)/2)}{\Gamma(j/2)} \frac{\Gamma((j+\lambda)/2)}{\Gamma((j+d-p+\lambda)/2)} \qquad (4\text{-}6\text{-}4)$$

We shall now apply these results to Poisson flat networks in $\mathbf{R}^d$. For further investigation in the case of Poisson hyperplanes, see also Chapter 6.

The Random Measure Associated with a Poisson Flat Network

Let A be a k-dimensional Poisson flat network ($0<k<d$), ψ be the functional $K \to \psi(K) = -\log P(\{A \cap K = \varnothing\})$ on $\mathcal{K}$, and G be the positive measure on $\mathcal{S}_k$ (see Proposition 3-5-1) such that

$$\psi(K) = \int_{\mathcal{S}_k} \mu_{d-k}(\Pi_S \perp K) G(dS) \qquad (K \in \mathcal{K}) \qquad (3\text{-}5\text{-}1)$$

If $V \in \mathcal{V}_k$ is a k-dimensional variety in $\mathbf{R}^d$, identified with the euclidean space $\mathbf{R}^k$, we may define a measure μ_V on $\mathbf{R}^d$ by putting, for any continuous function φ with compact support in $\mathbf{R}^d$,

$$\int_{\mathbf{R}^d} \varphi(x)\mu_V(dx) = \int_V \varphi(x)\mu_k(dx)$$

(μ_k is the Lebesgue measure on $V = \mathbf{R}^k$). The network A is the union of a Poisson process $\mathcal{Q}$ in $\mathcal{V}_k$, and $\mathcal{Q}$ is locally finite (i.e., each compact set K in $\mathbf{R}^d$ a.s. hits only a finite number of the varieties belonging to $\mathcal{Q}$), so that the measure $\nu^A = \sum_{V \in \mathcal{Q}} \mu_V$ is a.s. defined for each $\mathcal{Q}$ or, which is the same, for each A. Explicitly, for any continuous function φ with compact support in $\mathbf{R}^d$ and almost every A,

$$\int \nu^A(dx)\varphi(x) = \sum_{V \in \mathcal{Q}} \int \mu_V(dx)\varphi(x)$$

For a given function $\varphi \in \mathbf{C}\kappa$, the mapping $V \to \int \mu_V(dx)\varphi(x)$ is continuous on $\mathcal{V}_k$, so that $A \to \int \nu^A(dx)\varphi(x)$ is a random variable. For $\varphi \geqslant 0$, we have $\int \nu^A(dx)\varphi(x) \geqslant 0$ a.s., so that the functional $\varphi \to E(\int \nu^A(dx)\varphi(x))$ is positive and linear on the space $\mathbf{C}_{\mathcal{K}}(\mathbf{R}^d)$ (i.e., the space of the continuous functions with compact support). By the Riesz theorem, there exists a measure $\bar{\nu}$ on $\mathbf{R}^d$ such that $E(\int \nu^A(dx)\varphi(x)) = \int \bar{\nu}(dx)\varphi(x)$. But the RACS A is stationary, so that $\bar{\nu}$ is invariant under translation, that is, is proportional to the Lebesgue measure μ_d in $\mathbf{R}^d$. In other words, there exists a constant $\nu \geqslant 0$ such that $\bar{\nu}(dx) = E(\nu^A(dx)) = \nu\mu_d(dx)$. We shall say that ν *is the density of the k – dimensional Poisson flat network.*

Let us compute the density ν. Let B be the unit ball, so that $E(\nu^A(B)) = \nu b_d$, and use the results concerning the random secant A' of B (Section 3-5) for computing $E(\nu^A(B))$. When the isotropy of B is taken into account, the random variable $\mu_k(A' \cap B)$ is independent of the direction $S \in \mathcal{S}_k$ of the random secant A', and it follows from the Crofton formula that $E(\mu_k(A' \cap B)) = b_d / b_{d-k}$.

On the other hand, let N be the random number of the linear varieties $\in \mathcal{Q}$ hitting B, so that $E(N) = \psi(B) = b_{d-k}\gamma$ by (3-5-1′), with $\gamma = \int G(dS)$. Then we have $E(\nu^A(B)) = b_{d-k}\gamma \times (b_d / b_{d-k}) = \gamma b_d$, and thus

$$\nu = \int_{\mathcal{S}_k} G(dS) \tag{4-6-5}$$

The Covariance Measure

With the stationary random measure ν^A is also associated a (stationary) covariance measure C on $\mathbf{R}^d$, that is,

$$E\left[\left(\int \nu^A(dx)\varphi(x)\right)^2\right]=\int C(dh)g(h)$$

for any $\varphi\in\mathbf{C}_{\mathcal{K}}(\mathbf{R}^d)$ and $g=\varphi*\check{\varphi}$ defined by $g(h)=\int\varphi(x)\varphi(x+h)\,dx$. Let us give an explicit expression of this covariance measure C, but *only in the isotropic case*, that is, $G=\nu\varpi_k$.

Lemma 4-6-1. *Let C be an isotropic measure on $\mathbf{R}^d$ (i.e., invariant under rotation), B be the unit ball, and, for each $r\geqslant 0$, g_r be the function on $\mathbf{R}^d$ defined by*

$$g_r(h)=\int_{\mathbf{R}^d}1_{rB}(x)1_{rB}(x+h)\mu_d(x)$$

Then C is determined if the function $r\to\int g_r(h)C(dh)$ is known.

Proof. Denote as $\tilde{C}$ the measure on $\mathbf{R}^+$ induced from C by the mapping $x\to|x|$ from $\mathbf{R}^d$ into $\mathbf{R}^+$, so that the mapping $C\to\tilde{C}$ is one to one from the space of the isotropic measures on $\mathbf{R}^d$ onto the space of the measures on $\mathbf{R}_+$, and write $g_r(\rho)$ instead of $g_r(h)$, $\rho=|h|$. By Proposition 4-3-1, the function $\rho\to g_r(\rho)$ admits at any $\rho>0$ the derivative $-\gamma_r(\rho)=b_{d-1}(r^2-\rho^2/4)^{(d-1)/2}$ if $\rho\leqslant 2r$ and $-\gamma_r(\rho)=0$ if $\rho>2r$. Put $F(\rho)=\tilde{C}([0,\rho])$ for any $\rho>0$. Then $\int_0^\infty g_r(\rho)\tilde{C}(d\rho)=\int_0^\infty F(\rho)\gamma_r(\rho)=b_{d-1}\int_0^{2r}(r^2-\rho^2/4)^{(d-1)/2}F(\rho)\,d\rho$. Put $x=r^2$ and $\xi=\rho^2/4$, so that

$$\int_0^\infty g_r(\rho)\tilde{C}(d\rho)=b_{d-1}\int_0^x(x-\xi)^{(d-1)/2}\xi^{-1/2}F\left(2\sqrt{\xi}\,\right)d\xi.$$

By a Laplace transform, the lemma follows.

Lemma 4-6-2. *With the same notations, the isotropic measure C is*

$C(dh)=\sum_0^{d-1}(a_p/|h|^p)\mu_d(dh)+a_d\delta(dh)$ *if and only if*

$$\int g_r(h)C(dh)=\sum_0^{d-1}a_p\frac{2^{1+d-p}d}{(d-p)(1+d-p)}\frac{b_db_{2d-p}}{b_{1+d-p}}r^{2d-p}+a_db_dr^d$$

Proof. This follows, by simple calculations, from Lemma 4-6-1 and $g_r'(\rho)=-b_{d-1}(r^2-\rho^2/4)^{(d-1)/2}$.

If A is an isotropic k-dimensional Poisson flat network, the covariance measure C associated with the random measure ν^A is also isotropic, and Lemma 4-6-2 enables us to compute its explicit expression. If A' is the random secant of the ball rB, $r>0$, we have $E(\mu_k(A'\cap rB))=(b_d/b_{d-k})r^k$, and

$$E\left[\left(\mu_k(A'\cap rB)\right)^2\right]=\frac{(d-k)(b_k)^2}{r^{d-k}}\int_0^r (r^2-\rho^2)^k\rho^{d-k-1}\,d\rho$$

$$=\frac{(b_k r^k)^2 b_{d+k}}{b_{2k}b_{d-k}}$$

The random variable $\nu^A(rB)$ is the sum of N independent random variables equivalent to $\mu_k(A'\cap rB)$, where N is Poisson and $E(N)=\psi(rB)$ $=\nu b_{d-k}r^{d-k}$. It follows that

$$E\left[\left(\nu^A(rB)\right)^2\right]=\left[\psi(rB)E(\mu_k(A'\cap rB))\right]^2+\psi(rB)E\left[\left(\mu_k(A'\cap rB)\right)^2\right]$$

$$=\left(\nu b_d r^d\right)^2+\nu\left[\frac{b_{d-k}b_{d+k}(b_k)^2}{b_{2k}b_{d-k}}\right]r^{d+k}$$

Taking into account the relationship

$$b_{2k}=(1+k)b_k b_{1+k}2^{-1-k}$$

and making simple calculations, we obtain from Lemma 4-6-2 the following explicit expression of the (noncentered) covariance measure: If $k>0$,

$$C=\nu^2\mu_d+\nu\left(\frac{kb_k}{db_d}\right)\left(\frac{1}{|h|^{d-k}}\right)\mu_d \tag{4-6-6}$$

and if $k=0$ (i.e., A is a Poisson point process with density ν in $\mathbf{R}^d$),

$$C=\nu^2\mu_d+\nu\delta \tag{4-6-7}$$

(δ is the Dirac measure, and μ_d the Lebesgue measure in $\mathbf{R}^d$). We may rewrite these results as follows.

Proposition 4-6-1. *Let A be a k-dimensional Poisson flat network isotropic in $\mathbf{R}^d$ $(0<k<d)$, $G=\nu\varpi_k$ the measure associated with A by (3-5-1), $K\in\mathcal{K}$ a*

compact set, $V(K)$ its volume, and g_K the function defined by $g_K(h)=\int 1_K(x)\,1_K(x+h)\,dx$. Then the k-volume $v^A(K)$ of the intersection $A\cap K$ has expectation $E(v^A(K))=vV(K)$ and variance

$$D^2(v^A(K))=v\left(\frac{kb_k}{db_d}\right)\int_{\mathbf{R}^d}\frac{g_K(h)}{|h|^{d-k}}\,dh$$

Corollary. *If A' is the corresponding random secant of K, and K is convex, the first two moments of the random k-volume $\mu_k(A'\cap K)$ are*

$$E(\mu_k(A'\cap K))=\frac{b_d}{b_{d-k}}\frac{W_0(K)}{W_k(K)}$$

$$E\left[(\mu_k(A'\cap K))^2\right]=\frac{kb_k}{db_{d-k}W_k(K)}\int\frac{g_K(h)}{|h|^{d-k}}\,dh$$

Proof. We have

$$E(\mu_k(A'\cap K))=(\psi(K))^{-1}E(v^A(K))$$

and

$$E\left[(\mu_k(A'\cap K))^2\right]=(\psi(K))^{-1}D^2(v^A(K))$$

with

$$\psi(K)=v\int\mu_{d-k}(\Pi_S\perp K)\varpi_k(dS)=v(b_{d-k}/b_d)W_k(K),$$

and the corollary follows.

4-7. THE MINKOWSKI MEASURES

We will now give a local meaning to the Minkowski functionals W_k, $k=0,1,\dots,d$, by associating with each of them a mapping $K\to W_k^K$ from $C(\mathcal{K}')$ into the space $\mathfrak{M}_c^+$ of the positive measure with compact supports in $\mathbf{R}^d$. It is known that the space $\mathfrak{M}_c$ of the measures with compact supports is the (topological) dual of $\mathbf{C}(\mathbf{R}^d)$ (i.e., the space of the continuous functions on $\mathbf{R}^d$ topologized by the compact convergence). In the sequel, $\mathfrak{M}_c$ will always be provided with its weak topology; this may be defined by the corresponding convergence: $\mu=\lim\mu_n$ in $\mathfrak{M}_c$ if $\int\mu\varphi=\lim\int\mu_n\varphi$ for any

$\varphi \in \mathbf{C}(\mathbf{R}^d)$,which is equivalent to two conditions: (*a*) $\{\mu_n\}$ is weakly convergent toward μ, and (*b*) there exists a fixed compact set $K_0 \in \mathcal{K}$ such that Supp $\mu_n \subset K_0$ for any $n>0$ (Supp μ_n is the support of μ_n).

In a second step, the definition of the Minkowski measures will be extended to the space $C(\mathcal{F}')$, that is, the space of the nonempty closed convex sets, by a mapping $F \to W_k^F$ from $C(\mathcal{F}')$ into the space of the Radon measures on $\mathbf{R}^d$. We recall that a Radon measure is a linear and continuous functional defined on the space $\mathbf{C}_{\mathcal{K}}(\mathbf{R}^d)$ of the continuous functions with compact supports provided with its usual topology. (For this topology, $\varphi = \lim \varphi_n$ in $\mathbf{C}_{\mathcal{K}}$ if (*a*) $\{\varphi_n\}$ is uniformly converging toward φ, and (*b*) there exists a fixed compact set enclosing the support of φ_n for any $n>0$.) Finally, these various definitions will be extended to the convex ring $\mathfrak{S}$, that is, the class closed under finite union generated by $C(\mathcal{K})$.

We shall use the following lemmas.

Lemma 4-7-1. *The mapping $\mu \to$ Supp μ from $\mathfrak{M}_c$ into $\mathcal{K}$ is l.s.c..*

Proof. Let G be an open set and $\mu \in \mathfrak{M}_c$. The support K of the measure μ hits G if and only if there exists a function $\varphi \in \mathbf{C}(\mathbf{R}^d)$ the support of which is included in G and $|\int \varphi(x)\mu(dx)|>0$, that is, $K \in \mathcal{K}_G$ if and only if μ belongs to the set

$$\left\{ \mu:\ \exists \varphi \in \mathbf{C},\quad \text{Supp}\, \varphi \subset G, \quad \left|\int \varphi\mu\right| > 0 \right\}$$

which is open in $\mathfrak{M}_c$. Thus the mapping $\mu \to$Supp μ is l.s.c.

Lemma 4-7-2. *The mapping $K \to \partial K$ is continuous on $C(\mathcal{K}')$.*

Proof. The mapping $K \to \partial K$ is l.s.c. by $K \supset \partial K$ and Proposition 1-2-4, Corollary 3. Let $\{K_n\} \subset C(\mathcal{K})$ be a sequence, $K = \lim K_n$, $\{K_{n_k}\}$ a subsequence, and, for each k, x_{n_k} a point such that $x_{n_k} \in \partial K_{n_k}$ and $\lim x_{n_k} = x$ in $\mathbf{R}^d$. Obviously, $x \in K$. On the other hand, for each k, there exists a closed half space $H \supset K_{n_k}$ such that $x_{n_k} \in \partial H_{n_k}$. Let $\{H, H'\}$ be an adherent point in $\mathcal{F} \times \mathcal{F}$ for the sequence $\{H_{n_k}, \partial H_{n_k}\}$. Obviously, H is a half space and $H' = \partial H$ its boundary. Thus $K \subset H$ and $x \in \partial H$. It follows that ∂H is a supporting hyperplane for K and that $x \in \partial K$. Thus, by Proposition 1-2-4, ∂ is u.s.c. on $C(\mathcal{K})$.

Lemma 4-7-3. *The mapping $(\varphi, K) \to \int_K \varphi(x)\,dx$ is continuous on $\mathbf{C}(\mathbf{R}^d) \times C(\mathcal{K}')$.*

Proof. If we have, for any $\alpha > 0$, $\varphi - \alpha \leqslant \varphi' \leqslant \varphi + \alpha$ [φ, $\varphi' \in \mathbf{C}(\mathbf{R}^d)$] and

$K \subset K' \oplus \alpha B$, $K' \subset K \oplus \alpha B$ [$K, K' \in C(\mathcal{K}')$, B is the unit ball], then

$$\int_K \varphi(x)\,dx \leqslant \int_{K'\oplus\alpha B} \varphi'(x)\,dx + \alpha\|\varphi'\|\,V(K'\oplus\alpha B)$$

$$\int_{K'} \varphi'(x)\,dx \leqslant \int_{K\oplus\alpha B} \varphi(x)\,dx + \alpha\|\varphi\|\,V(K\oplus\alpha B)$$

(V is the volume, and $\|\varphi\|$ and $\|\varphi'\|$ are the suprema of φ and φ' on $K\oplus aB$, for a real $a>\alpha$). It follows that

$$\|\varphi\|(V(K\oplus\alpha B)-V(K)+\alpha V(K\oplus\alpha B)) \leqslant \left|\int_K \varphi(x)\,dx - \int_{K'}\varphi'(x)\,dx\right|$$

$$\leqslant (\alpha+\|\varphi\|)(V(K'\oplus\alpha B)-V(K')+\alpha V(K'\oplus\alpha B))$$

When the Steiner formula (4-1-8) and the continuity of the Minkowski functionals are taken into account, the lemma follows.

We are now able to define the *Minkowski measure*, by using a generalized version of the Steiner formula.

Take $K\in C(\mathcal{K}')$. By the classical projection theorem, for any $x\in\mathbf{R}^d$, there exists a unique point $x'=\Pi_K x$ such that $|x-x'|=\operatorname{Inf}\{|x-y|, y\in K\}$. Put $\rho_K(x)=|x-x'|$. Obviously, $x\in K$ is equivalent to $\rho_K(x)=0$ or also to $x=\Pi_K x$, and the mapping α: $x\to(\Pi_K x, \rho_K(x))$ from $\mathbf{R}^d$ into $K\times\mathbf{R}_+$ is continuous. Let W be the measure on $K\times\mathbf{R}_+$ induced by this mapping α from the Lebesgue measure dx on $\mathbf{R}^d$. Clearly, $\alpha(K)=K\times\{o\}$, and $\alpha(K^c)=\partial K\times(\mathbf{R}_+\setminus\{o\})$, so that $W=1_K\,dx\delta_0+W'$, for a positive measure W' on $\partial K\times(\mathbf{R}_+\setminus\{o\})$. If the boundary ∂K is a twice continuously differentiable manifold, it is known that

$$W'(dx,d\rho)=\sum_{k=1}^{d}\binom{d}{k}W_k^K(dx)\,d\rho$$

where W_k^K is proportional to the surface measure [associated with the $(k-1)$-volume ∂K]. Moreover, for each $k=2,\ldots,d$, W_k^K is absolutely continuous with respect to W_1^K, and the corresponding density depends only on the curvature radii (see, e.g., Bonnesen and Fenchel, 1934). Thus, for any $\varphi\in\mathbf{C}(\mathbf{R}^d)$ and $r\geqslant 0$,

$$\int_{K\oplus rB}\varphi(\Pi_K x)\,dx=\int_K\varphi(x)\,dx+\sum_{k=1}^{d}\binom{d}{k}r^k\int\varphi(x)\,W_k^K(dx)$$

This relationship still holds for an arbitrary $K\in C(\mathcal{K}')$, as we will see from the following theorem.

Theorem 4-7-1. *Let K be a convex compact set and Π_K its projector. Then there exist $d+1$ measures $W_k^K \geqslant 0 (0 \leqslant k \leqslant d)$; the first of them is $W_0^K(dx) = 1_K(x)\,dx$, and the others are concentrated on the boundary ∂K, so that, for any $\varphi \in \mathbf{C}(\mathbf{R}^d)$ and $r \geqslant 0$,*

$$\int_{K \oplus rB} \varphi(\Pi_K x)\,dx = \sum_{k=0}^{d} \binom{d}{k} r^k \int \varphi(x) W_k^K(dx) \tag{4-7-1}$$

(B is the unit ball). Moreover, the Minkowski functionals W_k satisfy

$$W_k(K) = \int W_k^K(dx) \qquad (0 \leqslant k \leqslant d) \tag{4-7-2}$$

and the mappings $(\varphi, K) \to \int \varphi(x) W_k^K(dx)$ are continuous on the product space $\mathbf{C}(\mathbf{R}^d) \times C(\mathcal{K}')$.

Proof. We have already seen that the measures W_k^K exist and satisfy (4-7-1) if ∂K is a $(n-1)$-dimensional manifold twice continuously differentiable. But the class of the convex compact sets the boundaries of which satisfy this condition is dense in $C(\mathcal{K})$. Thus, for any $K \in C(\mathcal{K})$, there exists a sequence $\{K_n\} \subset C(\mathcal{K})$ such that $K = \lim K_n$, and for each $n > 0$ the measures $W_k^{K_n} = W_k{}^n$ exist and satisfy (4-7-1). On the other hand, the mappings $\varphi \to \varphi \circ \Pi_K$ and $K \to K \oplus rB$ are continuous on $\mathbf{C}(\mathbf{R}^d)$ and $C(\mathcal{K}')$, respectively. Thus, by Lemma 4-7-3, for any $r \geqslant 0$, we have

$$\int_{K \oplus rB} \varphi(\Pi_K x)\,dx = \lim \int_{K_n \oplus rB} \varphi(\Pi_{K_n} x)\,dx$$

By using (4-7-1) and taking finite differences with respect to r, it follows that $\lim n \int W_k{}^n(dx)\varphi(x)$ exists for each $\varphi \in \mathbf{C}(\mathbf{R}^d)$, and defines a positive linear functional on $\mathbf{C}(\mathbf{R}^d)$. Thus the measures W_k^K exist and satisfy (4-7-1). If $r = 0$, (4-7-1) yields $W_0^K(dx) = 1_K(x)\,dx$. If $k > 0$, Supp $W_k{}^n \subset \partial K_n$ implies Supp $W_k^K \subset \partial K$ by Lemmas 4-7-1 and 4-7-2. Relationship (4-7-2) follows from the Steiner formula (4-1-8) when (4-7-1) is written with $\varphi = 1$.

Finally, it is easy to verify that the mapping $(\varphi, K) \to \varphi \circ \Pi_K$ from $\mathbf{C}(\mathbf{R}^d) \times C(\mathcal{K}')$ into $\mathbf{C}(\mathbf{R}^d)$ is continuous. But the Minkowski sum $\oplus$ and the positive homotheties also are continuous, and by Lemma 4-7-3 we conclude that the mapping $(\varphi, K, r) \to \int_{K \oplus rB} \varphi(x)\,dx$ is continuous on the product space $\mathbf{C}(\mathbf{R}^d) \times C(\mathcal{K}') \times \mathbf{R}_+$. By taking finite differences with respect to r, it follows that the mapping $(\varphi, K) \to \int \varphi(x) W_k^K(dx)$ is also continuous for each k.

Let us now extend onto $C(\mathcal{F}')$ the mapping $K \to W_k^K$ we have just defined on $C(\mathcal{K}')$. We need the two following lemmas, the first of which emphasizes the local meaning of W_k^K.

Lemma 4-7-4. *Let K, $K' \in C(\mathcal{K}')$ and $G \in \mathcal{G}$ be such that $K \cap G = K' \cap G$. Then $1_G W_k^K = 1_G W_k^{K'}$, $0 \leqslant k \leqslant d$.*

Proof. There exists $\alpha > 0$ such that $G \ominus \rho B \neq \varnothing$ for $0 < \rho < \alpha$, and $G \ominus \rho B \uparrow G$ if $\rho \downarrow 0$. Obviously, $K \cap (G \ominus \rho B) = K' \cap (G \ominus \rho B)$. It follows that $(K \oplus \rho B) \cap (G \ominus \rho B) = (K' \oplus \rho B) \cap (G \ominus \rho B)$, for, if a point x belongs to the left-hand side [i.e., $(\rho B)_x \subset G$ and $x = y + \rho b$ for two points $y \in K$ and $b \in B$], we have $y = x - \rho b \in (\rho B)_x \subset G$ and thus $y \in K \cap G = K' \cap G$. This implies $x = y + \rho b \in K' \oplus \rho B$ and $x \in (K' \oplus \rho B) \cap (G \ominus \rho B)$, because $(\rho B)_x \subset G$. It follows that $(K \oplus \rho B) \cap (G \ominus \rho B) \subset (K' \oplus \rho B) \cap (G \ominus \rho B)$. Hence we have the required equality by exchanging the roles of K and K'.

Clearly, $\Pi_K x \in G \ominus \rho B$ implies $\Pi_{K'} x = \Pi_K x$. Let $\{\varphi_n\} \subset \mathbf{C}(\mathbf{R}^d)$ be a sequence such that $\varphi_n \uparrow 1_{G \ominus \rho B}$. By the above considerations, we have

$$\int_{K \oplus \rho B} \varphi_n(\Pi_K x)\psi(\Pi_K x)\,dx = \int_{K' \oplus \rho B} \varphi_n(\Pi_{K'} x)\psi(\Pi_{K'} x)\,dx$$

for any $\psi \in \mathbf{C}(\mathbf{R}^d)$. By (4-7-1) and monotone continuity, it follows that $1_{G \ominus \rho B} W_k^K = 1_{G \ominus \rho B} W_k^{K'}$, and thus $1_G W_k^K = 1_G W_k^{K'}$ for $\rho \downarrow 0$.

Lemma 4-7-5. *Let $G \in \mathcal{G}$ and $K_0 \in \mathcal{K}$ be such that $G \subset K_0$, and $\{F_n\} \subset \mathcal{F}$ be such that* $\lim F_n = F$ *in $\mathcal{F}$. Then* $(\underline{\lim}\,(F_n \cap K_0)) \cap G = F \cap G$.

Proof. By the upper semicontinuity of $\cap$, we have

$$(\underline{\lim}\,(F_n \cap K_0)) \subset \overline{\lim}\,(F_n \cap K_0) \subset F \cap K_0.$$

Conversely, take $x \in F \cap G$. Then there exists a sequence $n \rightarrow x_n \in F_n$ such that $x = \lim x_n$ in $\mathbf{R}^d$. For n large enough, x_n belongs to the neighborhood G of x, that is, $x_n \in F_n \cap G \subset F_n \cap K_0$. Thus $x \in \underline{\lim}\,(F_n \cap K_0)$, and $F \cap G \subset (\underline{\lim}\,(F_n \cap K_0)) \cap G$.

It is now possible to define the extension $F \rightarrow W_k^F$ on $C(\mathcal{F}')$ of the mapping $K \rightarrow W_k^K$. Note that W_k^F will be a positive Radon measure generally not bounded on $\mathbf{R}^d$.

Proposition 4-7-1. *For any $F \in C(\mathcal{F}')$ and $k = 0, 1, \ldots, d$, there exists a unique Radon measure $W_k^F \geqslant 0$ on $\mathbf{R}^d$ such that $1_G W_k^F = 1_G W_k^{K \cap F}$ for any $K \in C(\mathcal{K}')$, $G \in \mathcal{G}$ satisfying $K \cap G = F \cap G$. Moreover, $W_0^F(dx) = 1_F(x)\,dx$,* Supp *$W_k^F \subset \partial F (k = 1, 2, \ldots, d)$, and the mapping $(\varphi, F) \rightarrow \int \varphi(x) W_k^F(dx)$ is continuous on $\mathbf{C}_{\mathcal{K}}(\mathbf{R}^d) \times C(\mathcal{F}')$ for $k = 0, 1, \ldots, d$.*

Proof. Let $\{B_n\} \subset C(\mathcal{K}')$ be a sequence such that $B_n \subset \mathring{B}_{n+1}$ and $B_n \uparrow R^d$. With each $F \in C(\mathcal{F}')$, let us associate the sequences $n \to K_n = F \cap B_n$ and $n \to W_k{}^n = 1_{\mathring{B}_n} W_k^{K_n}$. By Lemma 4-7-4, it follows that $W_k{}^n = 1_{\mathring{B}_n} W_k{}^m$ if $m \geqslant n$. Thus there exists a σ-finite measure $W_k^F \geqslant 0$ on $\mathbf{R}^d$ such that $W_k^F = \lim \uparrow W_k^n$ for $n \uparrow \infty$, and it is easy to verify that W_k^F does not depend on the choice of the sequence $\{B_n\}$. If $K \in C(\mathcal{K}')$ and $G \in \mathcal{G}$ are such that $K \cap G = F \cap G$, we have $1_G W_k^F = 1_G W_k^K$ by Lemma 4-7-4, and this property may be taken as definition. In particular, $W_0^F = 1_F dx$ and Supp $W_k^F \subset \partial F$ for $f = 1, 2, \ldots, d$.

Now let $\{\varphi_n\} \subset \mathbf{C}_{\mathcal{K}}(\mathbf{R}^d)$ be a sequence such that $\lim \varphi_n = \varphi$ in $\mathbf{C}_{\mathcal{K}}(\mathbf{R}^d)$, $K_0 \in C(\mathcal{K})$ be a compact convex set including for any $n > 0$ the support of φ_n, and $\{F_n\} \subset C(\mathcal{F}')$ be a sequence such that $\lim F_n = F \in C(\mathcal{F}')$. In order to complete the proof, we must show that $\int \varphi(x) W_k^F(dx) = \lim \int \varphi_n(x) W_k^{F_n}(dx)$. Let G be an open convex set such that $K_0 \subset G \subset \overline{G} \in C(\mathcal{K}')$, and put $K_n = F_n \cap \overline{G}$, $K = F \cap \overline{G}$. By the first part of the proof, we have $1_G W_k^{F_n} = 1_G W_k^{K_n}$, $1_G W_k^F = 1_G W_k^K$, and thus also $\int \varphi_n W_k^{F_n} = \int \varphi_n W_k^{K_n}$, $\int \varphi W_k^F = \int \varphi W_k^K$. From $K_n \subset \overline{G} \in C(\mathcal{K}')$, it follows that $\int W_k^{K_n} \leqslant \int W_k^{\overline{G}}$ (because the Minkowski functionals are increasing), so that the sequence $\{W_k^{K_n}\}$ is dominated in $\mathfrak{M}_c^+$. Thus the sequence $n \to \int \varphi_n W_k^{K_n}$ also is dominated in $\mathbf{R}$, and admits an adherent point $u \in \mathbf{R}$. Then there exists a subsequence $\{n_p\}$ such that $\lim K_{n_p} = K'$ in $C(\mathcal{K}')$, and $\lim W_k^{K_{n_p}} = W_k^{K'}$ in $\mathfrak{M}_c$. By Lemma 4-7-5, it follows that $K' \cap G = K \cap G$, and thus, by Theorem 4-7-1 and Lemma 4-7-4, $u = \lim \int \varphi_{n_p} W_k^{K_{n_p}} = \int \varphi W_k^{K'} = \int \varphi 1_G W_k^{K'} = \int \varphi 1_G W_k^K = \int \varphi W_k^K$. It follows that u is the unique adherent point of the sequence $n \to \int \varphi_n W_k^{K_n}$, and $\lim \int \varphi_n W_k^{F_n} = \lim \int \varphi_n W_k^{K_n} = \int \varphi W_k^K = \int \varphi W_k^F$.

Corollary. *Let φ be a positive function l.s.c. on $\mathbf{R}^d$ (the values of which may be infinite). Then the mapping $F \to \int \varphi(x) W_k^F(dx)$ $(k = 0, 1, \ldots, d)$ is l.s.c. on $C(\mathcal{F}')$. In particular, the mapping $F \to W_k(F) = \int W_k^F(dx) \leqslant \infty$ is an l.s.c. extension on $C(\mathcal{F}')$ of the corresponding Minkowski functional previously defined on $C(\mathcal{K}')$.*

Proof. Let $\{\varphi_n\} \subset \mathbf{C}_{\mathcal{K}}(\mathbf{R}^d)$ be a sequence such that $\varphi_n \uparrow \varphi$, so that $\int \varphi_n W_k^F \uparrow \int \varphi W_k^F$. We have $\int \varphi W_k^F > a$ (a real) if and only if F belongs to the set $\cup n \{F' \colon F' \in C(\mathcal{F}'), \int \varphi_n W_k^{F'} > a\}$, which is open in $C(\mathcal{F})$ by the proposition. Thus $F \to \int \varphi W_k^F$ is l.s.c.

The Random Version of the Minkowski Measures

Let $A = (\mathcal{F}, \sigma_f, P)$ be an a.s. convex RACS. For each $k = 0, 1, \ldots, d$, the measure W_k^F is defined for P-almost every $F \in \mathcal{F}''$ (if $F = \varnothing$, we put

$W_k^F=0$). By proposition 4-7-1, there exists for any $\varphi\in\mathbf{C}_{\mathcal{K}}(\mathbf{R}^d)$ a random variable $\int W_k^A(dx)\ \varphi(x)$ P-a.e. defined by the mapping $F\to\int\varphi W_k^F$. Moreover, by the same proposition, $\varphi=\lim\varphi_n$ in $\mathbf{C}_{\mathcal{K}}(\mathbf{R}^d)$ implies $\int\varphi W_k^A$ $=\lim\int\varphi_n W_k^A$ a.s. In other words, W_k^A is a *positive random measure*.

For each $\varphi\in\mathbf{C}_{\mathcal{K}}(\mathbf{R}^d)$, $\quad E(|\int\varphi W_k^A|)<\infty$.

Proof. We may suppose that $\varphi\geqslant 0$. Let $K_0\in\mathcal{K}$ be the support of φ, $G\in\mathcal{G}$, and $K_0'\in C(\mathcal{K}')$ such that $K_0\subset G\subset K_0'$. By Proposition 4-7-1, we have a.s.

$$\int\varphi W_k^A=\int W_k^{A\cap K_0'}\ \varphi\leqslant\ (\mathrm{Sup}\ \varphi)(W_k(K_0'\cap A))\leqslant(\mathrm{Sup}\ \varphi)W_k(K_0')$$

for the Minkowski functionals are increasing. Thus $E(\int\varphi W_k^A)\leqslant(\mathrm{Sup}\ \varphi)$ $W_k(K_0')<\infty$.

It follows that the mapping $\varphi\to E(\int\varphi W_k^A)$ is a positive linear functional on $\mathbf{C}_{\mathcal{K}}(\mathbf{R}^d)$, that is, a Radon measure, which is called the *expectation measure* of the *Minkowski measure* W_k and denoted by $E(W_k^A)$ or $E(W_k)$ or simply $W_k(dx)$ if there is no ambiguity.

If ψ is a positive l.s.c. function on $\mathbf{R}^d$ (with finite or infinite values), there exists a sequence $\{\varphi_n\}\subset\mathbf{C}_{\mathcal{K}}(\mathbf{R}^d)$ such that $\varphi_n\uparrow\psi$ It follows that $\int\varphi_n W_k^A\uparrow\int\psi W_k^A$ a.s. and $E(\int\varphi_n W_k^A)\uparrow E(\int\psi W_k^A)$, that is, $\int\psi(x)E(W_k(dx))$ $=E(\int\psi(x)W_k^A(dx))$. In the same way, for any measurable positive function f, the relationship $\int fW_k^A=\mathrm{Inf}\ \{\int\psi W_k^A,\ \psi$ is l.s.c., $\psi\geqslant f\}$ a.s. yields $E(\int fW_k^A)=\int fE(w_k)$. In other words:

Proposition 4-7-2. *Let $A=(\mathcal{F},\sigma_f,P)$ be an a.s. convex RACS. Then, for any measurable function f, the integral $\int fW_k^A$ $(k=0,\ 1,\dots,d)$ is a random variable defined P-a.e. on $\mathcal{F}$, and the mapping $\varphi\to\int\varphi W_k^A$ is a positive random measure on $\mathbf{C}_{\mathcal{K}}(\mathbf{R}^d)$. Moreover, the mapping $\varphi\to E(\int\varphi W_k^A)$ is a positive Radon measure on $\mathbf{C}_{\mathcal{K}}(\mathbf{R}^d)$, called the expectation measure of the Minkowski measure W_k and denoted as $E(W_k)$ For any measurable positive function f on $\mathbf{R}^d$, $E(\int fW_k^A)=\int fE(W_k)$.*

Note. If the expectation measure $E(W_k)$ is absolutely continuous with respect to the Lebesgue measure, its density will often be called the *density of the Minkowski functional W_k* (with respect to the RACS A).

The Convex Ring $\mathfrak{S}$

We denote by $\mathfrak{S}$ the class closed under finite union generated by $C(\mathcal{K})$. The class $\mathfrak{S}$ is dense in (but not identical with) $\mathcal{K}(\mathbf{R}^d)$. In integral geometry, a functional χ on $\mathfrak{S}$, called the *Euler-Poincaré characteristic*, is defined as follows: $\chi(\varnothing)=0$, $\chi(K)=1$ for any $K\in C(\mathcal{K}')$, and, if $K\in\mathcal{K}$ admits the representation $K=\cup_{i=1}^{p} K_i$ for compact convex sets $K_1,\ldots,K_p\in C(\mathcal{K}')$,

$$\chi(K)=\sum_i \chi(K_i)-\sum_{i_1<i_2}\chi(K_{i_1}\cap K_{i_2})+\cdots$$

$$+(-1)^{p-1}\chi(K_1\cap\cdots\cap K_p) \qquad (4\text{-}7\text{-}3)$$

It can be shown (see, e.g., Hadwiger, 1957) that the value $\chi(K)$ defined by (4-7-3) does not depend on the choice of the particular representation $K=\cup K_i$, $K_i\in C(\mathcal{K}')$ used for a given $K\in\mathfrak{S}$, so that (4-7-3) actually defines a function on $\mathfrak{S}$, the values of which are positive or negative integers. Moreover, by the definition of (4-7-3), χ is additive on $\mathfrak{S}$ in the following sense:

$$\chi(K\cup K')+\chi(K\cap K')=\chi(K)+\chi(K') \qquad (K,K'\in\mathfrak{S})$$

Let B be the unit ball and $r\geqslant 0$. If $K=\cup K_i$, $K_1,\ldots,K_p\in C(\mathcal{K}')$, we have, for any point $x\in\mathbf{R}^d$,

$$\chi(K\cap(rB)_x)=\sum_i \chi(K_i\cap(rB)_x)-\sum_{i_1<i_2}\chi(K_{i_1}\cap K_{i_2}\cap(rB)_x)+\cdots$$

$$+(-1)^{p-1}\chi(K_1\cap\cdots\cap K_p\cap(rB)_x)$$

In other words, the function $x\to\chi(K\cap(rB)_x)$ is

$$\sum_i 1_{K_i\oplus rB}-\sum_{i_1<i_2}1_{(K_{i1}\cap K_{i2})\oplus rB}+\cdots$$

and hence is integrable, so that we have

$$\int\chi(K\cap(rB)_x)\,dx=\sum_i V(K_i\oplus rB)-\sum_{i_1<i_2}V\big((K_{i_1}\cap K_{i_2})\oplus rB\big)+\cdots$$

V denoting the volume. In this relationship, the left-hand side does not depend on the choice of the particular representation $K=\cup K_i$. By the Steiner formula (4-1-8), the right-hand side is a polynomial in r.

This may be written as follows:

$$\int \chi(K \cap (rB)_x)\,dx = \sum_{k=0}^{d} \binom{d}{k} r^k \overline{W}_k(K)$$

$$\overline{W}_k(K) = \sum_i W_k(K_i) - \sum_{i_1 < i_2} W_k(K_{i_1} \cap K_{i_2}) + \cdots \qquad \text{(4-7-4)}$$

$$+(-1)^{p-1} W_k(K_1 \cap \cdots \cap K_p)$$

and the values $\overline{W}_k(K)$, $k=0, 1,\dots,d$, do not depend on the choice of the representation $K = \cup K_i$. Thus, for each $k=0, 1,\dots,d$, the mapping $K \to \overline{W}_k(K)$ is *an extension on* $\mathfrak{S}$ *of the Minkowski functional* W_k previously defined on $C(\mathcal{K})$. In particular, $\overline{W}_0$ is the volume, $d\overline{W}_1$ the surface area, and $(1/b_d)\overline{W}_d$ the Euler-Poincaré characteristic itself.

These functionals are *additive* on $\mathfrak{S}$, that is,

$$\overline{W}_k(K \cup K') + \overline{W}_k(K \cap K') = \overline{W}_k(K) + \overline{W}_k(K') \qquad (K, K' \in \mathfrak{S})$$

Also note that the *Crofton formula* (4-1-12) still holds for $\mathfrak{S}$, because it holds for each intersection $K_{i_1} \cap \dots \cap K_{i_n} \in C(\mathcal{K})$ in the right-hand side of (4-7-4). By taking $k'=k$ in (4-1-12), we have $\overline{W}_k^k(K \cap S_s) = b_k \chi(K \cap S_s)$. Hence we have the following formula:

$$\overline{W}_k^d(K) = \frac{b_d}{b_{d-k}} \int_{\mathfrak{S}_k} \varpi_k(dS) \int_{S^\perp} \chi(K \cap S_s)\,ds$$

which may also be used as a definition.

Note that these extensions $\overline{W}_k$, $k=0, 1,\dots,d$, on $\mathfrak{S}$ are not the only possible ones. For instance, $K \to (b_d/b_{d-k}) \int_{\mathfrak{S}_k} \mu_{d-k}(\Pi_S \perp K) \varpi_k(dS)$ is an extension on the entire space $\mathcal{K}$ which does not coincide with $\overline{W}_k$ on $\mathfrak{S}$. Let us now see another extension which also holds, not only for the functionals, but also for the Minkowski measures.

Extension of the Minkowski Measures

In order to extend to $\mathfrak{S}$ the Minkowski measures themselves, let us first give a definition. Let x be a point in $\mathbf{R}^d$ and $K \in \mathcal{K}$. Then a point $x' \in K$ is said to be a *projection* of x into K if there exists an open set G such that

$x' \in G$ and $|x-y|>|x-x'|$ for any $y \in G \cap K$, $y \neq x'$, and we denote by $\Pi_K(x)$ *the set of all projections of x into K*.

If $K \in \mathfrak{S}$, the set $\Pi_K(x)$ is *finite*; and, if $K = \cup K_i$, $K_1, \ldots, K_n \in C(\mathcal{K}')$ is a representation of K, any $x' \in \Pi_K(x)$ is the (unique) projection $\Pi_{K_i} x$ of x into K for at least one index $i (i = 1, 2, \ldots, n)$.

Proof. There exists an index i_0 such that $x' \in K_{i_0}$, since $x' \in \cup K_i$. If $x \in K_{i_0}$, $x = x'$ and the statement is true, because $\Pi_{K_{i0}} x = x = x'$. Suppose that $x \notin K_{i_0}$ and put $x_{i_0} = \Pi_{K_{i0}} x$. If $x' \neq x_{i_0}$, the segment $[x_{i_0}, x']$ is included in K_{i_0}, and for any open set $G \ni x'$ the segment $[x_{i_0}, x']$ includes a point $y \in G$, $y \neq x'$ such that $|x-y| \leqslant |x-x'|$. But this is impossible, by $x' \in \Pi_K(x)$. Thus $x' = x_{i_0}$.

Conversely, the point $x_i = \Pi_{K_i} x (i = 1, 2, \ldots, n)$ belongs to $\Pi_K(x)$ if and only if one of two conditions is satisfied:

a. The point $x = x_i$, that is, $x \in K_i$.

b. The point $x \neq x_i$, and there exists an open neighborhood G of x_i such that $|x-y| > |x-x_i|$ strictly for any point $y \neq x_i$, $y \in K \cap G$. Obviously, condition *b* is equivalent to the following:

b′. The point $x \neq x_i$, and the hyperplane $H \ni x_i$ is a supporting hyperplane for K in a local sense (i.e., there exists an open set $G' \ni x_i$ such that the point x and the set $K \cap G'$ are separated by H).

The set $\Pi_K(x)$ contains at most n distinct points. Denote by $x'_1, \ldots, x'_p$ $(0 \leqslant p \leqslant n)$ these distinct points in $\Pi_K(x)$, and put $x_i = \Pi_{K_i} x$, $x_{i_1, \ldots, i_k} = \Pi_{K_{i1} \cap \cdots \cap K_{ik}} x$ $(0 < k \leqslant n,\ 0 < i_1 < i_2 < \cdots < i_k \leqslant n)$. Also let F_i be the function defined by $F_i(x) = 1$ if $x_i = \Pi_{K_i} x \notin K_j$ for $j \neq i$, and by $F_i(x) = 0$ otherwise. In the same way, for $0 < i_1 < i_2 < \cdots < i_k \leqslant n$, put $F_{i_1, \ldots, i_k}(x) = 1$ if the points $x_{i_1}, \ldots, x_{i_k}$ are in $K_{i_1} \cap \cdots \cap K_{i_k}$ and none of them is in K_j, $j \neq i_1, \ldots, i_k$, and $F_{i_1, \ldots, i_k}(x) = 0$ otherwise. Obviously, $F_{i_1, \ldots, i_k}(x) = 1$ implies $x_{i_1} = \cdots = x_{i_k} = x_{i_1, \ldots, i_k}$.

Then for $i = 1, 2, \ldots, n$ the point $x_i = \Pi_{K_i} x$ belongs to $\Pi_K(x)$ if an only if $F_i(x) = 1$ or $F_{i_1, \ldots, i_k}(x) = 1$ for a sequence $0 < i_1 < \cdots < i_k \leqslant n$ containing the index i.

Proof. Suppose, for instance, that $i = n$. If $F_{i_1, \ldots, i_k, n}(x) = 1$ $(0 < i_1 < i_2 < \cdots < i_k < n)$ and $x \in K_{i_1} \cap \cdots \cap K_{i_k} \cap K_n$, obviously $x = x_n = x_{i_1, \ldots, i_k, n} \in \Pi_K(x)$. If $F_{i_1, \ldots, i_k}(x) = 1$ and $x \notin K_{i_1} \cap \cdots \cap K_{i_k} \cap K_n$, then $x \neq x_n$ and $x_{i_1} = \cdots = x_{i_k} = x_n = x_{i_1, \ldots, i_k, n}$. Then let H_n be the hyperplane orthogonal to the segment $[x, x_n]$ and such that $x_n \in H_n$. H_n is a supporting hyperplane for $K_{i_1}, K_{i_2}, \ldots, K_{i_k}$ and K_n, because $x_{i_1} = \cdots = x_{i_k} = x_n = x_{i_1, i_2, \ldots, i_k, n}$, and satisfies condition *b′* because $x_n \notin K_j$ if $j \neq i_1, i_2, \ldots, n$. Thus $x_n \in \Pi_K(x)$.

Conversely, let x' be a point in $\Pi_K(x)$ and $i_1,\dots,i_k$ be the indices i such that $x' \in K_i$. This implies $x' = x_{i_1} = \cdots = x_{i_k}$ and $x' \neq x_j$ if $j \neq i_1,\dots,i_k$, that is, $F_{i_1,\dots,i_k}(x) = 1$.

In other words:

Proposition 4-7-3. *With the above definitions, the indicator of the set* $\Pi_K(x)$ *is*

$$1_{\Pi_K(x)} = \sum_i F_i(x) 1_{\{x_i\}} + \sum_{i_1<i_2} F_{i_1,i_2}(x)\, 1_{\{x_{i1,\,i2}\}} + \cdots$$

and the number of the distinct points in $\Pi_K(x)$ *is*

$$n(K,x) = \sum_i F_i(x) + \sum_{i_1<i_2} F_{i_1,\,i_2}(x) + \cdots$$

Let us denote by $\Pi_K(x;r) \subset \Pi_K(x)$ the subset of $\Pi_K(x)$ defined by

$$\Pi_K(x;r) = \{x' :\ x' \in \Pi_K(x),\quad |x-x'| \leqslant r\} = \Pi_K(x) \cap B_r(x)$$

(where $r \geqslant 0$). The indicator of the set $\Pi_K(x;r)$ is $1_{\Pi_K}(x) 1(rB)_x$, and the number $n(K,r,x)$ of the distinct points in $\Pi_K(x,r)$ is

$$n(K,r,x) = \sum_i F_i(x) 1_{(rB)_x}(x_i) + \sum_{i_1<i_2} F_{i_1,\,i_2}(x) 1_{(rB)_x}(x_{i_1,\,i_2}) + \cdots \qquad (4\text{-}7\text{-}5)$$

More generally, for any $\varphi \in \mathbf{C}(\mathbf{R}^d)$, we have

$$\sum_{x' \in \Pi_K(x,r)} \varphi(x') = \sum_i F_i(x) 1_{(rB)_x}(x_i)\varphi(x_i) + \sum_{i_1<i_2} F_{i_1,i_2}(x) 1_{(rB)_x}(x_{i_1,i_2})\varphi(x_{i_1,i_2}) + \cdots \qquad (4\text{-}7\text{-}6)$$

It is not difficult to verify that the functions $F_{i_1,\dots,i_k}$ are measurable, and it follows that $x \to \sum_{x' \in \Pi_K(x,r)} \varphi(x')$ is also measurable.

By integrating in $\mathbf{R}^d$, we get

$$\int\left(\sum_{x' \in \Pi_K(x,r)} \varphi(x')\right) dx = \sum_i \int_{K_i \oplus rB} F_i(x)\varphi(\Pi_{K_i} x)\, dx + \sum_{i_1<i_2} \int_{(K_{i_1} \cap K_{i_2}) \oplus rB} F_{i_1,i_2}(x)\varphi(\Pi_{K_{i1} \cap K_{i2}} x)\, dx + \cdots$$

The left-hand side of this relationship does not depend on the choice of the particular representation $K = \cup K_i$. By (4-7-1), the right-hand side is a polynomial in r, giving the explicit formulae

$$\int \left(\sum_{x' \in \Pi_K(x,r)} \varphi(x') \right) dx = \sum_{k=0}^{d} \binom{d}{k} r^k \int W_k^K(dx)\varphi(x)$$

$$W_k^K = \sum_i F_i W_k^{K_i} + \sum_{i_1 < i_2} F_{i_1 i_2} W_k^{K_{i_1} \cap K_{i_2}} + \cdots \qquad \text{(4-7-7)}$$

The measures W_k^K that we have so defined do not depend on the choice of the representation $K = \cup K_i$, because the left-hand side of the first relationship in (4-7-7) does not depend on it. If $\varphi = 1$, the integrand is simply the number $n(K,r,x)$ of distinct points in $\Pi_K(x,r)$. By putting $W_k(K) = \int W_k^K(dx)$, the general formula then yields

$$\int n(K;r;x)\,dx = \sum_{k=0}^{d} \binom{d}{k} r^k W_k(K) \qquad \text{(4-7-8)}$$

Thus, by the Steiner formula (4-1-8), $K \to W_k(K)$ is an *extension on $\mathfrak{S}$ of the Minkowski functionals* previously defined on $C(\mathcal{K})$. Note that generally $W_k(K) \neq \overline{W}_k(K)$, so that the present extension does not coincide with the extension $\overline{W}_k$ defined by (4-7-4).

More generally, if $K \in C(\mathcal{K}')$, the right-hand side of (4-7-7) is simply $\int_{K \oplus rB} \varphi(\Pi_K x)\,dx$, so that the measures W_k^K are identical with the Minkowski measures (Theorem 4-7-1). Thus the mapping $K \to W_k^K$ defined by (4-7-7) actually *is an extension* on the convex ring $\mathfrak{S}$ of the analogous mapping we used for defining the Minkowski measures on $C(\mathcal{K})$.

If $k = o$, we have $W_0^K = 1_K dx$, and this measure is associated with the volume. If $k = 1$, W_1^K is (up to a multiplicative constant) the measure associated with the surface area. For $k = d$, the measure W_d^K generalizes the notion of total curvature, and the corresponding Minkowski functional $K \to W_k(K) = \int W_k^K(dx)$ is closely related to the convexity number we shall now define.

The Convexity Number

For any unit vector $u \in S_0$ and $r \in \mathbf{R}$, denote by $H(u,r)$ the hyperplane $H(u,r) = \{x\colon\ x \in \mathbf{R}^d,\ \langle x,u \rangle = r\}$ orthogonal to u and including the point ru. Then let K be a compact set in $\mathfrak{S}$, and C_i, $i = 1, 2, \ldots, p$, be the (nonempty) connected components of $K \cap H(u,r)$ ($p < \infty$, because $K \in \mathfrak{S}$). For a given index i_0 ($0 < i_0 \leqslant p$), we say that the connected component C_{i_0} is an *entering set* in K [with respect to u and $H(u,r)$] if there exists an open and connected set $G \supset C_{i_0}$ such that $G \cap C_i = \varnothing$ for $i \neq i_0$ and $G \cap K \cap H(u, r-\epsilon) = \varnothing$ for any $\epsilon > 0$ small enough. Denote by $f(r)$ the number ($\leqslant p$) of the entering sets associated with $H(u,r)$. If $u \in S_0$ is fixed, this number $f(r) \neq 0$ only for a *finite* number of reals $r_1, r_2, \ldots$, so that we may put $\nu(K,u) = \Sigma_r f(r)$. We shall say that $\nu(K,u)$ is *the convexity number of $K \in \mathfrak{S}$ with respect to the unit vector u*. By definition, the *mean convexity number $\nu(K)$* (or simply the convexity number) will be the rotation average of $\nu(K,u)$, that is, $\nu(K) = \int_{S_0} \nu(K,u)\, \varpi(du)$, ϖ denoting the unique invariant probability on S_0.

Proposition 4-7-4. *Let K be a compact set in $\mathfrak{S}$, $\nu(K)$ its convexity number, and $W_d(K)$ the value at K of the Minkowski functional defined by* (4-7-8) *for $k = d$. Then*

$$\nu(K) = \frac{1}{b_d} W_d(K) \tag{4-7-9}$$

Proof. Suppose first that $K \in C(\mathcal{K}')$ is a compact set, the boundary ∂K of which is very regular. Then the measure W_d^K is absolutely continuous with respect to the surface measure W_1^K, and its density is proportional to the total curvature on ∂K. If ϖ is the unique invariant probability on S_0, and $B \subset \partial K$ is a borelian set, $(1/b_d) W_d^K(B)$ is the ϖ-measure on S_0 of the set $\{u$: u is the unit vector of the external normal to ∂K at a point $x \in B\}$. In other words, $(1/b_d) W_d^K(B)$ is the probability for the hyperplane class $H(-u,r)$, $r \in \mathbf{R}$, the unit vector $-u$ of which is random on (S_0, ϖ) to have its entering point on K in the borelian set B. This interpretation holds as long as the normal at each point $x' \in \partial K$ is uniquely defined, and in particular for any $K \in C(\mathcal{K}')$ such that $K = K_0 \oplus \epsilon B$ for a $K_0 \in C(\mathcal{K}')$ and $\epsilon > 0$ (B is the unit ball), because the entering set in K of the random class $H(-u,r)$, $r \in R$, is ϖ-a.s. a single point.

Take now $K \in \mathfrak{S}$ such that $K = \cup (K_i \oplus \epsilon B) = (\cup K_i) \oplus \epsilon B$ for $i = 1, 2, \ldots, n$, $K_i \in C(\mathcal{K}')$ and $\epsilon > 0$. Let u be a unit vector, and $0 < i_1 < \cdots < i_k \leqslant n$. Then the entering point x' in $(K_{i_1} \oplus \epsilon B) \cap \cdots \cap (K_{i_k} \oplus \epsilon B)$ with respect to the unit vector $u \in S_0$ belongs to an entering set in K if and only if $F_{i_1, \ldots, i_k}(x) = 1$, and then this entering set is ϖ-a.s. a single point, that is, $\{x\}$. Thus the probability for the entering point in $(K_{i_1} \oplus \epsilon B) \cap \cdots \cap (K_{i_k} \oplus \epsilon B)$ with respect to the random unit vector u to be also an entering point in K

is

$$\frac{1}{b_d}\int F_{i_1,\ldots,i_k}(x)\,W_d^{(K_{i1}\oplus\epsilon B)\cap\cdots\cap(K_{ik}\oplus\epsilon B)}(dx)$$

By (4-7-7), it follows that $b_d\nu(K)=\int W_d^K(dx)=W_d(K)$, and (4-7-9) holds.

Finally, let K be an arbitrary compact set in $\mathfrak{S}$, and $K=\cup K_i$, $i=1, 2,\ldots,n$, $K_i\in C(\mathcal{K})$, be a representation of $K\in\mathfrak{S}$. For any $\epsilon>0$, put $K_\epsilon=K\oplus\epsilon B$, $K_i(\epsilon)=K_i\oplus\epsilon B$. By the first part of the proof, we have

$$b_d\nu(K_\epsilon)=W_d(K_\epsilon)\qquad(\epsilon>0)\tag{4-7-10}$$

Then $\epsilon\downarrow 0$ implies $\nu(K_\epsilon)\uparrow\nu(K)$, for, if $u\in S_0$ is a given unit vector, it is not too difficult to verify that the mapping $K\to\nu(K,u)$ is l.s.c.on $\mathfrak{S}$, and $\epsilon\to\nu(K_\epsilon,u)$ is a nonincreasing function on $\mathbf{R}_+$. Thus $\epsilon\downarrow 0$ implies $\nu(K_\epsilon,u)\uparrow\nu(K,u)$ and $\nu(K_\epsilon)=\int\nu(K_\epsilon,u)\varpi(du)\uparrow\int\nu(K,u)\varpi(du)=\nu(K)$ by monotone continuity.

On the other hand, $\epsilon\downarrow 0$ also implies $W_d(K_\epsilon)\uparrow W_d(K)$, for, if $x'\in\partial K$ is a projection on K of a point $x\neq x'$, $x'+\epsilon(x-x')/|x-x'|$ is also a projection of x on $K\oplus\epsilon B$ for ϵ small enough, and conversely. It follows that $\epsilon\downarrow 0$ implies $n(K_\epsilon,r-\epsilon,x)\uparrow n(K,r,x)$, and by (4-7-8) the desired implication holds. From (4-7-10), we conclude that $b_d\nu(K)=W_d(K)$.

The Generalized Random Minkowski Measures

Let us now return to the probabilistic version of the theory. In the following preliminary results, the convex ring $\mathfrak{S}$ is provided with the relative myope topology. For any $x\in\mathbf{R}^d$ and $K\in\mathfrak{S}$, the set $\Pi_K(x)$ of the projections of x into K is *finite*, and thus compact in $\mathbf{R}^d$. More precisely:

Lemma 4-7-6. *The mapping* $(x,K)\to\Pi_K(x)$ *from* $\mathbf{R}^d\times\mathfrak{S}$ *into* $\mathcal{K}$ *is l.s.c. In the same way, the mapping* $(x,K,r)\to\Pi_K(x)\cap\mathring{B}_r(x)$ *from* $\mathbf{R}^d\times\mathfrak{S}\times(\overline{\mathbf{R}}_+\setminus\{o\})$ *into* $\mathcal{K}$ *is l.s.c.* ($\mathring{B}_r(x)$ *is the open ball with center x and radius* $r>0$, *and* $\overline{\mathbf{R}}_+$ *is the extended half line* $[0,\infty]$).

Proof. Let $\{x_n\}\subset\mathbf{R}^d$ and $\{K_n\}\subset\mathfrak{S}$ be two sequences such that lim $x_n=x$ in $\mathbf{R}^d$ and lim $K_n=K\in\mathfrak{S}$ in $\mathcal{K}$. By the inclusion $\Pi_{K_n}(x_n)\subset K_n$ and Proposition 1-2-4, the first statement will hold if $\Pi_K(x)\subset\underline{\lim}\,\Pi_{K_n}(x_n)$. Let x' be a point in $\Pi_K(x)$. We shall prove that $x'\in\underline{\lim}\,\Pi_{K_n}(x_n)$.
a. Suppose that first $x'=x$, and thus $x\in K$. For each $n>0$, let x'_n be one of the points minimizing $|x_n-y|$, $y\in K_n$. In particular, $x'_n\in\Pi_{K_n}(x_n)$. It is easy to verify that $x=\lim x'_n$. Thus from $x=x'$ and $x'_n\in\Pi_{K_n}(x_n)$, we conclude that $x'\in\underline{\lim}\,\Pi_{K_n}(x_n)$.

b. Suppose now that $x' \neq x$. Thus there exists an open neighborhood G of x' such that

$$(b) \qquad |x-y| > |x-x'| \quad \text{for any } y \in G \cap K, \quad y \neq x'$$

Suppose that $x' \notin \underline{\lim}\ \Pi_{K_n}(x_n)$. Then there exists a subsequence $n_k \to \Pi_{K_{nk}}(x_{n_k})$ and a closed ball $B_\epsilon(x') \subset G$ such that

$$(b') \qquad B_\epsilon(x') \cap \Pi_{K_{nk}}(x_{n_k}) = \varnothing$$

for any $k > 0$. By $x' \in K'$, there exists a sequence $\{y_{n_k}\}$ such that $x' = \lim y_{n_k}$, $y_{n_k} \in K_{n_k}$; and, for k large enough, it follows that $K_{n_k} \cap B_\epsilon(x') \neq \varnothing$. Then let y'_{n_k} be one of the points minimizing $|x_{n_k} - y|$, $y \in K_{n_k} \cap B_\epsilon(x')$, that is, y_{n_k} is a projection of x_{n_k} into $K_{n_k} \cap B_\epsilon(x') \in \mathfrak{S}$. It follows that $y'_{n_k} \notin \mathring{B}(x')$, for, if so, y_{n_k} would also be a projection of x_{n_k} into K_{n_k} itself, and this is impossible by (b'). Thus $y'_{n_k} \in \partial B_\epsilon(x') \cap K_{n_k}$ and $|x_{n_k} - y'_{n_k}| < |x_{n_k} - y_{n_k}|$. Now let y_0 be an adherent point of the sequence $\{y'_{n_k}\}$. Then $y_0 \in \partial B_\epsilon(x')$, that is, $y_0 \neq x'$, and $|x - y_0| \leqslant |x - x'|$. But this is impossible by (b). We conclude that $x' \in \underline{\lim}\ \Pi_{K_n}(x_n)$, and $(x, K) \to \Pi_K(x)$ is l.s.c.

c. It remains to prove the second statement. Let $\{x_n, K_n, r_n\}$ be a sequence converging toward $\{x, K, r\}$ in $\mathbf{R}^d \times \mathfrak{S} \times (\overline{\mathbf{R}}_+ \setminus \{o\})$, and $x' \in \Pi_K(x) \cap \mathring{B}_r(x)$. By the first statement, there exists a sequence $\{x'_n\} \subset \mathbf{R}^d$ such that $x' = \lim x'_n$, and $x'_n \in \Pi_{K_n}(x_n)$ for n large enough. But $r = \lim r_n$ in $\overline{\mathbf{R}}_+ \setminus \{o\}$ and $|x - x'| < r$ imply $|x_n - x'_n| < r_n$ for n large enough, and thus $x'_n \in \Pi_{K_n}(x_n) \cap \mathring{B}_{r_n}(x_n)$. It follows that $x' \in \underline{\lim}\ (\Pi_{K_n}(x_n) \cap \mathring{B}_{r_n}(x_n))$, and the lemma is proved.

For any $K \in \mathfrak{S}$, $r \in \mathbf{R}_+ \setminus \{o\}$, $x \in \mathbf{R}^d$, and any function $\varphi \geqslant 0$ on $\mathbf{R}^d$, put

$$S(K, \varphi, r; x) = \sum_{x' \in \Pi_K(x) \cap B_r(x)} \varphi(x')$$

$$\mathring{S}(K, \varphi, r; x) = \sum_{x' \in \Pi_K(x) \cap \mathring{B}_r(x)} \varphi(x')$$

so that, by (4-7-7), the Minkowski measures W_k^K are defined by

$$\int S(K, \varphi, r; x)\, dx = \sum \binom{d}{k} r^k \int W_k^K(dx) \varphi(x) \qquad (4\text{-}7\text{-}7')$$

Proposition 4-7-5. *For any l.s.c. function* $\varphi \geqslant 0$, *the mapping* $(K, r) \to \int S(K, \varphi, r; x)\, dx$ *is l.s.c. on* $\mathfrak{S} \times (\overline{\mathbf{R}}_+ \setminus \{o\})$.

Proof. If $K\in\mathfrak{S}$ and $r>0$, the set of points x such that $\Pi_K(x)\cap\partial\bar{B}_r(x)\neq\varnothing$ is negligible for the Lebesgue measure, so that

$$\int S(K,\varphi,r;x)\,dx=\int\mathring{S}(K,\varphi,r;x)\,dx$$

On the other hand, by Lemma 4-7-6, the mapping $(K,r)\to\mathring{S}(K,\varphi,r;x)$ is l.s.c. on $\mathfrak{S}\times(\bar{\mathbf{R}}_+\backslash\{o\})$ for any fixed $x\in\mathbf{R}^d$. Thus, if $\{K_n,r_n\}\subset\mathfrak{S}\times(\bar{\mathbf{R}}_+\backslash\{o\})$ is a sequence converging toward $(K,r)\in\mathfrak{S}\times(\bar{\mathbf{R}}_+\backslash\{o\})$, it follows that

$$\mathring{S}(K,\varphi,r;x)\leqslant\underline{\lim}\;\mathring{S}(K_n,\varphi,r_n;x)$$

and, by the Fatou-Lebesgue lemma:

$$\int\mathring{S}(K,\varphi,r;x)\,dx\leqslant\underline{\lim}\int\mathring{S}(K_n,\varphi,r_n;x)\,dx$$

The proposition follows.

Corollary. *For any function $\varphi\geqslant 0$ l.s.c. on $\mathbf{R}^d$ and $k=0,1,\dots,d$, the mapping $K\to\int W_k^K(dx)\varphi(x)$ from $\mathfrak{S}$ into $\bar{\mathbf{R}}_+$ is measurable (with respect to the σ-algebra induced on $\mathfrak{S}$ by σ_f).*

Proof. By Proposition 4-7-5 and formula (4-7-7′), the mapping $(K,r)\to\Sigma\binom{d}{k}r^k\int\varphi W_k^K$ is measurable. The corollary follows by taking finite differences with respect to r.

Let us now extend the mapping $F\to W_k^F$ to the following class, $\mathfrak{S}_f\subset\mathfrak{F}$:

Definition 4-7-1. *We denote by $\mathfrak{S}_f$ the class of the closed sets $F\in\mathfrak{F}$ such that $F=\cup F_i$ for a locally finite family $\{F_i, i\in I\}\subset C(\mathfrak{F})$ (i.e., each F_i, $i\in I$, is closed and convex, and, for any $K\in\mathfrak{K}$, $F_i\cap K=\varnothing$ for any $i\in I$ except at most a finite number).*

For any point $x\in\mathbf{R}^d$, the set $\Pi_F(x)$ of its projections into a given $F\in\mathfrak{S}_f$ is defined in exactly the same way as for the case $F\in\mathfrak{S}$. In particular, $\Pi_F(x)$ is locally finite and thus closed, and $\Pi_F(x)\cap\mathring{B}_r(x)$ is finite $(0<r<\infty)$. Thus, for any $\varphi\in\mathbf{C}_{\mathfrak{K}}(\mathbf{R}^d)$, the integral $\int S(F,\varphi,r;x)\,dx$ is finite, and (4-7-7′) always holds. Note that the corresponding Radon measures W_k^F $(k=0,1,\dots,d)$ are no longer necessarily bounded. Lemma 4-7-6 and Proposition 4-7-3 also hold, so that the mapping $F\to\int\varphi W_k^K$ from $\mathfrak{S}_f$ into $\mathbf{R}_+$ is measurable for any positive function $\varphi\in\mathbf{C}_{\mathfrak{K}}(\mathbf{R}^d)$. Furthermore:

Lemma 4-7-7. *$\mathfrak{S}_f$ is a measurable subset in $\mathfrak{F}$.*

Proof. Let B_n be the closed ball with center o and radius n integer >0, and $\mathfrak{S}_{m,n} \subset \mathfrak{F}$ be the class of the closed sets $F \in \mathfrak{F}$ such that $F \cap B_n$ is the union of $m' \leqslant m$ compact convex sets. Then $\mathfrak{S}_{m,n}$ is measurable as the inverse image of a closed subset of $\mathfrak{K}(B_n)$ under the u.s.c. mapping $F \to F \cap B_n$. It follows that $\mathfrak{S}_f = \cap_n \cup_m \mathfrak{S}_{m,n} \in \sigma_f$.

If $F = \varnothing$, we put $W_k^F = 0$. Taking the preceding results into account, we may state:

Theorem 4-7-2. *For any $F \in \mathfrak{S}_f$ and $k = 0, 1, \ldots, d$, there exists a Radon measure $W_k^F \geqslant 0$ on $\mathbf{R}^d$ such that, for any l.s.c. function $f \geqslant 0$ and $r > 0$,*

$$\int S(F,f,r;x)\,dx = \sum_{k=0}^{d} \binom{d}{k} r^k \int W_k^F(dx) f(x)$$

and $W_0^F = 1_F dx$, Supp *$W_k^F \subset \partial F$ for $k = 1, \ldots, d$. Moreover, the mapping $(F,r) \to \int S(F,f,r;x)\,dx$ is l.s.c., and for each $k = 0, 1, \ldots, d$ the mapping $F \to \int f(x) W_k^F(dx)$ is measurable on $\mathfrak{S}_f$.*

Corollary. *Let A be a RACS a.s. in $\mathfrak{S}_f$, that is, $P(A \in \mathfrak{S}_f) = 1$. Then, for each $k = 0, 1, \ldots, d$, there exists a random measure $W_k^A \geqslant 0$ defined for almost every $F \in \mathfrak{F}$ and any $\varphi \in \mathbf{C}_{\mathfrak{K}}(\mathbf{R}^d)$ by the mapping $F \to \int \varphi(x) W_k^F(dx)$.*

Note. If $E(\int \varphi W_k^A) < \infty$ for any $\varphi \in \mathbf{C}_{\mathfrak{K}}^+(\mathbf{R}^d)$ (i.e., φ is a continuous positive function with a compact support in $\mathbf{R}^d$), the mapping $\varphi \to E(\int \varphi W_k^A)$ is a positive linear functional on $\mathbf{C}_{\mathfrak{K}}$, and thus there exists a Radon measure $E(W_k) \geqslant 0$, called the expectation measure of A, such that $E(\int \varphi W_k^A) = \int \varphi E(W_k)$. However, the condition $E(\varphi W_k^A) < \infty$ is not necessarily fulfilled.

Moreover, if the RACS A is *stationary* and the preceding condition is satisfied, the expectation measure $E(W_k)$ is invariant under translation, that is, is proportional to the Lebesgue measure. Thus there exists a number $w_k \geqslant 0$, called the *density of the Minkowski functional* W_k (with respect to A) such that $E(W_k(dx)) = w_k\,dx$. If the condition $E(\int \varphi(x) W_k^A(dx)) < \infty$ is not fulfilled for any $\varphi \in \mathbf{C}_{\mathfrak{K}}^+$, it is not difficult to see that $E(\int \varphi(x) W_k^A(dx)) = \infty$ for any $\varphi \in \mathbf{C}_{\mathfrak{K}}^+(\mathbf{R}^d)$, $\varphi \neq 0$, and we may put $w_k = \infty$.

Further Generalization

If $F \in \mathfrak{F}$ is an arbitrary closed set, not necessarily in $\mathfrak{S}_f$, it is still possible to define positive and linear mappings $\varphi \to W_k^F(\varphi)$ $(k = 0, 1, \ldots, d)$ from $\mathbf{C}_{\mathfrak{K}}^+$ into $\bar{\mathbf{R}}_+$ such that, for any $\varphi \in \mathbf{C}_{\mathfrak{K}}^+$, $F \to W_k^F(\varphi)$ is an extension of the mapping $F \to \int \varphi(x) W_k^F(dx)$ previously defined on $\mathfrak{S}_f$. But this extension is by no mean the only possible one (see, e.g., the book by Federer, 1969). Also note that $W_k^F(\varphi)$ may be infinite, so that generally the mapping $\varphi \to W_k^F(\varphi)$ is not a Radon measure. First, let us prove a lemma.

Lemma 4-7-8. *Let Ω be a LCS space, and f a mapping from Ω into the extended line $\bar{\mathbf{R}} = [-\infty, +\infty]$. Put $\gamma = \{(\omega, x)\colon \omega \in \Omega, x \in \bar{R}, x < f(\omega)\}$ and $\Gamma = \{(\omega, x)\colon \omega \in \Omega, x \in \bar{\mathbf{R}}, x \leqslant f(\omega)\}$. Then:*

a. The mapping f is l.s.c. (respectively, u.s.c.) if and only if γ(resp. Γ) is open (resp. closed) in $\Omega \times \bar{\mathbf{R}}$.

b. The mapping f admits a smallest u.s.c. upper bound and a largest l.s.c. lower bound.

c. Let f' be a l.s.c. (resp. u.s.c.) mapping from a subspace Ω' dense in Ω into $\bar{\mathbf{R}}$. Then f' admits a largest l.s.c. (resp. a smallest u.s.c.) extension f on Ω. Moreover, for any $\omega \in \Omega$, $\omega \not\in \Omega'$, there exists a sequence $\{\omega_n\} \subset \Omega'$ such that $\omega = \lim \omega_n$ and $f(\omega) = \lim f(\omega_n)$.

Proof. It is sufficient to prove the statements concerning u.s.c. functions (by the change $f \to -f$, $-\complement\Gamma$ is substituted for γ, and conversely).

a. Suppose that f is u.s.c., and let $\{\omega_n, x_n\} \subset \Omega \times \bar{\mathbf{R}}$ be a sequence such that $\lim\,(\omega_n, x_n) = (\omega, x)$ and $(\omega_n, x_n) \in \Gamma$, that is, $x_n \leqslant f(\omega_n)$ for any $n > 0$. This implies $x \leqslant \underline{\lim}\, f(\omega_n) \leqslant \overline{\lim}\, f(\omega_n) \leqslant f(\omega)$, for f is u.s.c., and $(\omega, x) \in \Gamma$. Thus Γ is closed.

Conversely, let Γ be a closed set in $(\Omega \times \bar{\mathbf{R}})$, and $\{\omega_n\} \subset \Omega$ be a sequence such that $\lim \omega_n = \omega$ in Ω. From $(\omega_n, f(\omega_n)) \in \Gamma$, it follows that $(\omega, \overline{\lim}\, f(\omega_n)) \in \Gamma$, because Γ is closed, that is, $\overline{\lim}\, f(\omega_n) \leqslant f(\omega)$. Thus f is u.s.c.

b. A function f is determined by its "subgraph" $\Gamma_f = \{(\omega, x), \omega \in \Omega, x \in \bar{\mathbf{R}}, x \leqslant f(\omega)\} \subset \Omega \times \bar{\mathbf{R}}$, and $f \leqslant g$ is equivalent to $\Gamma_f \subset \Gamma_g$. Thus a given function f admits a smallest u.s.c. upper bound, the subgraph of which is the closure $\bar{\Gamma}_f$ of the subgraph Γ_f.

c. Let $\Gamma' = \{(\omega', x), \omega' \in \Omega', x \in \bar{\mathbf{R}}, x \leqslant f(\omega')\}$ be the subgraph of an u.s.c. function $f'\colon \Omega' \to \bar{\mathbf{R}}$, so that Γ' is closed in $\Omega' \times \bar{\mathbf{R}}$. Thus there exists a smallest closed set $\Gamma \subset \Omega \times \bar{\mathbf{R}}$ such that $\Gamma \cap (\Omega' \times \bar{\mathbf{R}}) = \Gamma'$, that is, the closure $\bar{\Gamma}'$ of Γ' in $\Omega \times \bar{\mathbf{R}}$. If $\omega \in \Omega$, the set $\{x\colon (\omega, x) \in \bar{\Gamma}'\}$ is nonempty, for Ω' is dense in Ω. By putting $f(\omega) = \operatorname{Sup}\,\{x\colon (\omega, x) \in \bar{\Gamma}'\}$, f is the smallest u.s.c. extension of f'. If $\omega \not\in \Omega'$, that is, $(\omega, f(\omega)) \not\in \Gamma'$, there exists a sequence $\{(\omega_n, x_n)\} \subset \Gamma'$ such that $(\omega, f(\omega)) = \lim\,(\omega_n, x_n)$, because $\bar{\Gamma}'$ is the closure of Γ', and (a fortiori) $f(\omega) = \lim f(\omega_n)$.

The class $\mathfrak{S}_f$ encloses the class $\mathcal{G}$ of the finite subsets in $\mathbf{R}^d$, and thus is dense in $\mathcal{F}$ by Theorem 1-2-2, Corollary 2. For any l.s.c. function $f \geqslant 0$, the mapping

$$(F,r)\to\int S(F,f,r;x)\,dx=\sum_{k=0}^{d}\binom{d}{k}r^k\int W_k^F(dx)f(x)$$

from $\mathfrak{S}_f\times(\bar{\mathbf{R}}_+\setminus\{o\})$ into $\bar{\mathbf{R}}_+$ is l.s.c., as we have seen, so that by the lemma it admits a largest l.s.c. extension on $\mathfrak{F}\times(\bar{\mathbf{R}}_+\setminus\{o\})$. In other words:

Proposition 4-7-6. *For any $k=0, 1,\dots,d$, there exists a mapping $(F,\varphi)\to W_k^F(\varphi)$ from $\mathfrak{F}\times\mathbf{C}_{\mathfrak{K}}^+$ into $\bar{\mathbf{R}}_+$ such that:*

a. For any $F\in\mathfrak{F}$, the mapping $\varphi\to W_k^F(\varphi)$ is positively linear. We have $W_0^F(\varphi)=\int_F\varphi(x)\,dx$ and $W_k^F(\varphi)=0$ $(k=1,\dots,d)$ if Supp $\varphi\cap\partial F=\varnothing$. *Moreover, there exists a sequence $\{F_n\}\subset\mathfrak{S}_f$ such that $F=\lim F_n$, $W_k^F(\varphi)=\lim W_k^{F_n}(\varphi)$.*

b. $(F,\varphi)\to W_k^F(\varphi)$ is an extension of the mapping $(F,\varphi)\to\int W_k^F(dx)\varphi(x)$ from $\mathfrak{S}_f\times\mathbf{C}_{\mathfrak{K}}^+$ into $\mathbf{R}_+$.

c. For each $k=0, 1,\dots,d$, and $\varphi\in\mathbf{C}_{\mathfrak{K}}^+$, the mapping $F\to W_k^F(\varphi)$ is measurable, and $(F,r)\to\sum\binom{d}{k}r^kW_k^F(\varphi)$ is the largest l.s.c. extension of the mapping $(F,r)\to\sum\binom{d}{k}r^kW_k^F(\varphi)$ previously defined on $\mathfrak{S}_f\times\mathbf{R}_+\setminus\{o\}$.

By this proposition, the corresponding random functionals $W_k^A(\varphi)$ also exist for any RACS A. In particular, $\varphi\to dW_1^A(\varphi)$ is a generalized version of the random surface measure of the RACS A, and yields a notion of specific surface area if A is stationary. Unfortunately, as mentioned above, this extension is not the only possible one.

CHAPTER 5

Semi-Markovian Racs in $\mathbf{R}^d$

The semi-markovian property examined in this chapter is closely related to convexity. Although there exist semi-markovian RACS which are not infinitely divisible, we shall study chiefly the infinitely divisible semi-markovian random closed sets—for brevity, SMIDRACS. The first section gives general definitions and properties and then characterizes the SMIDRACS in terms of Poisson processes on $C(\mathcal{F}')$, so that this class of RACS may be identified with a class of C-additive Choquet capacities. If $d=1$, the stationary semi-markovian RACS are simple two-state renewal processes. In the third section, we give a more general example, that is, the boolean model with compact convex grains, or (which is the same) Poisson processes concentrated on $C(\mathcal{K}')$, and the corresponding parameters are interpreted in the stationary case. Section 5-4 gives a general characterization of stationary SMIDRACS. It turns out that any stationary SMIDRACS is the union of random cylinders the bases of which are boolean models with convex grains in suitable subspaces, so that the more general stationary smidracs may be constructed from two wimple prototypes: Poisson networks and boolean models. In particular, a RACS is a Poisson network if and only if it is stable, stationary, and semi-markovian. Covariances and linear granulometries are investigated; in particular, it turns out that the function $h \to Q(h)$ is always positive definite on $\mathbf{R}^d$.

5-1. THE SEMI-MARKOVIAN PROPERTY

If C, K, and $K' \in \mathcal{K}$, we say that K and K' are *separated* by C if, for any $x \in K$ and $x' \in K'$, the segment $[x,x'] = \{(1-\lambda)x + \lambda x'; 0 \leqslant \lambda \leqslant 1\}$ hits the compact set C (see Definition 4-2-1). In particular, if K or K' is empty, K and K' are separated by any $C \in \mathcal{K}$. On the other hand, if $C = \varnothing, K$ and K' are separated by C only if K or K' is empty. Under these conditions, we use the following definition:

Definition 5-1-1. *Let A be a RACS and Q its functional on $\mathcal{K}$, defined by $Q(K) = P(\mathcal{F}^K)$. Then A is said to be semi-markovian if*

$$Q(K \cup K' \cup C)Q(C) = Q(K \cup C)Q(K' \cup C) \qquad (5\text{-}1\text{-}1)$$

for any K, K', and $C \in \mathcal{K}$ such that K and K' are separated by C.

Let us interpret (5-1-1). If $Q(C) = 0$, (5-1-1) always holds. If $Q(C) \neq 0$, (5-1-1) is satisfied if and only if *the RACS $A \cap K$ and $A \cap K'$ are independent conditional upon $\mathcal{F}^C$ being true* (i.e., conditional upon $A \cap C = \varnothing$).

Proof. Let A' be the RACS A conditional upon $\mathcal{F}^C$, and P' the corresponding probability on σ_f, that is, $P'(\mathcal{A}) = P'(\mathcal{A} \cap \mathcal{F}^C)/Q(C)$, $\mathcal{A} \in \sigma_f$. In particular, $Q' : K_1 \to P'(\mathcal{F}^{K_1}) = Q(K_1 \cup C)/Q(C)$ is the associated functional on $\mathcal{K}$. Thus (5-1-1) may be rewritten $Q'(K \cup K') = Q'(K)Q'(K')$. Let $K_1, K'_1 \in \mathcal{K}$ be such that $K_1 \subset K$ and $K'_1 \subset K'$. A fortiori, K_1 and K'_1 are separated by C, and (5-1-1) implies $Q'(K_1 \cup K'_1) = Q'(K_1)Q'(K'_1)$, that is,

$$P'(\{A' \cap K \in \mathcal{F}^{K_1}\} \cap \{A' \cap K' \in \mathcal{F}^{K'_1}\})$$

$$= P'(A' \cap K \in \mathcal{F}^{K_1}) \times P'(\{A' \cap K' \in \mathcal{F}^{K'_1}\})$$

From (2-3-5), it follows that the RACS $A' \cap K$ and $A' \cap K'$ are independent.

Note also the following simple result:

Proposition 5-1-1. *Let A be a semi-markovian RACS and $K_0 \in C(\mathcal{K}')$. Then the dilatation $A \oplus \check{K}_0$ is a semi-markovian RACS.*

Proof. If K and $K' \in \mathcal{K}$ are separated by $C \in \mathcal{K}$, and $K_0 \in C(\mathcal{K}')$, it is not difficult to see that $K \oplus K_0$ and $K' \oplus K_0$ are separated by $C \oplus K_0$. On the other hand, the functional associated with $A \oplus \check{K}_0$ is $Q_{K_0} : K \to Q_{K_0}(K) = Q(K \oplus K_0)$. Thus, by (5-1-1), it follows that

$$Q_{K_0}(K \cup K' \cup C)Q_{K_0}(C) = Q((K \cup K' \cup C) \oplus K_0)Q(C \oplus K_0)$$

$$= Q((K \oplus K_0) \cup (K' \oplus K_0) \cup (C \oplus K_0))Q(C \oplus K_0)$$

$$= Q_{K_0}(K \cup C)Q_{K_0}(K' \cup C)$$

and $A \oplus \check{K}_0$ is semi-markovian.

In the sequel, it will be convenient to put $\psi = -\log Q$, so that, by (5-5-1), the semi-markovian property is characterized by

$$\psi(K \cup K' \cup C) + \psi(C) = \psi(K \cup C) + \psi(K' \cup C) \qquad (5\text{-}1\text{-}2)$$

If K, K', and $K \cup K' \in C(\mathcal{K})$, we know, from Proposition 4-2-1, that K and K' are separated by their intersection $K \cap K'$, and in this case (5-1-2) yields $\psi(K \cup K') + \psi(K \cap K') = \psi(K) + \psi(K')$. In other words, *the functional $\psi = -\log Q$ associated with a semi-markovian RACS is C-additive on $C(\mathcal{K})$.*

Moreover, if the semi-markovian RACS A is *stationary* and *isotropic*, its functional ψ is invariant under the euclidean displacements in $\mathbf{R}^d$. Thus, by Theorem 4-1-1, there exist constants $\beta_i \geq 0, i = 1, 2, \ldots, d$, such that

$$\psi(K) = -\log Q(K) = \sum_{i=1}^{d} \beta_i W_i(K) \qquad [K \in C(\mathcal{K})] \qquad (5\text{-}1\text{-}3)$$

(W_i is the corresponding Minkowski functional).

A semi-markovian RACS is not necessarily infinitely divisible with respect to $\cup$. We shall see a simple counterexample in Section 5-2. Nevertheless, the main purpose of the present chapter is to investigate the infinitely divisible case, that is, the SMIDRACS, because here the semi-markovian property is equivalent to the C-additivity. More precisely:

Theorem 5-1-1. *Let A be an IDRACS without fixed points, and $\psi = -\log Q$ be its functional* [*i.e.*, $\psi(K) = -\log P(\mathcal{F}^K), K \in \mathcal{K}$]. *Then the three following conditions are equivalent*:

a. A is semi-markovian.

b. The functional ψ is C-additive on $C(\mathcal{K})$.

c. A is the union of a Poisson process $\mathcal{Q}$ on $C(\mathcal{F}')$.

Proof. We have already seen that condition a implies b. Suppose that condition b is true. By Proposition 3-2-1, the IDRACS A is the union of a Poisson process $\mathcal{Q}$ on $\mathcal{F}'$, associated with a σ-finite measure θ such that $\theta(\mathcal{F}_K) = \psi(K)$. Let $\{B_n\} \subset \mathcal{K}$ be a sequence such that $B_n \subset \mathring{B}_{n+1}, B_n \uparrow \mathbf{R}^d$. For n large enough, we may suppose that $\psi(B_n) > 0$ (if not, $A = \emptyset$ a.s., and condition c holds), and $K \to \psi(K \cap B_n)/\psi(B_n)$ is a Choquet capacity associated with a probability P_n on σ_f. But ψ is C-additive, and so $K \to \psi(K \cap B_n)/\psi(B_n)$ is also. Thus, by Theorem 4-2-1, P_n is concentrated on $C(\mathcal{F}')$. Then the σ-finite measure θ is also concentrated on $C(\mathcal{F}')$, and condition c is true.

Finally, with the same notation, if condition c is true, Theorem 4-2-1 yields $\psi((K'\cup K\cup C)\cap B_n)+\psi(C\cap B_n)=\psi((K\cup C)\cap B_n)+\psi((K'\cup C)\cap B_n)$ if K and K' are separated by C. By putting $n\uparrow\infty$, (5-1-2) holds and A is a SMIDRACS.

By this theorem, the Poisson networks that we studied in Section 3-5 are SMIDRACS. We shall give more general examples, and first investigate the case $d=1$, that is, the semi-markovian RACS on the euclidean line. Note that, if A is a semi-markovian RACS (respectively, a SMIDRACS) and V a linear variety with dimension $k<d$ (identified with the euclidean space $\mathbf{R}^k$), then $A\cap V$ is a semi-markovian RACS (resp., a SMIDRACS) in $\mathbf{R}^k$, which is called the semi-markovian RACS (resp., the SMIDRACS) *induced* by A on V. This is so because (5-1-1) obviously is still satisfied by $A\cap V$ for K, K', and C included in V. In particular, the results concerning the unidimensional case may be applied to the semi-markovian RACS induced on a line in $\mathbf{R}^d$ by an arbitrary stationary semi-markovian RACS.

5-2. STATIONARY SEMI-MARKOVIAN RACS ON **R**

Let A be a *stationary* semi-markovian RACS (not necessarily infinitely divisible) on the euclidean line $\mathbf{R}$. Then the probability $Q([x,x+h])$ for the segment $[x,x+h]=\{x+\lambda h, 0\leqslant\lambda\leqslant 1\}$ not to hit A depends not on the point $x\in\mathbf{R}$, but only on h, and will simply be denoted by $Q(h)$. Clearly, the semi-markovian property implies

$$qQ(h_1+h_2)=Q(h_1)Q(h_2) \qquad [h_1,h_2\geqslant 0, q=Q(o)] \tag{5-2-1}$$

For the segments $[o,h_1]$ and $[h_1,h_1+h_2]$ are separated by the point h_1. Moreover, $h\to Q(h)$ is nonincreasing on R_+, so that

$$Q(h)=q\exp(-\theta h) \tag{5-2-2}$$

$[q=Q(o)\geqslant 0$, and θ is a constant $\geqslant 0]$. If $q=0, A=\mathbf{R}$ a.s.; we shall suppose that $q>0$. If $\theta=0, Q(h)=q$ is constant, and $P(A=\varnothing)=q, P(A=\mathbf{R})=p=1-q$. Thus we suppose that $\theta>0$. If $p=1-q=0$, it is not difficult to show, by the semi-markovian property, that A is a *Poisson point process* on $\mathbf{R}$. In the sequel, we suppose that $0<q<1$.

According to the notations defined in Section 2-7, we denote by $P(h)$ the probability for a segment $[x,x+h]$ to be contained in A, by $h\to C(h)$ the covariance, that is, $C(h)=P(x\in A, x+h\in A)$, and by $p-F(h)=P(x\in A_{[o,h]})$ the probability for a given point $x\in\mathbf{R}$ to belong to the opening of A by $[o,h]$. In other words, $p-F(h)$ is the probability of the event {there exists a segment with length $\geqslant h$ contained in A and containing x}. In Section 2-7, we saw that $P(h)$ admits a derivative $P'(h)$ at any $h>0$, and

$$p-F(h)=P(h)-hP'(h) \qquad (h>0) \tag{5-2-3}$$

It is not obvious that a derivative from the right $P'(o)$ always exists, but this turns out to be true. If $0\in A$, put $L=\text{Sup}\{h:\ h\geqslant 0,[0,h]\subset A\}$. Conditionally, if $0\in A$, the random variable L admits the law $P(L\geqslant h)=(1/p)$ $P(h)$. This law cannot admit an atom at O, for, if $\alpha=P(L=0), 2p\alpha$ is the probability for 0 to belong to the boundary ∂A, and, if $\alpha\neq 0$, the number N of the boundary points in $[0,1]$ admits the expectation $E(N)=\infty$. But (5-2-2) and the semi-markovian property imply $E(N)\leqslant 2\theta$ (i.e., $E(N)$ is smaller than twice the expectation of the number in $[0,1]$ of the points belonging to a Poisson process the density of which is θ), and $\theta=\infty$ implies $Q(h)=0$ for any $h>0$, and thus $q=Q(o)=0$, because Q is l.s.c.

In other words, there exists a density function $\varpi>0$ such that

$$P(h)=p\int_h^\infty \varpi(h)\,dh \tag{5-2-4}$$

Conditionally, if $0\notin A$, we may define the two random variables $L_0=\text{Sup}\{x:\ [o,x]\cap A=\varnothing\}$ and $L_1=\text{Sup}\{y:\ [L_0,L_0+y]\subset A\}$. In particular, $L_1=0$ if the connected component of A enclosing L_0 is one point. These two variables are independent, by the semi-markovian property. By (5-2-2) the law of L_0 is exponential, with the density $\theta\exp(-\theta h)$ $(h\geqslant 0)$. Put $F_1(x)=P(L_1<x)$. In order to compute $F_1(x), x>0$, consider the event $\{x\notin A, x+h\in A$, and only one boundary point belongs to the interval $[x,x+h]\}$, the probability of which is $qP(L_0+L_1\geqslant h)$. By changing the orientation of the x-axis, we see that the same event also admits the probability $pP(L_0+L_1'\geqslant h)$, where L_1' is the random variable the density of which is $\varpi(x)$. It follows that

$$q\theta\int_0^h(1-F_1(x))\exp(-\theta(h-x))\,dx=p\int_0^h\varpi(x)\exp(-\theta(h-x))\,dx$$

and thus

$$1-F_1(x)=\frac{p}{q\theta}\,\varpi(x) \tag{5-2-5}$$

In particular, the random variable L_1 has a *finite expectation* that we shall denote by m_1:

$$m_1 = \int_0^{\infty} (1 - F_1(x))\, dx = \frac{p}{q\theta}$$

By putting $m_0 = 1/\theta = E(L_0)$, this relationship implies $p = m_1/(m_0 + m_1), q = m_0/(m_0 + m_1)$, and also $\theta q = 1/(m_0 + m_1)$.

The probability for a small segment $(x, x+dx)$ to contain an entering point is $\theta q\, dx$. Thus the number of entering points enclosed in a unit segment admits the expectation $q\theta = 1/(m_0 + m_1)$. For $x \downarrow 0$, the left-hand side of (5-2-5) converges toward $P(L_1 > 0) = 1 - \beta, \beta = P(L_1 = 0)$ denoting an eventual atom at O for the law F_1. By (5-2-4), we also have

$$\frac{p - P(h)}{h} = \frac{\theta q}{h} \int_0^h (1 - F_1(x))\, dx$$

and $(1 - F_1(x)) \uparrow (1 - \beta)$ if $x \downarrow 0$. Thus $P(h)$ actually admits at $h = 0$ a derivative from the right, say,

$$P'_+(o) = \theta q(1 - \beta) = -\theta q \quad P(L_1 > 0)$$

Note that $P(h)$ and $Q(h)$ admit at $h = 0$ the same derivative from the right if and only if F_1 is without an atom at 0. Generally, $Q'_+(o) = -\theta q$ and $P'_+(o) = -\theta q(1 - \beta)$ are not equal.

The stationary semi-markovian RACS A is entirely defined if the parameter θ and the law F_1 are given, for, by the semimarkovian property, A, considered as a stochastic process, starts from scratch each time that the point x leaves the set A (see Feller 1966, pp. 365 ff.). In other words, A is a stationary two-state renewal process on **R**, or a stationary partition of **R** by consecutive segments. The random lengths are independent and alternatively admit the law F_1 and the exponential law $\theta \exp(-\theta h)$.

Conversely, such a renewal process is ergodic and may be made stationary by a suitable choice of the initial law (and so is a stationary semi-markovian RACS) if and only if the law F_1 admits a finite expectation $m_1 = \int_0^{\infty} (1 - F_1(x))\, dx < \infty$. If so, $m_0 = 1/\theta, p = m_1/(m_0 + m_1), Q(h) = q \exp(-\theta h)$, and $P(h) = q\theta \int_h^{\infty} (1 - F_1(x))\, dx$.

According to the terminology of Section 2-1, the law F_1 is the *linear number granulometry* of the RACS A. The length-weighted linear granulometry is given by the probability $p - F(h)$ for a given point x to belong to a segment length $\geqslant h$ enclosed in A (i.e., $x \in A_{[o,h]}$). By (5-2-3) and (5-2-5), we get $F(dh) = h\theta q F_1(dh) = (ph/m_1)F_1(dh)$. This result emphasizes the length-weighted character of the granulometry F, for "each" linear intercept has its frequency weighted by a coefficient proportional to its length h.

Let us now examine the *covariance* $C(h) = P(\{x \in A\} \cap \{x+h \in A\})$. It is related to $P(h)$ by a convolution equation, for the event $\{0 \in A$ and $h \notin A\}$, the probability of which is $p - C(h)$, is equivalent to $\{L_1' < h$ and $h \notin A\}$, where $L_1' = \operatorname{Sup}\{x\colon [o,x] \subset A\}$. By the semi-markovian property, it follows that

$$p - C(h) = -\frac{1}{q}\int_0^h P'(x)(1-2p+C(h-x))\,dx \qquad \text{(5-2-6)}$$

or (which is the same)

$$p - C(h) = \theta\int_0^h (1-F_1(x))(1-2p+C(h-x))\,dx$$

The covariance $C(h)$ admits at $h=0$ the *derivative from the right* $C'_+(o) = P'_+(o) = -\theta q(1-\beta)$, for, by the obvious inequalities $p \geqslant C(h) \geqslant P(h)$, we have $0 \leqslant p - C(h) \leqslant p - P(h)$ and, by substituting in (5-2-6),

$$\frac{p-P(h)}{h} \geqslant \frac{p-C(h)}{h} \geqslant \frac{\theta}{h}\int_0^h (1-F_1(x))(1-2p+P(h-x))\,dx$$

from which it follows that $\lim(p-P(h))/h = \lim(p-C(h))/h$ for $h \downarrow 0$.

The semi-markovian RACS A is not necessarily infinitely divisible. For instance, suppose that F_1 is the law of the random variable $L_1 = l$ a.s., where l is a given constant $\geqslant 0$. Then the corresponding stationary semi-markovian RACS cannot be infinitely divisible.

Suppose that A is a stationary SMIDRACS. We shall see in the sequel (but this is obvious in the present case) that A is the union of closed segments, the random lengths of which are independent and admit the same law G. Moreover, the origins (e.g., from the left) of these segments are a Poisson point process with density θ. It follows that

$$q = \exp(-\theta\mu) = \exp\left[-\theta\int_0^\infty (1-G(x))\,dx\right]$$

μ denoting the expectation associated with the law G of these "primary" segments. In particular, $A = \mathbf{R}$ a.s. if $\mu = \infty$. Suppose that $\mu < \infty$. Then it is easy to show that

$$Q(h) = q\exp(-\theta h)$$

By comparison with (5-2-2), it follows that the parameter θ occurring in (5-2-2) actually is identical with the density of the Poisson germs (which was also denoted by θ). Concerning the covariance, we easily find

$$1-2p+C(h)=q\exp\left[-\theta\int_0^h(1-G(x))\,dx\right] \tag{5-2-7}$$

By comparing (5-2-6) with (5-2-7), we conclude that the number granulometry F_1 and the "primary granulometry" G are related by an integral equation, which can always be solved (at least theoretically) by a Laplace transform. In particular, the probability $P(l)$ for a segment with length l to be covered by the union A of the primary segments may be computed by this procedure (this is the well-known covering problem). Conversely, by solving (5-2-6), we can find the covariance $C(h)$ associated with a given number granulometry F_1. But generally, if F_1 is an arbitrary law, $h\to -C'(h)/(1-2p+C(h))$ is not a decreasing function, and there exists no primary law G such that the corresponding number granulometry is F_1 (see, e.g., the above counterexample, i.e., F_1 is the law of a random variable a.s. $=$). It would be interesting to find under what conditions this problem can be solved, that is, when a given law F_1 is the number granulometry of a SMIDRACS, but we shall not examine this question.

5-3. THE BOOLEAN MODELS WITH CONVEX GRAINS

If the SMIDRACS A in Theorem 5-1-1 is the union of a Poisson process on $C(\mathcal{K}')$, that is, is associated with a σ-finite measure θ concentrated on $C(\mathcal{K}')$, we say that A is a *boolean model with convex (primary) grains*, for A a.s. is a locally finite union of compact convex sets (the "primary grains"), that is, a given $K\in\mathcal{K}$ a.s. hits only a finite number among these primary grains. The present section is devoted to this class of SMIDRACS, and particular attention is paid to the stationary case. (This was partially investigated in Section 3-2, but it is better here to start from scratch.)

More generally, let A be the IDRACS equivalent to the union of a Poisson process in $\mathcal{K}'$, that is, a boolean model with compact primary grains (no longer necessarily convex). Let θ be the corresponding σ-finite measure concentrated on $\mathcal{K}'$ (Proposition 3-2-1), and put $\psi(K)=\theta(\mathcal{F}_K), K\in\mathcal{K}$. We shall use Proposition 3-2-3 to characterize these IDRACS as follows: A is equivalent to the union $\cup A_{x_i}$, where the A_{x_i} are independent RACS a.s. compact, and the points $x_i\in A_{x_i}$ (i.e., the "primary germs") constitute a Poisson point process in $\mathbf{R}^d$.

In order to apply Proposition 3-2-1, we need a measurable mapping $K\to x(K)$ from $\mathcal{K}'$ into $\mathbf{R}^d$, so that it will be possible to identify a particular point in each $K \in \mathcal{K}'$. In Section 3-3, we used the center of the circumscribed ball. Now it is better to use the *lexicographic ordering* $\leqslant$ in $\mathbf{R}^d$, defined as follows: let x,y be two points in $\mathbf{R}^d$, and $x_i, y_i, i=1,2,\ldots,d$ be their coordinates. Then we put $x \leqslant y$ if $x_d < y_d$, or $x_d = y_d$ and $x_{d-1} < y_{d-1}$, or ..., or $x_d = y_d, x_{d-1} = y_{d-1}, \ldots, x_2 = y_2$ and $x_1 \leqslant y_1$.

For any $K \in \mathcal{K}'$, there exists a unique point $x(K) \in K$ such that $x(K) = \operatorname{Inf}\{x, x \in K\}$ in the sense of the lexicographic ordering, and it can easily be verified that the mapping $K \to x(K)$ from $\mathcal{K}'$ onto $\mathbf{R}^d$ is *continuous*. Now we may extend this mapping onto $\mathcal{F}'$ by putting, for instance, $x(F)=0$ for any $F \in \mathcal{F}' \setminus \mathcal{K}'$. Then the mapping $F \to x(F)$ is measurable on $\mathcal{F}'$ (and even θ-a.e. continuous, because θ is concentrated on $\mathcal{K}'$). On the other hand, for any $B \in \mathcal{K}', \theta(x^{-1}(B)) \leqslant \theta(\mathcal{F}_B) < \infty$, for θ is σ-finite and $x(F) \in F \theta$-a.e. on $\mathcal{F}'$. Thus the condition stated in Proposition 3-2-3 is fulfilled. It follows that there exist a σ-finite measure $\mu \geqslant 0$ on $\mathbf{R}^d$, and, for μ-almost every $x \in \mathbf{R}^d$, a RACS $A(x)$, the probability P_x of which on σ_f is measurable with respect to $x \in \mathbf{R}^d$, and such that

$$\theta(\mathcal{V} \cap x^{-1}(B)) = \int_B P_x(\mathcal{V}) \mu(dx) \tag{5-3-1}$$

for any $\mathcal{V} \in \sigma_f$ and any borelian set $B \subset \mathbf{R}^d$. Moreover, by Proposition 3-2-3, Corollary 1, for μ-almost every $x \in \mathbf{R}^d$, $A(x) \in \mathcal{K}'$ and $x = x(A) \in A(x) (P_x\text{-a.s.})$.

Now, if $\mathcal{Q}$ denotes the Poisson process on $\mathcal{K}'$, the union of which is the boolean model $A, x(\mathcal{Q})$ [i.e., the image of $\mathcal{Q}$ under the mapping $F \to x(F)$] is a *Poisson point process* on $\mathbf{R}^d$ associated with the σ-finite measure μ.

Proof. For any borelian set B in $\mathbf{R}^d$, the number $N(B)$ of the distinct points in $x(\mathcal{Q}) \cap B$, that is, the number of the closed sets belonging to $x^{-1}(B) \cap \mathcal{Q}$, is a Poisson random variable, and $E(N(B)) = \theta(x^{-1}(B)) = \mu(B)$ by (5-3-1). Moreover, if B_1 and B_2 are two disjoint borelian sets, $x^{-1}(B_1) \cap x^{-1}(B_2)$ is θ-negligible, so that $N(B_1)$ and $N(B_2)$ are independent. Thus $x(\mathcal{Q})$ is a Poisson point process associated with the measure μ.

Conversely, let μ be a positive σ-finite measure on $\mathbf{R}^d$ and, for μ-almost every $x \in \mathbf{R}^d, P_x$ be a probability on σ_f such that $P_x(\mathcal{K}')=1$ and, for any $\mathcal{V} \in \sigma_f$, the mapping $x \to P_x(\mathcal{V})$ is μ-measurable. Then the mapping $\mathcal{V} \to \theta(\mathcal{V}) = \int P_x(\mathcal{V}) \mu(dx)$ is a σ-finite measure concentrated on $\mathcal{K}'$ if and only if $\theta(\mathcal{F}_K) = \int P_x(\mathcal{F}_K) \mu(dx) < \infty$ for any $K \in \mathcal{K}$. It is no longer necessary to suppose that $x(A(x)) \in A(x) (P_x\text{-a.s.})$, for the only goal of the particular mapping $K \to x(K)$ used above was to associate a determined point $x(K)$ with each $K \in \mathcal{K}'$, and any other measurable mapping would have worked as well. Then, by the definition, the union A of the Poisson process $\mathcal{A}$ on $\mathcal{F}'$ associated with θ is a boolean model with compact grains, and we may state:

Proposition 5-3-1. *Let A be a RACS and* ψ: $K \to \psi(K) = -\log P(A \cap K = \emptyset)$ *be its functional on* $\mathcal{K}$. *Then the three following conditions are equivalent*:

a. The RACS A is a boolean model with compact grains.

b. There exist a class $\{A(x), x \in \mathbf{R}^d\}$ *of RACS a.s. compact and measurable with respect to x* [i.e., *for any* $K \in \mathcal{K}, x \to T_x(K) = P(A(x) \in \mathcal{F}_K)$ *is measurable*] *and a Poisson point process* Π *on* $\mathbf{R}^d$ *associated with a σ-finite measure μ such that* $\int T_x(K)\mu(dx) < \infty$ *for any* $K \in \mathcal{K}$, *and A is equivalent to* $\cup_{x \in \Pi} A(x)$.

c. For any $K \in \mathcal{K}, \psi(K) < \infty$ *and*

$$\psi(K) = \int T_x(K)\mu(dx) \tag{5-3-2}$$

for a σ-finite positive measure μ on $\mathbf{R}^d$ *and a μ-measurable class* $T_x, x \in \mathbf{R}^d$, *of Choquet capacities associated with RACS a.s. compact.*

By Theorems 4-2-1 and 5-1-1, the following corollary is obvious:

Corollary. *If one of the three preceding conditions is satisfied, A is a boolean model with convex grains if and only if, for μ-almost every* $x \in \mathbf{R}^d, A(x)$ *is a.s. convex, or (which is the same)* T_x *is C-additive on* $C(\mathcal{K}')$.

Suppose now that A is *stationary*, that is ψ is invariant under translations. Then, as we saw in Section 3-3, it follows from Proposition 3-2-3, Corollary 3, that μ is also invariant under translations, that is, is proportional to the Lebesgue measure on $\mathbf{R}^d$, and the Poisson point process associated with μ is stationary. Moreover, for μ-almost every x and $x' \in \mathbf{R}^d$, the RACS $A(x')$ and $A(x) \oplus \{x' - x\}$ are equivalent. In other words, each $A(x)$ is equivalent to the translate $A'_x = A' \oplus \{x\}$, where $A' = A(o) = A_0$. By (3-3-1), the functional ψ admits the representation

$$\psi(K) = aE\big(V(\check{A}_0 \oplus K)\big) = aE\big(V(A_0 \oplus \check{K})\big) \tag{5-3-3}$$

for a constant $a \geqslant 0$ (V is the volume).

In particular, A is a boolean model with convex grains if and only if the RACS A_0 in (5-3-3) is a.s. convex and $E(V(A_0 \oplus \check{K})) < \infty, K \in \mathcal{K}$. In the rest of this section, we suppose that these conditions are satisfied and we examine their implications.

Boolean Models Induced on the Lines

Let A be a stationary boolean model with convex grains. Obviously, if L is a one-dimensional linear variety in $\mathbf{R}^d$, that is, a line, and $s \in \mathbb{S}_1$ is its direction, $A \cap L$ is also a boolean model with convex grains in L identified with $\mathbf{R}^1$, and its probability depends only on the direction $s \in \mathbb{S}_1$. We denote by $A(s)$ a RACS equivalent to $A \cap L$, and say that $A(s)$ is the boolean model induced by A on $s \in \mathbb{S}_1$. From Section 5-2, it follows that $A(s)$ is determined if the density $\theta(s)$ of the primary germs on s and the primary granulometry G_s are given. Note that the primary grains are convex, that is, are line segments the random lengths of which admit the law G_s.

Let us first compute $\theta(s)$, and also denote as s one of the unit vectors of the line $s \in \mathbb{S}_1$. Let $h \in \mathbf{R}^d, h \neq 0$, be a vector the direction of which is s, that is, $s = h/|h|$, and $Q(h)$ be the probability for the segment $\{x + \lambda h, 0 \leqslant \lambda \leqslant 1\}$ to be disjoint of A. By (5-3-3), we have $Q(h) = \exp(-aV(A_0 \oplus \bar{h}))$ (with $\bar{h} = \{\lambda h, 0 \leqslant \lambda \leqslant 1\}$). For any $K \in C(\mathcal{K}), V(K \oplus \bar{h}) = V(K) + |h|\, \mu_{d-1}(\Pi_s \perp K), \mu_{d-1}$ denoting the Lebesgue measure in $\mathbf{R}^{d-1}$, and also, by (4-5-5), $V(K \oplus \bar{h}) = V(K) + |h| \int_{\mathbb{S}_{d-1}} |s^\perp, S^\perp| G^K_{d-1}(dS)$, where G^K_{d-1} is the surface measure associated with $K \in C(\mathcal{K})$. When taking an a.s. convex and compact RACS A_0 instead of $K, \mu_{d-1}(\Pi_s \perp A_0)$ is a random variable and $G^{A_0}_{d-1}$ is a random measure. In other words, $Q(h) = q\exp[-\theta(s)|h|]$ with

$$\theta(s) = aE(\mu_{d-1}(\Pi_s \perp A_0)) = a \int_{\mathbb{S}_{d-1}} |s^\perp, S| G_{d-1}(dS) \qquad (5\text{-}3\text{-}4)$$

where $G_{d-1} = E(G^{A_0}_{d-1})$ is the expectation measure associated with the random surface measure $G^{A_0}_{d-1}$. From Theorem 4-5-1, it follows that the function $h \to Q(h) = q\exp(-|h|\theta(h/|h|))$ is *positive definite on* $\mathbf{R}^d$.

Consider now the *covariance* $h \to C(h)$, that is, $1 - 2p + C(h) = P(\{x \notin A\} \cap \{x + h \notin A\}) = \exp(-\psi(\{x; x+h\}))$. By (5-3-3), $\psi(\{x, x+h\}) = a[2E(V(A_0)) - E(V(A_0 \cap A_0 \oplus \{h\}))]$. In Section 4-3, we associated with any $K \in C(\mathcal{K})$ the function $h \to g_K(h) = \int 1_K(x) 1_K(x+h)\,dx$. By taking A_0 instead of K, we get a random function $h \to g_{A_0}(h)$, and we denote by g its expectation, that is, by (4-3-1),

$$g(h) = E(g_{A_0}(h)) = E(V(A_0 \cap A_0 \oplus \{h\}))$$

Thus, taking into account $q = \exp[-aE(V(A_0))] = P(x \notin A)$, we can express the covariance $C(h)$ by the following formula:

$$C(h) = p - q + q\exp[-a(g(o) - g(h))] \qquad (5\text{-}3\text{-}5)$$

On comparing with (5-2-7), we see that the linear granulometry G_s of the primary convex grains induced on s is defined by

$$\theta(s)\int_0^r (1-G_s(\xi))\,d\xi = a(g(o)-g(rs)) \qquad (r\geqslant 0) \qquad (5\text{-}3\text{-}6)$$

Proposition 4-3-1 enables us to interpret this result. If the convex RACS A_0 has its interior a.s. nonempty, we have, a.s.,

$$\gamma_{A_0}(o;s)=\mu_{d-1}(\Pi_s \perp A_0)=\lim\left(\frac{g_{A_0}(o)-g_{A_0}(rs)}{r}\right)$$

as $r\downarrow 0$. By (5-3-4) and (5-3-6), it follows that $\lim(1-G_s(\xi))=1$ for $\xi\downarrow 0$, that is, the law G_s is without an atom at o. In particular, the covariance C and the function Q admit for each $x\in\mathbb{S}_1$ the same derivative from the right along this direction x, that is,

$$\lim_{r\downarrow 0}\frac{C(o)-C(rs)}{r}=\lim_{r\downarrow 0}\frac{Q(o)-Q(rs)}{r}=q\theta(s)$$

These relationships no longer hold if the probability $\alpha=P(\mathring{A}_0=\varnothing)$ for A_0 to have its interior empty is strictly >0. In this case, we denote by A_0' and A_1' the RACS A_0 conditional if $\mathring{A}_0=\varnothing$ and $\mathring{A}_0\neq\varnothing$, respectively. Then $g_{A_0'}=0$, because A_0' has its interior empty, and for, any $r\geqslant 0$, we have $g(rs)=(1-\alpha)E(g_{A_1'}(rs))$. If $r\downarrow 0$, it follows that $(1-\alpha)E(\mu_{d-1}\Pi_s\perp A_1')$ $=\lim(g(o)-g(rs))/r$. In the same way, $\theta(s)=\alpha\theta_0(s)+(1-\alpha)\theta_1(s)$, with $\theta_0(s)=aE(\mu_{d-1}(\Pi_s\perp A_0'))$ and $\theta_1(s)=aE(\mu_{d-1}(\Pi_s\perp A_1'))$. Then, if $r\downarrow 0$ in (5-3-6), we see that the law G_s admits at 0 an atom $\beta(s)>0$ such that

$$1-\beta(s)=(1-\alpha)\frac{\theta_1(s)}{\theta(s)} \qquad (5\text{-}3\text{-}7)$$

In this case, the derivatives from the right along the direction s of the covariance C and the function Q are, respectively, at $h=0$, $(1-\beta(s))q\theta(s)$ and $q\theta(s)$. Let us summarize these results.

Proposition 5-3-2. *Let A be a stationary boolean model with convex grains. Then the SMIDRACS induced on the lines with direction $s\in\mathbb{S}_1$ are characterized by* (5-3-4) *and* (5-3-6). *In particular, the function $h\to Q(h)$ is positive definite on $\mathbf{R}^d$. The functions Q and C admit the same tangential cone at $h=0$ if and only if the primary convex compact RACS A_0 has a.s. its interior nonempty, or if and only if the induced granulometries G_s are without* *an atom at the origin for any* $s \in \mathbb{S}_1$. *If* $\alpha = P(\mathring{A}_0 = \varnothing) \neq 0$ *and* $\theta_1(s)$ *is the density induced by the RACS* A'_1 *equivalent to* A_0 *conditional upon* $\mathring{A}_0 \neq \varnothing$, *the induced primary granulometry* G_s *admits at the origin an atom* $\beta(s)$ *given by* (5-3-7).

Note that, if the boolean models induced on the lines are known [i.e., if $\theta(s)$ and the primary granulometry G_s are known for any $s \in \mathbb{S}_1$], the probability P on σ_f associated with the boolean model A with convex grains in $\mathbf{R}^d$ is *not* determined. For instance, if A is a boolean model and A' a Poisson point process independent of A, A and $A \cup A'$ induce on the lines the same boolean models.

The Densities of the Minkowski Functionals

If K is the ball rB with radius $r \geqslant 0$, it follows from the Steiner formula and (5-3-3) that

$$\psi(rB) = a \sum_{k=0}^{d} \binom{d}{k} r^k E(W_k(A_0)) \tag{5-3-8}$$

The expectations $E(W_k(A_0))$ of the Minkowski functionals for the a.s. convex compact primary RACS A_0 are important parameters of the boolean model A itself. If A is *isotropic*, we may substitute for A_0 its isotropized RACS (see Section 4-2), that is, suppose A_0 itself to be isotropic. Then (5-3-8) may be applied to any $K \in C(\mathcal{K})$, for, by Proposition 4-2-4, we have

$$\psi(K) = \frac{a}{b_d} \sum_{k=0}^{d} \binom{d}{k} E(W_k(A_0)) W_{d-k}(K) \qquad [K \in C(\mathcal{K})]$$

so that the restriction of ψ to $C(\mathcal{K})$ (but not ψ itself on $\mathcal{K}$) is determined if the $E(W_k(A_0)), k = 0, 1, \ldots, d$, are known.

In the general case, the stationary boolean model A with convex grains is not isotropic. But $A \in \mathfrak{S}_f$ a.s., and the random measures $W_k^A, k = 0, 1, \ldots, d$, admit expectations $W_k = E(W_k^A)$ which are Radon measures on $\mathbf{R}^d$ (Theorem 4-7-2, corollary), such that, for any $\varphi \in \mathbf{C}_{\mathcal{K}}(\mathbf{R}^d)$ and $r > 0$,

$$E\left[\int \left(\sum_{x' \in \Pi_A(x,r)} \varphi(x')\right) dx\right] = \sum_{k=0}^{d} \binom{d}{k} r^k \int \varphi(x) W_k(dx) \tag{5-3-9}$$

On the other hand, A is stationary, so that the measures W_k are invariant under translations, that is, proportional to the Legesgue measure on $\mathbf{R}^d$. In other words, there exist $d+1$ constants $w_0, w_1, \ldots, w_d$ such that $W_k(dx) = w_k\,dx$. We shall say that these numbers w_k are the *densities of the Minkowski functionals*, with respect to the boolean model A.

Clearly, $w_0 = E(1_A(x)) = P(x \in A) = p$. In the same way, $dw_1 = E(W_1^A(rB))/(b_d r^d)$ if B is the unit ball, and dw_1 is the expectation *of the specific surface area of A*.

Let us now show that w_d/b_d can be interpreted as a *specific convexity number* and

$$\frac{w_d}{b_d} = aq \tag{5-3-10}$$

Proof. Let $s \in S_0$ be a unit vector, and $S = s^{\perp} \in \mathbb{S}_{d-1}$. With a suitable choice of the coordinate axes, the mapping $K \to x(K)$ based on the lexicographic ordering (see the beginning of the present section) is such that the hyperplane $S_{x(K)}$ parallel to S and including $x(K)$ is a supporting plane for K, and K is enclosed in the half space containing the point $x(K)+s$. For ϖ-almost every $s \in S_0$, the entering set of the hyperplanes with direction S in the a.s. convex compact RACS A_0 is a.s. a point set, and the entering points of these hyperplanes into the primary convex grains $A'(x_i)$ may be identified with the Poisson germs x_i. Then, if B is a borelian set, the number of the entering points in the $A'(x_i)$ which are contained in B admits the expectation $aV(B)$. On the other hand, conditional upon x_i being a Poisson germ, the probability for x_i not to be enclosed in any other primary grain (or for x_i to be an entering point in A itself) is q. Thus the number of entering points in A enclosed in B admits (for ϖ-almost every direction $s \in \mathbb{S}_1$) the expectation $aqV(B)$. On the other hand (see Proposition 4-7-4), this number admits (with respect to s) the rotation average $(1/b_d)E(W_d^A(B)) = (w_d/b_d)V(B)$, and (5-3-10) follows. In particular, w_d/b_d is the expectation of the number of entering points per unit volume, with respect to a random direction $s \in \mathbb{S}_1$. Thus w_d/b_d really is a specific convexity number.

More generally, for $k = 1, 2, \ldots, d$, it is possible to show that the density w_k of the Minkowski functional W_k is related to the expectation $E(W_k(A_0))$ of the same functional with respect to the primary convex grain A_0 by

$$w_k = aqE(W_k(A_0))$$

Boolean Models Induced on Linear Varieties

Suppose now that the compact set K in (5-3-3) is *Steiner* and hence, in particular, $K=\check{K}$. By Theorem 4-5-2, there exist $d-1$ measures $G_k^{\ K}, k=1,2,\ldots,d-1$, such that (4-5-7) holds for any $K'\in C_0(\mathcal{K}')$. Thus, by (5-3-3), we have for, any $r\geqslant 0$,

$$\psi(rK)=a\left[r^d\mu_d(K)+\sum_{k=1}^{d-1}r^{d-k}\int_{\mathbb{S}_{d-k}}E(\mu_k(\Pi_S\perp A_0))G_{d-k}^K(dS)\right.$$
$$\left.+E(\mu_d(A_0))\right] \qquad (5\text{-}3\text{-}11)$$

This relationship enables us to compute the expectation of the Minkowski functionals for the boolean model induced on any linear variety in $\mathbf{R}^d$. Let $S_p\in\mathbb{S}_p$ be the direction of a p-dimensional variety, and take $K=B\cap S_p$, that is, the unit ball in S_p. The measure $G_k^{\ B\cap S_p}$ is concentrated on the set $\mathbb{S}_k(S_p)$ of the k-dimensional subspaces in $S_p(0<k\leqslant p)$, and is invariant under the rotations in S_p. Thus it is proportional to the invariant probability $\varpi_k^{\ S_p}$ on $\mathbb{S}_k(S_p)$. By applying the corollaries of Theorem 4-5-2, we get explicitly

$$G_k^{\ B\cap S_p}=\frac{\binom{p}{k}b_p}{b_{p-k}}\varpi_k^{\ S_p}$$

(and $=0$ if $k>p$). On substituting in (5-3-11), it follows that

$$\psi(r(B\cap S_p))=a\left[E(\mu_d(A_0))+\sum_{k=1}^{p}\binom{p}{k}\frac{b_p}{b_{p-k}}r^k\right.$$
$$\left.\times\int_{\mathbb{S}_k(S_p)}E(\mu_{d-k}(\Pi_S\perp A_0)\varpi_k^{\ S_p}(dS))\right]$$

On the other hand, in the space S_p identified with $\mathbf{R}^p$, the RACS $A_p=A\cap S_p$ induced by A is still a stationary boolean model with convex grains. Thus there exists a RACS $A_0(S_p)$, a.s. convex and compact in $\mathbf{R}^p$, such that

$$\psi(K)=E\big[a(S_p)V(A_0(S_p)\oplus K)\big]$$

for any $K\in\mathcal{K}, K\subset S_p$. On taking $K=r(B\cap S_p)$ and identifying with the above expression of $\psi(r(B\cap S_p))$, it follows that

$$a(S_p)E\big[W_k{}^p(A_0(S_p))\big]=a\left(\frac{b_p}{b_{p-k}}\right)\int_{\mathbb{S}_k(S_p)}E(\mu_{d-k}(\Pi_S\perp A_0))\varpi_k{}^{S_p}(dS)$$

$(p\leqslant k)$. The index p in $W_k{}^p$ signifies that the corresponding Minkowski functional is taken in the euclidean space $\mathbf{R}^p$. If $k=p$, this formula becomes $a(S_p)=aE(\mu_{d-p}(\Pi_{S_p^\perp}A_0))$. If $k=0$, we get $a(S_p)E[\mu_p(A_0(S_p))]=aE(\mu_d(A_0))$. Then the density $a(S_p)$ of the germ point Poisson process and the expectations $E[W_k{}^p(A_0(S_p))]$ of the Minkowski functionals of the boolean model induced on $S_p\in\mathbb{S}_p$ are

$$a(S_p)=aE\big(\mu_{d-p}(\Pi_{S_p^\perp}A_0)\big)$$

$$E\big[W_k{}^p(A_0(S_p))\big]$$

$$=\frac{b_p}{b_{p-k}E\big(\mu_{d-p}(\Pi_{S_p^\perp}A_0)\big)}\int_{\mathbb{S}_k(S_p)}E(\mu_{d-k}(\Pi_S\perp A_0))\varpi_k{}^{S_p}(dS)$$

and, in particular, if $k=0$,

$$E\big[\mu_p(A_0(S_p))\big]=\frac{E(\mu_d(A_0))}{E\big(\mu_{d-p}(\Pi_{S_p^\perp}A_0)\big)}$$

Also, we can see that the unidimensional models induced on a given line $L\subset S_p$ by A and $A_p=A\cap S_p$ are obviously identical. Thus, by (5-3-6), for any $s\in\mathbb{S}_1, s\subset S_p$, we have a $g(rs)=a(S_p)g(S_p;rs)$, where

$$g(S_p;rs)=E\big(\textstyle\int 1_{A_0(S_p)}(x)1_{A_0(S_p)}(x+h)\mu_p(dx)\big).$$

By taking into account $\int_{S_p}g(S_p;h)\mu_p(dh)=E\{[\mu_p(A_0(S_p))]^2\}$, the order -2 moment of the p-volume of the primary induced grain $A_0(S_p)$ is

$$E\left\{\left[\mu_p(A_0(S_p))\right]^2\right\} = \frac{b_p}{E\left(\mu_{d-p}(\Pi_{S_p^\perp}A_0)\right)}\int_0^\infty r^{p-1}\,dr\int_{\mathbb{S}_1(S_p)} g(r,s)\,\varpi_1^{S_p}(ds)$$

If the boolean model A is *isotropic*, the various preceding formulae are greatly simplified, for, as stated above, we may suppose the primary grains A_0 and $A_0(S_p)$ themselves to be isotropic in $\mathbf{R}^d$ and $\mathbf{R}^p$, respectively, and, moreover, the parameters related to $A_0(S_p)$ are constants independent of $S_p \in \mathbb{S}_p$. Moreover, A is isotropic and the definition in (4-1-3) through (4-1-5) of the Minkowski functionals implies $E(\mu_{d-k}(\Pi_S \perp A_0)) = (b_{d-k}/b_d)E(W_k(A_0))$. By putting $a_p = a(S_p)$, $W_k = E(W_k(A_0))$ and $W_k{}^p = E[W_k{}^p(A_0(S_p))]$, formulae (5-3-12) yield in the isotropic case

$$a_p = a\left(\frac{b_{d-p}}{b_d}\right)W_p$$

$$W_k{}^p = \frac{b_p b_{d-k}}{b_{p-k}b_{d-p}}\,\frac{W_k}{W_p}$$

As far as the first two moments of the induced primary grains $A_0(S_r)$ are concerned, we have, in the same way,

$$E\left[\mu_p(A_0(S_p))\right] = \frac{b_d W_0}{b_{d-p}W_p}$$

$$E\left\{\left[\mu_p(A_0(S_p))\right]^2\right\} = \frac{b_p b_d}{b_{d-p}W_p}\int_0^\infty r^{p-1}g(r)\,dr$$

5-4. THE STATIONARY SMIDRACS

Let A be a stationary SMIDRACS. Obviously, if A admits fixed points, $A = \mathbf{R}^d$ a.s. In the sequel, we suppose that A is without fixed points. By Theorem 5-1-1, A is the union of a Poisson process $\mathcal{Q}$ on $\mathcal{F}'$ associated with a σ-finite measure θ concentrated on $C(\mathcal{F}')$ and invariant under translations. In particular, $\psi(K) = \theta(\mathcal{F}_K)$ for any $K \in \mathcal{K}$. Our present goal is to characterize the structure of such a measure θ.

a. By the preceding section, the restriction of θ to $C(\mathcal{K}')$ is associated with a boolean model with convex grains, and there exists an a.s. compact and convex RACS A_0 such that $\theta(\mathcal{F}_K \cap C(\mathcal{K}')) = aE(V(A_0 \oplus \check{K}))$. It remains to study the restriction of the measure θ to $C(\mathcal{F}')\backslash C(\mathcal{K}')$.

b. Suppose now that θ is concentrated on $C(\mathcal{F}')\backslash C(\mathcal{K}')$, and let $\mathcal{T}$ be the set of the *closed convex cones* in $\mathbf{R}^d$. $\mathcal{T}$ is a compact subset of $C(\mathcal{F}')$, and we suppose that it is provided with its borelian tribe σ_t, that is, the σ-algebra induced on $\mathcal{T}$ by σ_f.

We say that a cone $T \in \mathcal{T}$ is *dominated* by $F \in C(\mathcal{F}')$ if there exists a translate of T contained in F. If T is dominated by $F \in C(\mathcal{F}')$, clearly $T_x = T \oplus \{x\} \subset F$ for any $x \in F$, and thus $T \oplus F \subset F$ and $F = T \oplus F$. Conversely, the cones $T \in \mathcal{T}$ dominated by F are characterized by this equality $F = T \oplus F$.

There exists a cone T such that $F = T \oplus F$ for a given $F \in C(\mathcal{F}')$ if and only if F is not compact, that is, $F \in C(F')\backslash C(\mathcal{K}')$. The union $T(F) = \cup \{T\colon\ T \in \mathcal{T}, T \oplus F = F\}$ of all the cones $T \in \mathcal{T}$ which satisfy this equality is still a convex closed cone, as can easily be shown, and $T(F) \oplus F = F$. In other words, $T(F)$ is the largest convex closed cone dominated by F. In order to complete the definition of the mapping $F \to T(F)$, we put $T(F) = \varnothing$ for $F \in C(\mathcal{K}')$. Then the mapping $T\colon\ C(\mathcal{F}') \to \mathcal{T} \cup \{\varnothing\}$ is u.s.c. and thus measurable, and, moreover, obviously invariant under translations. Then we may apply Proposition 3-2-2, for the measure θ is concentrated on $C(\mathcal{F}')\backslash C(\mathcal{K}')$ and invariant under translations. Thus there exists a probability τ on $\mathcal{T}$, and, for τ-almost every $T_0 \in \mathcal{T}$, a σ-finite positive measure θ_{T_0} invariant under translations, concentrated on $C(\mathcal{F}')\backslash C(\mathcal{K}')$, and such that $\theta_{T_0}(\{T(F) \neq T_0\}) = 0$ and

$$\theta(\mathcal{V}) = \int_{\mathcal{T}} \theta_T(\mathcal{V})\tau(dT) \qquad (\mathcal{V} \in \sigma_f) \tag{5-4-1}$$

c. Now we will show that this probability τ on $\mathcal{T}$ is actually *concentrated on the set* $\mathcal{S} = \cup_{k=1}^{d} \mathcal{S}_k$ *of the subspaces in* $\mathbf{R}^d$. Let u be a unit vector, $L \in \mathcal{S}_1$ the straight line $L = \{\lambda u, \lambda \in R\}$, and $L^+ = \{\lambda u, \lambda \geqslant 0\}$. For any $x \in \mathbf{R}^d$ and $\lambda \geqslant 0$, the inclusion $L^+ \oplus \{x - \lambda u\} \subset F$ implies $L^+ \oplus \{x\} \subset F$. But θ is invariant under translations, and $\theta(\{L^+ \oplus \{x - \lambda u\} \subset F\}) = \theta(\{L^+ \oplus \{x\} \subset F\})$. Thus these two events are equal θ-a.e. on $C(\mathcal{F}')$. It follows that $\{L^+ \subset T(F)\} = \{L \subset T(F)\}$ θ-a.e. for any $L \in \mathcal{S}_1$, and thus $\theta(\{T(F) \not\in \mathcal{S}\}) = 0$. In other words, τ is concentrated on $\mathcal{S} = \cup \mathcal{S}_k$.

d. For $k = 1, 2, \ldots, d$, denote now by τ_k the restriction of τ to $\mathcal{S}_k$. Thus, for τ_k-almost every $S \in \mathcal{S}_k$, there exists a σ-finite measure $\theta_S \geqslant 0$, concentrated on $C(\mathcal{F}')$, invariant under translations, and such that $\theta_S(\{T(F) \neq S\}) = 0$. In other words, θ_S is concentrated on the set of cylinders

$K \oplus S, K \in C(\mathcal{K}'), K \subset S^{\perp}$($S^{\perp}$ is the subspace in $\mathbf{R}^d$ orthogonal to $S \in \mathbb{S}_k$). Then the results of Section 5-3 may be applied to θ_S: there exists a constant $a_k(S) \geq 0$ and an a.s. compact and convex RACS A_S in the space $S^{\perp}$ (identified with $\mathbf{R}^{d-k}$) such that, for any $K \in \mathcal{K}$,

$$\theta_S(\mathcal{F}_K) = a_k(S) E(\mu_{d-k}(A_S \oplus \Pi_{S^\perp} K)) \qquad (5\text{-}4\text{-}2)$$

On the other hand, by (5-4-1), $\psi(K) = \sum_{k=1}^{d} \int_{\mathbb{S}_k} \theta_S(\mathcal{F}_K) \tau_k(dS)$ for any $K \in \mathcal{K}$. If $K = rB$ is the ball with radius $r \geq 0$, the Steiner formula yields

$$\mu_{d-k}(A_S \oplus r\Pi_{S^\perp} B) = \sum_{p=0}^{d-k-1} \binom{d-k}{p} r^p W_p^{d-k}(A_S) + b_{d-k} r^k$$

and the mapping $S \to \theta_S(\mathcal{F}_B) = a_k(S) E(\mu_{d-k}(A_S \oplus \Pi_{S^\perp} rB))$ is measurable for any $r \geq 0$, so that the mapping $S \to a_k(S)$ itself is measurable. Thus there exists a positive measure $G_k = a_k \tau_k$ on $\mathbb{S}_k$, and the functional ψ associated with A admits the representation

$$\psi(K) = \sum_{k=1}^{d} \int_{\mathbb{S}_k} E(\mu_{d-k}(A_S \oplus \Pi_{S^\perp} K)) G_k(dS)$$

If $k = d$, $\mathbb{S}_d = \{\mathbf{R}^d\}$ is a one-point set, and $S^{\perp} = \{o\}$ for any $S \in \mathbb{S}_d$, so that $\int_{\mathbb{S}_k} E(\mu_{d-k}(A_S \oplus \Pi_{S^\perp} K)) G_k(dS)$ is the constant $a_d = \theta(\{\mathbf{R}^d\})$, and $P(\{A \neq \mathbf{R}^d\}) = \exp(-a_d)$.

Taking into account an eventual component of θ on $C(\mathcal{K}')$, we may summarize as follows the preceding results.

Theorem 5-4-1. *A RACS A is a stationary SMIDRACS if and only if the functional $\psi = -\log Q$ is $< \infty$ on $\mathcal{K}$ and admits the representation*

$$\psi(K) = a_0 \psi_0(K) + \sum_{k=1}^{d-1} \int_{\mathbb{S}_k} \psi_S^k(K) G_k(dS) + a_d \qquad (K \in \mathcal{K}) \quad (5\text{-}4\text{-}3)$$

where a_0, a_d are constants ≥ 0, $G_k, k = 1, \ldots, d-1$, is a positive measure on $\mathbb{S}_k$, $\psi_0(K) = E(\mu_d(A_0 \oplus \check{K}))$ for an a.s. compact and convex RACS A_0, and the mappings $(S, K) \to \psi_S^k(K), k = 1, 2, \ldots, d-1$, are such that:

a. For any $K \in \mathcal{K}, S \to \psi_S^k(K)$ is G_k-measurable.

b. For G_k almost every $S \in \mathbb{S}_k$, there exists an a.s. convex and compact RACS $A_S \subset S^{\perp}$ such that

$$\psi_S^k(K) = E\big(\mu_{d-k}(A_S \oplus \Pi_{S^\perp} \check{K})\big) \qquad (5\text{-}4\text{-}4)$$

We may interpret these results by saying that A is equivalent to the union $\cup A_k$ of $d+1$ stationary independent SMIDRACS such that A_0 is a stationary boolean model with convex grains, A_d is a.s. $\varnothing$ on $\mathbf{R}^d$, and, for each $k=1,2,\ldots,d-1$, the SMIDRACS A_k is the union of cylinders the bases of which are $(d-k)$-dimensional boolean models with convex grains. If the RACS A_S in (5-4-3) are single points, A_k is a k-dimensional Poisson flat network. In other words, the Poisson networks and the boolean models are the two prototypes from which any stationary SMIDRACS may be constructed.

Theorem 5-4-2. *A RACS A is equivalent to a Poisson flat network if and only if it is stable, stationary, and semi-markovian.*

Proof. The "only if" part follows from Section 3-5. Conversely, let A be stable (and thus infinitely divisible), stationary, and semi-markovian. In particular, A is a stationary SMIDRACS, and (5-4-3) holds. Take $K=rB(r\geqslant 0, B$ is the unit ball). For $k=0$, it follows from the Steiner formula that

$$\psi_0(rB)=\sum_{p=0}^{d}\binom{d}{p}r^pE\left(W_p^{\,d}(A_0)\right)$$

In the same way, for $0<k<d$ and $S\in\mathbb{S}_k, A_S\oplus\Pi_S\perp rB=A_S\oplus(rB)\cap S^{\perp})$, and by applying the Steiner formula in $S^{\perp}$ (identified with $\mathbf{R}^{d-k}$) we obtain

$$\mu_{d-k}(A_S\oplus\Pi_S\perp rB)=\sum_{p=0}^{d-k}\binom{d-k}{p}r^pE\left(W_p^{\,d-k}(A_S)\right)$$

In other words, $\psi(rB)$ is a polynomial with respect to r, say,

$$\psi(rB)=\sum_{p=0}^{d}B_pr^p$$

with the following coefficients:

$$B_0=a_0E(\mu_d(A_0))+\sum_{k=1}^{d-1}\int_{\mathbb{S}_k}E\left(W_p^{\,d-k}(A_S)\right)G_k(dS)+a_d$$

$$B_p=\binom{d}{p}a_0E\left(W_p^{\,d}(A_0)\right) \qquad (5\text{-}4\text{-}5)$$

$$+\sum_{k=1}^{d-p}\binom{d-k}{p}\int_{\mathbb{S}_k}E\left(W_p^{\,d-k}(A_S)\right)G_k(dS)$$

On the other hand, A is stable, and $\psi(rB)=r^{\alpha}\psi(B)$ for a real $\alpha \geqslant 0$, by Theorem 3-4-1. If $\alpha=0, \psi(rB)$ is constant, and A is a.s. $\varnothing$ or $\mathbf{R}^d$, that is, is a d-dimensional Poisson network. If $\alpha>0$, (5-4-5) implies $\alpha=p$ for an integer p such that $0<p\leqslant d$. Then $B_{p'}=0$ for $p'\neq p$ implies

$$E\left(W_{p'}{}^{d}(A_0)\right)=0 \quad \text{and} \quad \int_{\mathbb{S}_{d-p'}} E\left(W_{p'}{}^{d-p'}(A_S)\right)G_{d-p'}(dS)=0$$

If $p=d$, it follows that A_0 is a point set, $a_d=0$, the RACS $A_S, S\in\mathbb{S}_k$ ($k=1,2,\ldots,d-1$), are empty for G_k-almost every $S\in\mathbb{S}_k$, and A is a Poisson point process. If $p<d, a_0=a_d=0$, and the $A_S, S\in\mathbb{S}_{p'}$, are empty if $p\neq p'$ and are single points if $p'=p$. Then, by Proposition 3-5-1, A is a $(d-p)$-dimensional Poisson flat network.

Corollary. *A stationary SMIDRACS A is a Poisson flat network if and only if $\psi(rB)=r^{\alpha}\psi(B)$ for a real $\alpha\geqslant 0$. If so, α is an integer $p, 0\leqslant p\leqslant d$, and A is a $(d-p)$-dimensional Poisson flat network.*

In view of the applications, we shall now study the parameters mentioned in Section 5-3 for the case of a stationary SMIDRACS.

SMIDRACS Induced on the Lines

It follows from Section 5-2 that the stationary SMIDRACS induced on the lines in $\mathbf{R}^d$ are entirely determined if the functions Q and C are given. Using the same notations as in Theorem 5-4-1, we find, for $q=Q(o)$,

$$q=\exp\left[-a_0-\sum_{k=1}^{d-1}\int_{\mathbb{S}_k}E(\mu_{d-k}(A_S))G_k(dS)-a_d\right]$$

For any $h\in\mathbf{R}^d, h\neq 0$, we denote by $u=h/|h|\in S_0$ the unit vector associated with h, by $r=|h|$ its modulus, and by $s=\{\lambda u, \lambda\in R\}\in\mathbb{S}_1$ the one-dimensional subspace generated by h or u. Then $Q(h)=q\exp(-|h|\,\theta(h/|h|))$ for a symmetric function $\theta\geqslant 0$ on the unit sphere S_0, which may be identified with a function $s\rightarrow\theta(s)$ on $\mathbb{S}_1$. Let $\bar{u}$ be the segment $\{\lambda u, 0\leqslant\lambda\leqslant 1\}$. It follows from Theorem 5-4-1 that

$$\psi(r\bar{u})=a_0E(\mu_d(A_0\oplus r\bar{u}))+\sum_{k=1}^{d-1}\int_{\mathbb{S}_k}E(\mu_{d-k}(A_S\oplus\Pi_{S^{\perp}}r\bar{u}))G_k(dS)+a_d$$

For $S\in\mathbb{S}_k$, let $v=\Pi_S\perp u/|\Pi_S\perp u|$ be the unit vector associated with $\Pi_S\perp u$, so that $r\Pi_S\perp u=r|s,S^\perp|\bar{v}$, and $G_{d-k-1}^{A_S}$ be the random surface measure on $\mathbb{S}_{d-k-1}(S)$ associated with $A_S\subset S^\perp$. Then, by (4-5-5),

$$\mu_{d-k}(A_S\oplus r\Pi_S\perp\bar{u})=\mu_{d-k}(A_S)+r|x,S^\perp|\int_{\mathbb{S}_{d-k-1}(S^\perp)}|\sigma,v^\perp|G_{d-k-1}^{A_S}(d\sigma)$$

Note also that $|\sigma,v^\perp|=|\sigma^\perp\cap S^\perp,v|$, and hence $|s,S^\perp||\sigma,v^\perp| =|s,\sigma^\perp\cap S^\perp|=|s^\perp,\sigma\oplus S|$. Let now $G_{d-1}^{A_S}$ be the measure induced from $G_{d-k-1}^{A_S}$ by the mapping $\sigma\to\sigma\oplus S$ from $\mathbb{S}_{d-k-1}(S^\perp)$ into $\mathbb{S}_{d-1}$. We may write

$$\mu_{d-k}(A_S\oplus r\Pi_S\perp\bar{u})=\mu_{d-k}(A_S)+r\int_{\mathbb{S}_{d-1}}|s^\perp,\sigma|G_{d-1}^{A_S}(d\sigma)$$

Now denote by $F_{d-1}^k(S;d\sigma)=E(G_{d-1}^{A_S}(d\sigma))$ the expectation measure associated with the random measure $G_{d-1}^{A_S}$. After an easy calculation, we obtain

$$\psi(r\bar{u})=-\log q+a_0r\int_{\mathbb{S}_{d-1}}|s^\perp,\sigma|F_{d-1}^0(d\sigma)$$

$$+r\sum_{k=1}^{d-1}\int_{\mathbb{S}_k}G_k(dS)\int_{\mathbb{S}_{d-1}}|s^\perp,\sigma|F_{d-1}^k(d\sigma)$$

Hence the desired expression of the function θ is

$$\theta(s)=\int_{\mathbb{S}_{d-1}}|s^\perp,\sigma|F_{d-1}(d\sigma) \tag{5-4-6}$$

with a positive measure F_{d-1} on $\mathbb{S}_{d-1}$ defined by

$$F_{d-1}(.)=a_0F_{d-1}^0(.)+\sum_{k=1}^{d-1}\int_{\mathbb{S}_k}F_{d-1}^k(S;.)G_k(dS)$$

From this representation, it follows (exactly as in the case of the boolean models with convex grains) that the function $h\to Q(h)$ is *positive definite* on $\mathbf{R}^d$.

Concerning the *covariance* C, we have, for any $h\in\mathbf{R}^d, 1-2p+C(h) =P(\{x\notin A\}\cap\{x+h\in A\})=\exp(-\psi(\{o,h\}))$, and, after a calculation already made in the case of a boolean model, we get

$$\psi_S{}^k(\{o,h\})=E[\mu_{d-k}(A_S\cup(A_S\oplus\{\Pi_S\perp h\}))]$$
$$=2E(\mu_{d-k}(A_S))-g_S(h)$$

where $g_S(h) = E(\int_{S\perp} 1_{A_S}(x) 1_{A_S}(x+h)\,dx)$, and in the same way,

$$\psi^0(\{o,h\}) = a_0[2E(\mu_d(A_0))] - g_0(h)$$

$$g_0(h) = E\left(\int 1_{A_0}(x) 1_{A_0}(x+h)\,dx\right)$$

Hence we have

$$1 - 2p + C(h) = q\exp[-(g(o) - g(h))]$$

$$g(h) = a_0 g_0(h) + \sum_{k=0}^{d-1} \int_{\mathbb{S}_k} g_S(h) G_k(dS)$$

Exactly in the same way as with the boolean model with convex grains, it may be verified that the functions C and Q admit the same tangential cone at $h=0$ if and only if A_0 has a.s. a nonempty interior and for each $k = 1,2,\ldots,d-1$, A_S has a.s. a nonempty interior in $S^\perp$ for G_k-almost every $S \in \mathbb{S}_k$.

The Densities of the Minkowski Functionals

Let B be the unit ball, $r \geqslant 0$, and take $K = rB$ in (5-4-3). For any $S \in \mathbb{S}_k$, it follows from the Steiner formula that

$$E(\mu_{d-k}(A_S \oplus r\Pi_{S\perp} B)) = \sum_{p=0}^{d-k} \binom{d-k}{p} r^p E(W_p^{d-k}(A_S))$$

and thus

$$\psi(rB) = \sum_{p=0}^{d} B_p r^p$$

with the coefficients B_p already written in (5-4-5). Then it may be shown that the coefficients B_p are related to the densities of the Minkowski functionals w_p in the same way as in the boolean model case, say,

$$w_k = \frac{qB_k}{\binom{d}{k}} \qquad (k = 1,2,\ldots,d)$$

For $k=d$, we have $B_d = a_0 b_d$. Thus the convexity number of the SMIDRACS A is $a_0 q$ and depends only on the boolean component.

SMIDRACS Induced on Linear Varieties

Let K be a Steiner compact set (so that $K=\check{K}$), $G_k{}^K, k=1,2,\dots,d-1$, the measure on $\mathbb{S}_k$ associated with K by Theorem 4-5-2, and compute $\psi(rK)$ by (5-4-3). From Theorem 4-5-2, Corollary 1, it follows, for $S\in\mathbb{S}_k$, that

$$\mu_{d-k}(A_S\oplus r\Pi_S\perp K)$$

$$=\mu_{d-k}(A_S)+\sum_{p=1}^{d-k} r^p\int_{\mathbb{S}_p}|\sigma,S^{\perp}|\,\mu_{d-k-p}(\Pi_{S^{\perp}\cap\sigma^{\perp}}A_s)\,G_p{}^K(d\sigma)$$

$$\mu_d(A_0\oplus rK)=\mu_d(A_0)+\sum_{p=1}^{d-1} r^p\int_{\mathbb{S}_p}\mu_{d-p}(\Pi_\sigma\perp A_0)\,G_p{}^K(d\sigma)+r^d\mu_d(K)$$

By substituting in (5-4-3), an explicit expression could be obtained for $\psi(K)$, but is not needed now. Suppose that K is contained in a subspace $S_j\in\mathbb{S}_j\,(0<j<d)$, and take $K=B\cap S_j$, that is, the unit ball in S_j. As we have already seen,

$$G_p{}^{B\cap S_j}=\binom{j}{p}\frac{b_j}{b_{j-p}}\varpi_p{}^{S_j}\qquad(p\leqslant j)$$

where $\varpi_p{}^{S_j}$ is the unique probability on $\mathbb{S}_p(S_j)$ invariant under rotations. By identifying with the expansion $\psi(r(B\cap S_j))=\Sigma_{p=0}^{j}B_p(S_j)r^p$, explicit expressions may be obtained for the densities of the Minkowski functionals of the induced SMIDRACS $A\cap S_j$. But we shall write only the formula concerning the *induced convexity number* $(1/b_j)w_j(S_j)$, and leave the proof to the reader:

$$\frac{1}{b_j}w_j(S_j)=qa_0(S_j)=qa_0E\big(\mu_{d-j}(\Pi_{S_j^{\perp}}A_0)\big)$$

$$+\sum_{k=1}^{d-j}\int_{\mathbb{S}_k}|S_j,S^{\perp}|E\big(\mu_{d-k-j}(\Pi_{S_j^{\perp}\cap S^{\perp}}A_S)\big)G_k(dS)$$

The Isotropic Case

If the stationary SMIDRACS A is *isotropic*, the preceding results are greatly simplified, for in this case the measures G_k in (5-4-3) are invariant under rotations, and for any $S\in\mathbb{S}_k$ the RACS A_S may be chosen isotropic in $S^{\perp}$. Then, for any $K \in C(\mathcal{K})$, it follows from Proposition 4-2-4 that

$$\psi_0(K) = \frac{1}{b_d} \sum_{p=0}^{d} \binom{d}{p} E(W_p{}^d(A_0))\, W_{d-p}^d(K)$$

$$\psi_S(K) = \frac{1}{b_{d-k}} \sum_{p=0}^{d-k} \binom{d-k}{p} E(W_p{}^{d-k}(A_k))\, W_{d-k-p}^{d-k}(\Pi_S \perp K)$$

(A_k denotes the a.s. convex compact RACS in R^{d-k} equivalent to A_S for any $S \in \mathbb{S}_k$.) On the other hand, G_k is invariant under rotations, and thus $G_k = a_k \varpi_k$ for a suitable constant $a_k \geqslant 0$, and it follows from (4-1-6) that

$$\int_{\mathbb{S}_k} W_{d-k-p}^{d-k}(\Pi_S \perp K)\varpi_k(dS) = \frac{b_{d-k}}{b_d} W_{d-p}^d(K)$$

Hence, in the isotropic case, the following relationships hold for any $K \in C(\mathcal{K})$:

$$\psi(K) = \frac{1}{b_d} \sum_{p=0}^{d} \beta_p W_{d-p}^d(K)$$

$$\beta_p = \sum_{k=0}^{d-p} a_k \binom{d-k}{p} E(W_p^{d-k}(A_k)) \tag{5-4-7}$$

In particular, $\beta_d = a_0 b_d$, and $q(\beta_d / b_d)$ is the specific convexity number [with $q = \exp(-\psi(\{o\})) = \exp(-\beta_0)$].

By (5-4-7) and Theorem 4-5-2, Corollary 6, it would not be difficult to calculate the corresponding parameters for the induced SMIDRACS, but we shall not do it here.

CHAPTER 6

Poisson Hyperplanes and Polyhedra

In this chapter, the preceding results are used for studying in a fairly detailed manner the Poisson hyperplane networks and the convex polyhedra they determine in $\mathbf{R}^d$. Many results are due to R.E. Miles; some others are new. First, we show that a Poisson hyperplane network is entirely characterized by an associated Steiner compact set Λ, and the network induced on a p-dimensional subspace S_p is characterized by the Steiner compact set $\Pi_{S_p}\Lambda$, that is, the projection of Λ onto S_p. The positive measures $G_k{}^{\Lambda}$ associated with Λ may also be used to characterize the (non-Poisson) networks of order $2,3,\ldots,d$, that is, the intersections of $2,3,\ldots,d$ hyperplanes belonging to the initial network. With the order-k network is associated the random measure $N_k(dx)$, such that $N_k(K)$ is the $(d-k)$-volume of the intersection of the k-network and a given compact set K. The corresponding expectation measure and also (but only in the isotropic case) the covariance measures are calculated. In the second section, the Poisson polyhedra are defined and characterized by a conditional invariance property, which exactly generalizes the well-known characteristic property of the exponential distribution. Also note the very important distinction between the number law and the measure law. Then various characteristics of the Poisson polyhedra are calculated: expectations of the Minkowski functionals, granulometries with respect to the unit ball, first moments of the volume, the surface area, the projection area, and so on. The laws of the volume, the surface area, the number of faces, and other parameters are not exactly known. Nevertheless the law of the volume V is simply related to the conditional expectation (with respect to V) of the surface area. In the isotropic case, the number of faces and the norm are also related. Finally, the particular case of isotropic Poisson polygons in $\mathbf{R}^2$ is examined.

6-1. STATIONARY POISSON HYPERPLANE NETWORKS

In this section, A is a stationary Poisson hyperplane network, that is, by Section 3-5, A is an IDRACS the functional ψ of which admits the representation

$$\psi(K)=\int_{\mathbb{S}_{d-1}}\mu_1(\Pi_{S^\perp}K)\lambda(dS) \qquad (K\in\mathcal{K}) \tag{6-1-1}$$

for a measure $\lambda\geqslant 0$ on $\mathbb{S}_{d-1}$ (μ_1 is the Lebesgue measure on $\mathbf{R}$, and $\Pi_{S^\perp}K$ the projection of K into $S^\perp\in\mathbb{S}_1$).

Let G_1 be the positive measure on $\mathbb{S}_1$ induced from λ by the mapping $S\to S^\perp$ from $\mathbb{S}_{d-1}$ onto $\mathbb{S}_1$, so that (6-1-1) may be rewritten as $\psi(K)=\int\mu_1(\Pi_S K)G_1(dS)$. With this measure G_1 is associated the *Steiner convex set* Λ, whose supporting function $r_\Lambda\in\mathcal{R}_1$ is defined by

$$r_\Lambda(u)=\tfrac{1}{2}\int_{\mathbb{S}_{d-1}}|u^\perp,S|\lambda(dS)=\tfrac{1}{2}\int_{\mathbb{S}_1}|u,S'|G_1(dS') \tag{6-1-2}$$

(for any $u\in S_0$ identified with the corresponding one-dimensional subspace $u\in\mathbb{S}_1$).

Conversely, by Theorem 4-5-1, the measure λ is determined if the supporting function $r_\Lambda\in\mathcal{R}_1$ is given. In other words, the mapping $\lambda\to\Lambda$ is one to one from the Poisson hyperplane networks onto $\mathcal{R}_1$.

This geometrical representation of a Poisson network by its associated Steiner convex set Λ leads to a very easy characterization of the networks induced on linear varieties, for, if u is a unit vector and $\bar{u}=\{\lambda u, 0\leqslant\lambda\leqslant 1\}$, we may write $r_\Lambda(u)=\frac{1}{2}\psi(\bar{u})$. But, for any $S_p\in\mathbb{S}_p$ $(0<p<d)$, the functional ψ_{S_p} associated with the network induced on S_p is $\psi_{S_p}(K)=\psi(K)$, $K\in\mathcal{K}$, $K\subset S_p$. Hence, if $\Lambda(S_p)$ is the Steiner set associated with the induced network and u a unit vector in S_p, we have $r_{\Lambda(S_p)}(u)=r_\Lambda(u)$. In other words, *the Steiner convex set associated with the induced network $A\cap S_p$ is the projection of Λ into S_p*, that is, $\Lambda(S_p)=\Pi_{S_p}\Lambda$. It follows that, once a property of the network A is explicitly determined as a function of Λ, the corresponding property of the induced network $A\cap S_p$ may be obtained without further calculations by changing d into p and Λ into $\Pi_{S_p}\Lambda$.

From this point of view, the isotropic case deserves special attention, for, if A is isotropic, λ is proportional to the invariant measure ϖ on $\mathbb{S}_{d-1}$, say $\lambda=\lambda_d\varpi$ with a constant $\lambda_d=\int\lambda(dS)=\int G_1(dS)=dW_{d-1}(\Lambda)/b_{d-1}$ (Theorem 4-5-2, Corollary 4), and Λ is the *ball* with radius a given by

$$a=\frac{b_{d-1}}{db_d}\int\lambda(dS) \tag{6-1-3}$$

Then the induced network $A \cap S_p$ is characterized by the convex set $\Lambda(S_p) = \Pi_{S_p}(\Lambda)$, that is, *the ball with the same radius a* in S_p. In other words, the initial network and all the induced networks are completely defined by the single parameter a. For this reason, all the results that we shall obtain in the isotropic case will be presented in terms of a, and they will automatically hold for the induced network, simply by changing d into p. For instance, the definition formula (6-1-1) may be rewritten as

$$\psi(K) = a\left(\frac{2d}{b_{d-1}}\right) W_{d-1}(K) = \frac{2a}{b_{d-1}} N(K) \tag{6-1-4}$$

($N = dW_{d-1}$ is the norm in $\mathbf{R}^d$), and the induced network $A \cap S_p$ admits the functional ψ_p: $K \to (2a/b_{p-1}) N^p(K) = a(2p/b_{p-1}) W_{p-1}^p(K)$, where N^p is the norm in $\mathbf{R}^p$. If $K = B$ is the unit ball, it follows from (6-1-4) that

$$\psi(B) = \left(\frac{2db_d}{b_{d-1}}\right) a \tag{6-1-5}$$

In the general case, that is, if A is not isotropic, we have in the same way

$$\psi(B) = \frac{2dW_{d-1}(\Lambda)}{b_{d-1}} = \frac{2N(\Lambda)}{b_{d-1}} \tag{6-1-5$'$}$$

The Densities of the $(d-k)$-Volumes

Any p-dimensional linear variety V_p may be identified to $\mathbf{R}^p$ and associated with the measure $\mu_p^{V_p}$ defined on $\mathbf{R}^d$ by

$$\int_{\mathbf{R}^d} f(x) \mu_p^{V_p}(dx) = \int_{V_p} f(x') \mu_p(dx') \qquad [f \in \mathbf{C}_{\mathcal{K}}(\mathbf{R}^d)]$$

(μ_p is the Lebesgue measure on $\mathbf{R}^p$), and, conversely, V_p is determined if $\mu_p^{V_p}$ is given. The network A is locally finite and may be associated with the *random measure* ν_1 such that

$$\int_{\mathbf{R}^d} f(x) \nu_1(dx) = \sum_i \int_{H_i} f(x') \mu_{d-1}(dx') \qquad [f \in \mathbf{C}_{\mathcal{K}}(\mathbf{R}^d)]$$

where the H_i are the hyperplanes the union of which is A. In particular, for any $K \in \mathcal{K}$, the random variable $\nu_1(K) = \sum \mu_{d-1}(K \cap H_i)$ is the $(d-1)$-volume of $A \cap K$.

More generally, for any integer $k=2,3,\ldots,d$, we may consider the intersections $H_{i_1}\cap\cdots\cap H_{i_k}$ and say that their union is the *order-k network* A_k associated with A (A itself is the order-1 network). The order-k network is also locally finite and may be associated with the random measure ν_k on $\mathbf{R}^d$ defined by

$$\int f(x)\nu_k(dx)=\sum_{i_1<i_2\cdots<i_k}\int_{H_{i_1}\cap\cdots\cap H_{i_k}} f(x')\mu_{d-k}(dx') \qquad [f\in\mathbf{C}_{\mathcal{K}}(\mathbf{R}^d)]$$

Obviously, the order-k network is determined if the measure ν_k is known. For any $K\in\mathcal{K}$, $\nu_k(K)$ is the $(d-k)$-volume of $A_k\cap K$, that is, $\nu_k(K)=\sum\mu_{d-k}(K\cap H_{i_1}\cap\cdots\cap H_{i_k})$.

The order-k network is a stationary RACS, but it is not Poisson if $k>1$, and not even an IDRACS. In what follows, we shall determine the expectation measure $E(\nu_k(dx))$. By stationarity, this measure is invariant under translation. Hence $E(\nu_k(dx))=\nu_k\,dx$ with a constant $\nu_k\geqslant 0$, called the *density of the* $(d-k)$*-volume of* A, and $E(\nu_k(K))=\nu_k\mu_d(K)$, $K\in\mathcal{K}$. On the other hand, the stationary random measure is also associated with a *covariance measure* C_k on $\mathbf{R}^d$, such that

$$\int g_K(h)C_k(dh)=E\big[(\nu_k(K))^2\big]$$

for any $K\in\mathcal{K}$ $[g_K(h)=\int 1_K(x)1_K(x+h)\,dx]$. In the isotropic case, the covariance measure C_k itself is invariant under rotations and an explicit expression can be derived.

Calculation of ν_k, $k=1,2,\ldots,d$

In order to derive the expression for ν_k, we suppose first that exactly k hyperplanes of the network hit the ball RB with radius $R\geqslant 0$ and compute the corresponding conditional expectation of the $(d-k)$-volume $\mu_{d-k}(H_1\cap\cdots\cap H_k\cap RB)$, from which the value of ν_k easily follows.

a. Let $H_1,\ldots,H_k$ be the hyperplanes hitting RB. For any $i=1,2,\ldots,k$, we denote by $S_i\in\mathcal{S}_{d-1}$ the direction of H_i and by R_i the radius of the $(d-i)$-ball $RB\cap H_1\cap\cdots\cap H_i$ (if it is not empty). By (3-5-3), the law of S_1 is the probability $\lambda(dS_1)/\lambda_d$ $[\lambda_d=\int\lambda(dS)=dW_{d-1}(\Lambda)/b_{d-1}]$ on $\mathcal{S}_{d-1}$, and for a fixed $S_1\in\mathcal{S}_{d-1}$ the single point in $H_1\cap S_1^{\perp}$ is uniformly distributed on $RB\cap S_1^{\perp}$. Thus R_1 is equivalent to $\rho_1 R$, where $\rho_1=\sqrt{1-X^2}$, X uniformly distributed on $[0,1]$, and, moreover, ρ_1 and S_1 are independent.

In the same way, S_2 is independent of S_1 and ρ_1 and admits the law $\lambda(dS_2)/\lambda_d$. If S_2, S_1, and ρ_1 are fixed, the probability for H_2 to hit $H_1 \cap RB$ is $\rho_1|S_2^\perp, S_1|$, and if so the ball $RB \cap H_1 \cap H_2$ admits the radius $\rho_2 R_1 = \rho_1\rho_2 R$, where ρ_2 is independent of, and identically distributed as, ρ_1.

By a recurrence procedure, if $S_1, S_2, \ldots, S_{k-1}$, $R_1 = \rho_1 R$, $R_2 = \rho_1\rho_2 R, \ldots$, $R_{k-1} = \rho_{k-1}R_{k-2} = \rho_1\rho_2 \cdots \rho_{k-1}R$ are fixed, the direction S_k still admits the law $\lambda(dS_k)/\lambda_d$, and the probability for H_k to hit $RB \cap H_1 \cap \cdots \cap H_{k-1}$ is $\rho_1\rho_2 \cdots \rho_{k-1}|S_k^\perp, S_1 \cap \cdots \cap S_{k-1})$. If so, the ball $RB \cap H_1 \cap \cdots \cap H_k$ admits the radius $\rho_1\rho_2 \cdots \rho_k R$, where ρ_k is independent of, and identically distributed as, the ρ_i, $i = 1, 2, \ldots, k-1$.

It follows that *the probability* p_k *for the intersection* $H_I \cap \cdots \cap H_k$ *to hit* RB is

$$p_k = E(\rho_1{}^{k-1}\rho_2{}^{k-2} \cdots \rho_{k-1})E(V(S_1^\perp, S_2^\perp, \ldots, S_k^\perp))$$

where $V(S_1^\perp, S_2^\perp, \ldots, S_k^\perp)$ is the k-volume of a parallelotope constructed from k unit vectors the directions of which are $S_1^\perp, \ldots, S_k^\perp$ in $\mathbf{R}^d$, and the ρ_i, $i = 1, \ldots, k$, are k independent random variables equivalent to $\sqrt{1 - X^2}$, X uniformly distributed on $[0, 1]$. Although an explicit expression of p_k is not needed for the sequel, it can easily be derived. First, we have by simple calculations

$$E(\rho_1{}^{k-1}\rho_2{}^{k-2} \cdots \rho_{k-1}) = 2^{-k}b_k$$

In order to compute $E(V(S_1^\perp, S_2^\perp, \ldots, S_k^\perp))$, we shall use the measures $G_k = G_k{}^\Lambda$ associated by Theorem 4-5-2 with the Steiner convex set Λ. We have seen that G_k is induced from $(1/k!)V(S_1^\perp, \ldots, S_k^\perp)G_1(dS_1^\perp) \times \cdots \times G_1(dS_k^\perp)$ by the mapping $(L_1, L_2, \ldots, L_k) \to L_1 \oplus \cdots \oplus L_k$ (a.e. defined) from $(\mathbb{S}_1)^k$ into $\mathbb{S}_k$. In other words,

$$E(V(S_1^\perp, \ldots, S_k^\perp)) = \frac{k!}{(\lambda_d)^k} \int_{\mathbb{S}_k} G_k(dS)$$

By Theorem 4-5-2, Corollary 4, the integral $\int G_k(dS)$ is very simply related to the Minkowski functional value $W_{d-k}(\Lambda)$. Hence we have

$$E(V(S_2^\perp, \ldots, S_k^\perp)) = \frac{k!\binom{d}{k}}{b_{d-k}} \frac{W_{d-k}(\Lambda)}{(\lambda_d)^k} \qquad (6\text{-}1\text{-}6)$$

$$\lambda_d = \int_{\mathbb{S}_{d-1}} \lambda(dS) = \frac{dW_{d-1}(\Lambda)}{b_{d-1}}$$

Hence the probability p_k is

$$p_k = \frac{2^{-k}d!}{(d-k)!}\frac{b_k}{b_{d-k}}\frac{W_{d-k}(\Lambda)}{(\lambda_d)^k}$$

In the isotropic case, the preceding formulae reduce to

$$E(V(S_1^{\perp},S_2^{\perp},\dots,S_k^{\perp})) = \frac{d!}{(d-k)!}\frac{b_d}{b_{d-k}}\left(\frac{b_{d-1}}{db_d}\right)^k \tag{6-1-6'}$$

and

$$p_k = \frac{2^{-k}d!}{(d-k)!}\frac{b_d b_k}{b_{d-k}}\left(\frac{b_{d-1}}{db_d}\right)^k$$

Return now to the expectation of $\mu_{d-k}(H_1 \cap \cdots \cap H_k \cap RB)$, which is $b_{d-k}R^{d-k}(\rho_1\rho_2\cdots\rho_k)^{d-k}$ conditional upon $H_1 \cap \cdots \cap H_k \cap RB \neq \varnothing$. Thus

$$E(\mu_{d-k}(H_1 \cap \cdots \cap H_k \cap RB)) = b_{d-k}R^{d-k}E(\rho_1{}^{d-1}\rho_2{}^{d-2}\cdots\rho_k{}^{d-k})$$
$$\times E(V(S_1^{\perp},S_2^{\perp},\dots,S_k^{\perp}))$$

Simple calculations give $E(\rho_1{}^{d-1}\cdots\rho_k{}^{d-k}) = 2^{-k}b_d/b_{d-k}$, and thus by (6-1-6),

$$E(\mu_{d-k}(H_1 \cap \cdots \cap H_k \cap RB)) = \frac{2^{-k}d!}{(d-k)!}\frac{b_d}{b_{d-k}}\frac{W_{d-k}(\Lambda)}{(\lambda_d)^k}R^{d-k} \tag{6-1-7}$$

In the isotropic case, this formula reduces to

$$E(\mu_{d-k}(H_1 \cap \cdots \cap H_k \cap RB)) = \frac{2^{-k}d!}{(d-k)!}\frac{(b_d)^2}{b_{d-k}}\left(\frac{b_{d-1}}{db_d}\right)^k R^{d-k} \tag{6-1-7'}$$

Now, if $n \geqslant k$ hyperplanes hit the ball RB, the probability of which is $[(R\psi(B))^n/n!]\exp(-R\psi(B))$, the expectation of the $(d-k)$-volume is $\binom{n}{k}$ times the preceding one. By (6-1-5′) and

$$\sum_{n=k}^{\infty}\frac{x^n}{(n-k)!}\exp(-x) = x^k$$

it easily follows that

$$E(\nu_k(RB)) = \binom{d}{k} \frac{b_d}{b_{d-k}} W_{d-k}(\Lambda) R^d$$

But obviously $E(\nu_k(RB)) = \nu_k b_d R^d$, so that the *density* ν_k of $(d-k)$-volume is

$$\nu_k = \frac{\binom{d}{k}}{b_{d-k}} W_{d-k}(\Lambda) = \int_{\mathbb{S}_k} G_k(dS) \tag{6-1-8}$$

and, in particular, in the isotropic case

$$\nu_k = \binom{d}{k} \frac{b_d}{b_{d-k}} a^k \tag{6-1-8'}$$

If $k=d$, the order d network is made up of all the intersection points of d hyperplanes in A, called the *vertices* of the network A. Thus the specific vertex number, that is, the expectation ν_d of their number per unit volume, is

$$\nu_d = W_0(\Lambda) \tag{6-1-9}$$

or, in the isotropic case,

$$\nu_d = b_d a^d \tag{6-1-9'}$$

If $k<d$, we may also be interested in the directions of the k-dimensional varieties in the order-k network, and define the random measure $\nu_k(dx;dS)$ on $\mathbf{R}^d \times \mathbb{S}_k$ so that $\nu(K \times \mathbb{S})$ is the $(d-k)$-volume of the intersections of K and the varieties in A_k whose directions belong to $\mathbb{S} \subset \mathbb{S}_k$. From the preceding results, it easily follows that the corresponding expectation measure is $E(\nu_k(dx;dS)) = \nu_k dx \times G_k(dS)/\int G_k(dS)$. In other words, the directions of the order-k varieties in A_k admit a distribution law proportional to the measure G_k associated with the Steiner convex set Λ.

The Covariance Measures (Isotropic Case)

In the isotropic case, the covariance measure C_k associated with the order-k network is invariant under rotations, and thus, by Lemma 4-6-1, it is determined if the function $R \to \int C_k(dh) g_R(h)$ is known $[g_R(h) = \int 1_{RB}(x)$

$1_{RB}(x+h)dx]$. For this reason, we must first calculate $E[(\nu_k(RB))^2] = \int C_k(dh) g_R(h)$,

a. Suppose that exactly n hyperplanes $H_1,\ldots,H_n$ hit the ball $RB (n \geq k)$ and put

$$X_{i_1,\ldots,i_k} = \mu_{d-k}(H_{i_1} \cap \cdots \cap H_{i_k} \cap RB)$$

We already know the (conditional) expectation of $X_{i_1,\ldots,i_k}$. For n fixed, we have

$$(\nu_k(RB))^2 = \sum_{0<i_1<\cdots<i_k \leq n} \; \sum_{0<j_1<\cdots \leq n} X_{i_1,\ldots,i_k} X_{j_1,\ldots,i_k}$$

and we must calculate $E(X_{i_1,\ldots,i_k} X_{j_1,\ldots,j_k})$ for all the possible choices of the sequences $\{i_1,\ldots,i_k\}$ and $\{j_1,\ldots,j_k\}$ in $\{1,2,\ldots,n\}$. Let p be the number of indices appearing in both sequences, that is, there exist $2k-p$ hyperplanes $H_1,\ldots,H_p$, $H_{p+1},\ldots,H_k$, $H'_{p+1},\ldots,H'_k$ such that

$$X_{i_1,\ldots,i_k} = \mu_{d-k}(H_1 \cap \cdots \cap H_p \cap H_{p+1} \cap \cdots \cap H_k \cap RB)$$
$$X_{j_1,\ldots,j_k} = \mu_{d-k}(H_1 \cap \cdots \cap H_p \cap H'_{p+1} \cap \cdots \cap H'_k \cap RB)$$

Again let $S_1,\ldots,S_k$, $S'_{p+1},\ldots,S'_k$ be the directions of these hyperplanes, $\rho_1,\ldots,\rho_k$, $\rho'_{p+1},\ldots,\rho'_k$ be independent random variables equivalent to $\sqrt{1-X^2}$, X uniformly distributed on $[0,1]$, and put

$$\sigma = S_1 \cap \cdots \cap S_p \in \mathbb{S}_{d-p} \text{ (a.s.)}.$$

The calculation is closely analogous to the preceding one. The probability for $H_1 \cap \cdots \cap H_p \cap RB$ not to be empty is

$$\rho_1^{p-1} \rho_2^{p-2} \cdots \rho_{p-1} |S_2^{\perp}, S_1| \times \cdots \times |S_p^{\perp}, S_1 \cap \cdots \cap S_{p-1}|$$

Conditionally upon $H'_1 \cap \cdots \cap H_p \cap RB \neq \emptyset$, the probability for the two intersections $H_1 \cap \cdots \cap H_k \cap RB$ and $H_1 \cap \cdots \cap H'_k \cap RB$ not to be empty is

$$(\rho_1 \rho_2 \cdots \rho_p)^{2(k-p)} \rho'^{\,k-p-1}_{p+1} \cdots \rho'_{k-1} \rho^{k-p-1}_{p+1} \cdots \rho_{k-1} \times |\Pi_\sigma S_{p+1}^{\perp}|$$
$$\times |\Pi_\sigma S_{p+2}^{\perp}, \sigma \cap S_{p+1}| \times \cdots \times |\Pi_\sigma S_k^{\perp}, \sigma \cap S_{p+1} \cap \cdots \cap S_{k-1}|$$
$$\times |\Pi_\sigma S'^{\perp}_{p+1}| \times |\Pi_\sigma S'^{\perp}_{p+2}, \sigma \cap S'_{p+1}| \times \cdots \times |\Pi_\sigma S'^{\perp}_k, \sigma \cap S'_{p+1} \cap \cdots \cap S'_{k-1}|$$

Finally, conditionally upon these intersections not being empty, we have

$X_{i_1,\ldots,i_k}X_{j_1,\ldots,j_k}$

$$=(b_{d-k}R^{d-k})^2(\rho_1\rho_2\cdots\rho_p)^{2(d-k)}(\rho'_{p+1}\cdots\rho'_k)^{d-k}(\rho_{p+1}\cdots\rho_k)^{d-k}$$

Thus we must evaluate the expectation of the product:

$$(b_{d-k}R^{d-k})^2\rho_1{}^{2d-p-1}\rho_2{}^{2d-p-2}\cdots\rho_p{}^{2d-2p}(\rho_{p+1}\rho'_{p+1})^{d-p-1}\cdots(\rho_k\rho'_k)^{d-k}$$
$$\times|S_2^\perp,S_1|\times\cdots\times|S\perp_p,S_1\cap\cdots\cap S_{p-1}|\times|\Pi_\sigma S_{p+1}^\perp|\times|\Pi_\sigma S'^{\perp}_{p+1}|\times\cdots$$
$$\times|\Pi_\sigma S_k^\perp,\sigma\cap S_{p+1}\cap\cdots\cap S_{k-1}|\times|\Pi_\sigma S'^{\perp}_k,\sigma\cap S'_{p+1}\cap\cdots\cap S'_{k-1}|$$

By Section 4-6, it turns out that these various random variables are independent and [with the notations used in (4-6-2)] the above product is equivalent to

$$(b_{d-k}R^{d-k})^2\rho_1{}^{2d-p-1}\cdots\rho_p{}^{2d-2p}(\rho_{p+1}\rho'_{p+1})^{d-p-1}\cdots(\rho_k\rho'_k)^{d-k}$$
$$\times\prod_{d-p+1}^{d}Y_j\times\prod_{d-k+1}^{d-p}Y'_j\times\prod_{d-k+1}^{d-p}Y''_j$$

where Y_j, Y'_j, Y''_j are independent and distributed as in (4-6-2).

Taking into account $E(\rho^j)=b_{1+j}/2b_j$ and $E(Y_j)=(j/d)(b_{d-1}/b_d)\times(b_j/b_{j-1})$, and after doing some calculations, we finally obtain

$$E(X_{i_1,\ldots,i_k}X_{j_1,\ldots,j_k})=B(p,k)R^{2(d-k)}$$

$$B(p,k)=2^{p-2k}\frac{d!(d-p)!}{((d-k)!)^2}\frac{b_{2d-p}b_d(b_{d-p})^3}{b_{2d-2p}(b_{d-k})^2}\left(\frac{b_{d-1}}{db_d}\right)^{2k-p}$$

b. If n hyperplanes hit RB, there exist $\binom{n}{k}$ k-dimensional varieties associated with them, and exactly

$$\binom{n}{p}\binom{n-p}{k-p}\binom{n-k}{k-p}=n!/\left[((k-p)!)^2p!(n-2k+p)!\right]$$

possible pairs of the type

$$(H_1 \cap \cdots \cap H_p \cap H_{p+1} \cap \cdots \cap H_k, H_1 \cap \cdots \cap H_p \cap H'_{p+1} \cap \cdots \cap H'_k)$$

considered above. It follows that

$$E\left[(\nu_k(RB))^2\right]$$

$$= \sum_{p=0}^{k} \sum_{n=2k-p}^{\infty} \frac{n!\, B(p,k) R^{2(d-k)}}{p!(n-2k+p)!((k-p)!)^2} \frac{(R\psi(B))^n}{n!} \exp(-R\psi(B))$$

$$= \sum_{p=0}^{k} \frac{B(p,k)}{p((k-p)!)^2} (\psi(B))^{2k-p} R^{2d-p}$$

and thus, by (6-1-5),

$$E\left[(\nu_k(RB))^2\right]$$

$$= \sum_{p=0}^{k} \frac{d!(d-p)!}{p!((d-k)!(k-p)!)^2} \frac{b_{2d-p} b_d (b_{d-p})^3}{b_{2d-2p}(b_{d-k})^2} a^{2k-p} R^{2d-p} \tag{6-1-10}$$

c. By applying Lemmas 4-6-1 and 4-6-2, and after doing an easy calculation, it follows from (6-1-10) that

$$C_k(dh) = \sum_{p=0}^{k} \frac{(d-p)(d-1)!(d-p)!}{p!((d-k)!(k-p)!)^2} \left(\frac{b_{d-p}}{b_{d-k}}\right)^2 a^{2k-p} \frac{dh}{|h|^p} \tag{6-1-11}$$

if $k < d$, and if $k = d$

$$C_d(dh) = \sum_{p=0}^{d-1} \binom{d-1}{p} (b_{d-p})^2 a^{2d-p} \frac{dh}{|h|^p} + b_d a^d \delta(dh) \tag{6-1-12}$$

where δ is the Dirac measure. By comparing with (4-6-5) through (4-6-6), we conclude that the order-k network ($k \geqslant 2$) is not Poisson.

6-2. THE POISSON POLYHEDRA AND THE CONDITIONAL INVARIANCE

In order to obtain the most general class of Poisson polyhedra, we must extend a little the definition of the Poisson networks, by starting from a

Poisson point process on the product space $S_0 \times \mathbf{R}_+$, where S_0 is the unit sphere, rather than on $\mathbb{S}_1 \times \mathbf{R}$ as above. With each point $(u,r) \in S_0 \times \mathbf{R}_+$, let us associate the hyperplane $H(u,r) = \{x: \langle u,x \rangle = r\}$ in $\mathbf{R}^d$. The mapping H: $(u,r) \to H(u,r)$ from $S_0 \times \mathbf{R}_+$ into $\mathcal{F}'$ is continuous. Let $\lambda(dy)$ be a positive measure on the unit sphere S_0, dr be the Lebesgue measure on $\mathbf{R}_+$, and $\mathcal{Q} \subset \mathcal{F}'$ be the image under the mapping H of the Poisson point process in $S_0 \times \mathbf{R}_+$ associated with the product measure $\lambda(ds) \otimes dr$. We say that $\mathcal{Q}$ is a *Poisson network* (in a generalized sense). Let A be the union of the hyperplanes $H \in \mathcal{Q}$. Then A is a RACS, for the number $n(K)$ of the hyperplanes in $\mathcal{Q}$ hitting a compact set $K \in \mathcal{K}$, that is, the number of points of the Poisson process on $S_0 \times \mathbf{R}_+$ belonging to $H^{-1}(\mathcal{F}_K)$, is a Poisson random variable, and its expectation is

$$\psi(K) = \int_{H^{-1}(\mathcal{F}_K)} \lambda(du)\, dr \tag{6-2-1}$$

But $H^{-1}(\mathcal{F}_K)$ is compact in $S_0 \times \mathbf{R}_+$, as can easily be verified, so that $\psi(K) < \infty$. Thus the network $\mathcal{Q}$ is locally finite and A is a.s. closed in $\mathbf{R}^d$, that is, is a RACS.

Obviously, the RACS A is infinitely divisible and semi-markovian, for it may be considered as the union of a Poisson process in $C(\mathcal{F}')$. Note that the SMIDRACS A is *stationary* if and only if the measure λ on S_0 is *symmetric*.

Let $\mathcal{Q}$ be the family of hyperplanes such that $A = \cup \mathcal{Q}$, let x be a point not belonging to A, and, for each hyperplane $H \in \mathcal{Q}$, let $H(x)$ be the open half space containing x and limited by H. Clearly,

$$\Pi(x) = \cap \{H(x), H \in \mathcal{Q}\}$$

is an open and convex polyhedron. Hence the complementary set A^c is a.s. the union of open, convex, and disjoint polyhedra in $\mathbf{R}^d$. Moreover, we also have $P(0 \in A^c) = 1$. Thus one of these polyhedra, say Π_0, a.s. encloses the origin O and is called a *Poisson polyhedron*. Obviously, Π_0 is an open random set a.s. convex and enclosing O. Its closure $\overline{\Pi}_0$ is a RACS a.s. convex, the interior of which is a.s. identical to Π_0 itself. For any compact set $K \in \mathcal{K}$, we have $K \subset \Pi_0$ (i.e., $\Pi_0 \in \mathcal{G}_{K_o}$ with the notations defined in Chapter 1) if and only if the convex hull $C(\{o\} \cup K)$ and the Poisson network A are disjoint, that is, $A \in \mathcal{F}^{C(\{o\} \cup K)}$, and the probability of this event is $Q(C(\{o\} \cup K))$.

Thus the probability P on $(\mathcal{G}, \sigma_g)$ associated with the random open set Π_0 is entirely determined if the functional $K \to P(\mathcal{G}_K)$ is known for any $K \in C_0(\mathcal{K})$ (i.e., K is compact, convex, and encloses the point O), and is identical on $C_0(\mathcal{K})$ with the functional Q associated with the network A

itself. Explicitly, if r_K is the *supporting function* of $K \in C_0(\mathcal{K})$, the set $H^{-1}(\mathcal{F}_K)$ in $S_0 \times \mathbf{R}_+$ is $\{(u,r): r \leqslant r_K(u)\}$, and it follows from (6-2-1) that

$$\left.\begin{aligned} P(\mathcal{G}_K) &= Q(K) = \exp(-\psi(K)) \\ \psi(K) &= \int_{S_0} r_K(u)\lambda(du) \end{aligned}\right\} \qquad [K \in C_0(\mathcal{K})] \qquad (6\text{-}2\text{-}2)$$

This relationship, extended on the whole space $\mathcal{K}$ by putting

$$Q(K) = Q(C(\{o\} \cup K)) \qquad (K \in \mathcal{K}) \qquad (6\text{-}2\text{-}3)$$

may be used as a *definition* of the Poisson polyhedron Π_0. In particular, there is a one-to-one correspondence between the Poisson polyhedra and the positive Radon measures on the unit sphere S_0. If the measure λ on S_0 is *symmetric*, the corresponding Poisson polyhedron Π_0 may be associated with a *stationary* Poisson network A. If λ is proportional to the (unique) probability ϖ on S_0 invariant under rotations, the Poisson polyhedron is said to be *isotropic*. If so, the network A itself is necessarily stationary and isotropic.

The Conditional Invariance

We saw in Section 4-4 that the mapping $K \to r_K$, where r_K is the supporting function of $K \in C_0(\mathcal{K})$, is a homeomorphism from $C_0(\mathcal{K})$ onto the cone $\mathcal{R} \subset \mathbf{C}(S_0)$. Thus it follows from (6-2-2) that the functionals ψ and Q are continuous on $C_0(\mathcal{K})$, and by (6-2-3) they are even continuous on the whole space $\mathcal{K}$. On the other hand, the relationships $r_{K+K'} = r_K + r_{K'}$ and $r_{\alpha K} = \alpha r_K$ yield $\psi(K \oplus K') = \psi(K) + \psi(K')$ and $\psi(\alpha K) = \alpha\psi(K)$, $\alpha \geqslant 0$, for any K and $K' \in C_0(\mathcal{K})$. By (6-2-3), these relationships always hold if K and K' are arbitrary compact sets enclosing O. Thus we have

$$\left.\begin{aligned} Q(\alpha K) &= (Q(K))^{\alpha} \\ Q(K \oplus K') &= Q(K)Q(K') \end{aligned}\right\} \qquad (\alpha \geqslant 0,\ K,\ K' \in \mathcal{K}_0)$$

where Q: $K \to Q(K) = P(K \subset \Pi_0)$ is the functional associated by (6-2-2) and (6-2-3) with the Poisson polyhedron Π_0.

The *semigroup relationship* $Q(K \oplus K') = Q(K)Q(K')$ is related to a Markovian property that we will now examine, and turns out to yield a characterization of the Poisson polyhedra. First, note that $K \subset \Pi_0$ is equivalent to $0 \in \Pi_0 \ominus \check{K}$, and in the same way $K \oplus K' \subset \Pi_0$ is equivalent to $K' \subset \Pi_0 \ominus \check{K}$. Thus the semigroup relationship may be rewritten as

$$P(K' \subset \Pi_0 \ominus \check{K}) = P(K' \subset \Pi_0)P(K \subset \Pi_0)$$

On the other hand, $K' \subset \Pi_0 \ominus \check{K}$ implies $K \subset \Pi_0$, for $0 \in K'$, and $P(K \subset \Pi_0) = Q(K)$ never vanishes. Thus the semigroup relationship may be interpreted in terms of conditional probabilities, that is,

$$P(K' \subset \Pi_0 \ominus \check{K} | 0 \in \Pi_0 \ominus \check{K}) = P(K' \subset \Pi_0)$$

In other words, *conditionally upon O belonging to the erosion $\Pi_0 \ominus \check{K}$ of Π_0 by $K \in \mathcal{K}$, $\Pi_0 \ominus \check{K}$ is a Poisson polyhedron equivalent to Π_0 itself*. We shall state this property by saying that *Poisson polyhedra are conditionally invariant under erosions* by the compact sets $K \in \mathcal{K}_0$. This conditional invariance will be reinterpreted below in a slightly different way, that is, from the point of view of the number law.

Actually, the Poisson polyhedra are *characterized* by this conditional invariance property. More precisely, by putting $Q(K) = P(K \subset \Pi_0)$, the following statement holds:

Theorem 6-2-1. *A random open set Π_0 a.s. enclosing the point O is a Poisson polyhedron if and only if $Q(K \oplus K') = Q(K)Q(K')$ for any $K, K' \in \mathcal{K}_0$.*

Proof. We have just proved the "only if" part. Conversely, let Π_0 be an open random set such that $P(O \in \Pi_0) = 1$ and

$$Q(K \oplus K') = Q(K)Q(K') \qquad (K, K' \in \mathcal{K}_0) \qquad (6\text{-}2\text{-}4)$$

[with $Q(K) = P(K \subset \Pi_0)$]. In what follows, rB will denote the ball with center O and radius $r \geq 0$.

a. $Q(rB) \uparrow 1$ if $r \downarrow 0$, for the function Q always is l.s.c. on $\mathcal{K}$, and $\lim(rB) = \{o\}$ in $\mathcal{K}$ if $r \downarrow 0$.

b. The function Q is *continuous* on $\mathcal{K}$.

It is sufficient to verify the continuity on $\mathcal{K}_0$, for $Q(K) = Q(K \cup \{o\})$, and the union $\cup$ is continuous. Let $\{K_n\} \subset \mathcal{K}_0$ be a sequence such that $\lim K_n = K$ in $\mathcal{K}_0$, and $\epsilon > 0$. Thus $K \subset K_n \oplus \epsilon B$ and $K_n \subset K \oplus \epsilon B$ for n large enough. Then it follows from (6-2-4) that

$$Q(K) \geq Q(\epsilon B)\,\overline{\lim}\, Q(K_n), \qquad \underline{\lim}\, Q(K_n) \geq Q(K)Q(\epsilon B)$$

Taking statement *a* into account, we conclude that $Q(K) = \lim Q(K_n)$, and Q is continuous.

c. Let $K \in \mathcal{K}$ be a compact set, and $C_0 = C(\{o\} \cup K)$. Then $Q(K) = Q(C_0)$.

Let $C = C(K)$ be the convex hull of K. It is sufficient to prove that $Q(K) = Q(C)$, because $O \in \Pi_0$ a.s. By Proposition 1-5-7, there exists a real

$r>0$ such that $C\oplus rB\subset K\oplus rB\oplus\epsilon B$, and thus $Q(K)Q(\epsilon B)\leqslant Q(C)$ by (6-2-4). Then, from part *a*, it follows that $Q(K)\leqslant Q(C)$, and also $Q(K)=Q(C)$, because the reverse inequality is obvious.

d. The mapping Q is continuous on $\mathcal{K}$, so that (6-2-4) implies

$$Q(\alpha K)=(Q(K))^{\alpha} \qquad (\alpha\geqslant 0, K\in\mathcal{K})$$

Note that Q does not vanish on $\mathcal{K}$, for suppose that there exists $K_0\in\mathcal{K}$ such that $Q(K_0)=0$. For $\rho_0>0$ large enough, $K_0\subset\rho_0 B$ implies $Q(\rho_0 B)=(Q(B))^{\rho_0}=0$, and thus $Q(\rho B)=0$ for any $\rho>0$. But this is impossible by part *a*. Thus $Q(K)>0$ for any $K\in\mathcal{K}$, and the function $\psi=-\log Q$ is continuous on $\mathcal{K}$.

e. The function ψ is continuous, increasing, and positively linear on $\mathcal{K}$ and (a fortiori) on $C_0(\mathcal{K})=\mathcal{R}$. Thus it follows from Proposition 4-4-4 that there exists a Radon measure $\lambda\geqslant 0$ on the unit sphere S_0 such that

$$\psi(K)=\int_{S_0} r_K(u)\lambda(du) \qquad [K\in C_0(\mathcal{K})]$$

Taking into account part *c* and the definition of (6-2-2) through (6-2-3), we conclude that Π_0 is the Poisson polyhedron associated with the measure λ.

The Number Law

Now let λ be a positive *symmetric* measure on S_0, and A the *stationary* Poisson hyperplane network induced by the mapping H from the Poisson point process in $S_0\times\mathbf{R}_+$ associated with the product measure $\lambda(du)\otimes dr$. Then the complementary set A^c is the union of open convex and disjoint polyhedra Π_i, of which one, say Π_0, a.s. contains the origin O. From a statistical (and heuristic) point of view, we may consider the family Π_i as a "population," that is, assign the same weight to "each" of these individual polyhedron Π_i, and the corresponding probability law is said to be the *number law* of the Poisson polyhedron Π. Generally, Π is defined only up to a translation, and the corresponding probability is given on the quotient space $\mathcal{G}/\tau$, when τ is the translation equivalence, that is, $G\tau G'$ if G' is a translate of G. In particular, the condition $O\in\Pi$ is no longer required. On the other hand, each of the individual polyhedra Π_i may be assigned a weight proportional to its volume $V(\Pi_i)$, giving us the corresponding volume-weighted law called, for brevity, the *volume law*. The Poisson polyhedron Π_0, as defined above, fits the point of view of the volume law, for (roughly speaking) the point O will belong more often to a large polyhedron than to a small one. The reader is referred to Miles (1964–71) for a more rigorous treatment, based on an ergodic theorem. In fact, the following result has been proved by Miles. Let $F_0(d\Pi)$ be the probability of the Poisson polyhedron Π_0 as defined above, and $F(d\Pi)$ the corresponding number law. Then

$$F_0(d\Pi)=\frac{V(\Pi)}{E(V(\Pi))}F(d\Pi)$$

where V is the volume.

It will be interesting to reinterpret the conditional invariance defined above in terms of the number law. To do this, we will use some preliminary results. Since the volume $V(\Pi_0)$ plays an important role in the sequel, we will always suppose that Π_0 *is a.s. bounded*. It is easy to verify that Π_0 is a.s. bounded if and only if the linear varieties in A are not a.s. parallel to a fixed line or (which is the same) if the subspace spanned in $\mathbf{R}^d$ by the support of the measure λ on S_0 is equal to $\mathbf{R}^d$ itself. Another necessary and sufficient condition for Π_0 to be a.s. bounded is $W_0(\Lambda)>0$, where Λ is the Steiner convex set associated with the network A. By (6-1-9), this condition is also equivalent to $\nu_d>0$.

In order to characterize an open set up to a translation, we shall use the quotient space $\mathcal{G}/\tau$; and, if $\mathcal{G}_k\subset\mathcal{G}$ is the space of the bounded open set, we shall also use the quotient space $\mathcal{G}_k/\tau\subset\mathcal{G}/\tau$.

If G is a bounded open set and $c(G)$ the center of the ball circumscribing G, the mapping $G\to c(G)$ from $\mathcal{G}_k$ into $\mathbf{R}^d$ is measurable, as is easy to verify, and so also is the mapping u from $\mathcal{G}_k$ into itself, defined by

$$u(G)=G\oplus\{-c(G)\}$$

Clearly, u is invariant under translation and, for any $B\in u(\mathcal{G}_k)$, $u^{-1}(B)$ is the class of B modulo τ. In other words, we can identify $\mathcal{G}_k/\tau$ with $u(\mathcal{G}_k)$, and provide it with the corresponding σ-algebra. If P is a probability on $\mathcal{G}$ concentrated on $\mathcal{G}_k$ (i.e., the corresponding random open set Γ is a.s. bounded), and F is the probability on $u(\mathcal{G}_k)$ induced from P by u, F may also be considered as a probability on $\mathcal{G}/\tau$ concentrated on $\mathcal{G}_k/\tau$, that is, F is the law of the random open set Γ up to a translation. In this identification, the hypothesis that Γ is a.s. bounded is obviously essential.

With the measurability of u taken into account, the random open set Γ conditional upon $u(\Gamma)=B$, which will be denoted by $\Gamma_u(B)$, exists for F-almost every $B\in\mathcal{G}/\tau$. It follows from Proposition 2-3-2 that $\Gamma_u(B)$ is a.s. a translate B_x of B. In other words, $\Gamma_u(B)=B_x$ is characterized by the law $\varpi_B(dx)$ of the random translation vector $x\in\mathbf{R}^d$. If $\Gamma=\Pi_0$ is the Poisson polyhedron, we will see that, conditionally upon $u(\Pi_0)=B$, Π_0 is equivalent to B_x, where x is uniformly distributed on $\check{B}=\{y,-y\in B\}$. This means that, once B is fixed, the location of the origin O may be considered as uniformly distributed inside Π_0. This interpretation is analogous to the results obtained in Section 2-7 concerning the linear granulometry. In the following statement, the hypothesis that Γ is a.s. convex is probably not necessary, but it simplifies the proof.

Proposition 6-2-1. *Let Γ be a random open set, a.s. convex and bounded and containing a.s. the origin O. Let Q be the functional defined on $\mathcal{K}$ by $Q(K)=P(K\subset\Gamma)$. Then the three following conditions are equivalent:*

1. *For any $K\in\mathcal{K}_0$ and $h\in K$, $Q(K)=Q(K_{-h})$.*
2. *The functional Q admits the representation*

$$Q(K)=\int\frac{V((\check{B}\ominus K)\cap\check{B})}{V(B)}F(dB)\qquad(K\in\mathcal{K})\qquad(a)$$

for a unique probability F on $\mathcal{G}/\tau$, concentrated on the classes of convex and bounded open sets. In particular, if $K\in\mathcal{K}_0$,

$$Q(K)=\int\frac{V(\check{B}\ominus K)}{V(B)}F(dB)\qquad(K\in\mathcal{K}_0)\qquad(a')$$

3. *Conditionally upon $u(\Gamma)=B$, Γ is equivalent to the translate B_x, where x is uniformly distributed on $\check{B}$.*

Proof. If x is uniformly distributed on $\check{B}\in u(\mathcal{G}_k)$, we have, for any $K\in\mathcal{K}$, $P(K\subset B_x)=P(x\in\check{B}\ominus K)=V((\check{B}\ominus K)\cap\check{B})/V(B)$. Hence, if condition 3 is true and F is the probability on $u(\mathcal{G}_k)$ induced from P by the mapping u, relationship (a) is satisfied, and condition 2 is true. If O and h belong to $K\in\mathcal{K}$, we have $K\in\mathcal{K}_0$ and $K_{-h}\in\mathcal{K}_0$. Hence, by (a'), condition 2 implies $Q(K)=Q(K_{-h})$.

It remains to prove that condition 1 implies 3. Since $Q(\{o\})=1$, there exist compact neighborhoods K of O such that $Q(K)\neq 0$. Let $K\in\mathcal{K}_0$ be such that $Q(K)\neq 0$ and $h\in K$, so that we have $Q(K)=Q(K_{-h})$. For any $K'\in\mathcal{K}$, we have $K\cup K'\in\mathcal{K}_0$ and $h\in K\cup K'$. It follows that $Q(K_{-h}\cup K'_{-h})=Q(K\cup K')$ and

$$P(K'\subset\Gamma|K\subset\Gamma)=\frac{Q(K\cup K')}{Q(K)}=\frac{Q(K_{-h}\cup K'_{-h})}{Q(K_{-h})}=P(K'\subset\Gamma_h|K\subset\Gamma_h)$$

In other words, the random open set Γ conditional upon $K\subset\Gamma$ is equivalent to the translate $\Gamma_h=\Gamma\oplus\{h\}$ conditional upon $K\subset\Gamma_h$.

Now let u be the mapping $G \to u(G) \oplus \{-c(G)\}$ and F the probability induced from P by u on the space $u(\mathcal{G})$ identified with $\mathcal{G}/\tau$ (this identification is legitimate because Γ is a.s. bounded). The probability F is concentrated on the classes of convex and bounded open sets. The random open set $\Gamma_u(B)$ [i.e., Γ conditional upon $u(\Gamma) = B$] is defined for F-almost every $B \in \mathcal{G}/\tau$. From Proposition 2-3-2, it follows that $\Gamma_u(B)$ is a.s. convex and bounded and contains a.s. the origin O, and, since u is invariant under translation, $\Gamma_u(B)$ conditional upon $K_0 \subset \Gamma_u(B)$ is equivalent to $\Gamma_u(B) \oplus \{h\}$ conditional upon $K_0 \subset \Gamma_u(B) \oplus \{h\}$ for any $K_0 \in \mathcal{K}_0$ and $h \in K_0$. Let us denote by Q_B the functional associated with $\Gamma_u(B)$. We have $Q_B(\{o\}) = 1$, and thus $Q(\{o,h\}) > 0$ for $|h|$ small enough. Taking $K_0 = \{o,h\}$, we have for any $K \in \mathcal{K}_0$ such that $h \in K$,

$$\frac{Q_B(K)}{Q_B(\{o,h\})} = \frac{Q_B(K_{-h})}{Q_B(\{o,-h\})}$$

that is, $Q_B(K_{-h}) = f_B(h) Q_B(K)$ with a function f_B on a neighborhood of O defined by $f_B(h) = Q_B(\{o,-h\})/Q_B(\{o,h\})$.

Let us prove that $f_B = 1$. Take $h, h' \in K$ such that $h + h' \in K (K \in \mathcal{K}_0)$. We have $Q_B(K_{-h-h'}) = f_B(h+h') Q(K)$ and $Q_B(K_{-h-h'}) = f_B(h) Q_B(K_{-h'}) = f_B(h) f_B(h') Q_B(K)$. Hence $f_B(h+h') = f_B(h) f_B(h')$. This relationship and the measurability of f_B imply $f_B(h) = \exp(\langle b(B), h \rangle)$ for a vector $b(B) \in \mathbf{R}^d$ ($\langle\rangle$ is the scalar product). But, for any compact neighborhood K of O and any $h \in K$, we have $Q(K) = Q(K_{-h})$, that is,

$$Q(K) = \int \exp^{(-\langle b(B), h \rangle)} Q_B(K) F(dB)$$

This relationship holds for any h belonging to a given compact neighborhood K of O. By the properties of the Laplace transform, it follows that $b(B) = 0$ almost everywhere for the measure $Q_B(K) F(dB)$, that is, $\int (b(B))^2 Q_B(K) F(dB) = 0$. By choosing a sequence $\{K_n\}$ of compact neighborhoods of O such that $K_n \downarrow 0$, we have $Q_B(K_n) \uparrow Q_B(\{o\}) = 1$ and $\int (b(B))^2 F(dB) = 0$ by monotone continuity. Hence (a.s.) $b(B) = 0$, $f_B = 1$, and

$$Q_B(K_{-h}) = Q_B(K) \qquad (K \in \mathcal{K}_0, h \in K) \tag{b}$$

On the other hand, $\Gamma_u(B)$ is equivalent to a translate B_x of B. Let $\varpi_B(dx)$ be the probability of the random translation vector $x \in \mathbf{R}^d$. Since $O \in B_x$ a.s., that is, $x \in \check{B}$ a.s., ϖ_B is concentrated on $\check{B}$. Moreover, for any $K \in \mathcal{K}_0$ and $h \in K$, B_x conditional upon $K \subset B_x$ is equivalent to B_{x+h} conditional upon $K \subset B_{x+h}$. Hence $1_{\check{B} \ominus K}(x) \varpi_B(dx)/Q_B(K) = 1_{\check{B} \ominus K}(x)$ $\varpi_B(dx-h)/Q_B(K_{-h})$, that is, by (b),

$$1_{\check{B} \ominus K}(x)\, \varpi_B(dx-h) = 1_{\check{B} \ominus K}(x)\, \varpi_B(dx) \tag{c}$$

This relationship holds for any compact neighborhood K of O and any $h \in K$. Moreover, B is convex, and thus $\check{B} \ominus K$ is convex. It follows that $1_{\check{B}\ominus K}\varpi_B$ is proportional to the restriction of the Lebesgue measure λ to the open set $\check{B} \ominus K$, that is,

$$1_{\check{B}\ominus K}\varpi_B = a(K) 1_{\check{B}\ominus K}\lambda$$

But the coefficient $a(K)$ is independent of K. In fact, if $O \in K' \subset K$, $\check{B} \ominus K' \supset \check{B} \ominus K$ and thus $a(K) = a(K')$. It follows, for any $K'' \in \mathcal{K}_0$, that $a(K) = a(K \cap K'') = a(K'') = a$. Finally, if $\{K_n\}$ is a sequence of compact neighborhoods of O such that $K_n \downarrow \{o\}$, we have $1_{\check{B}\ominus K_n} \uparrow 1_{\check{B}}$, and thus $1_{\check{B}\ominus K_n}\varpi_B = a 1_{\check{B}\ominus K_n}\lambda$ implies $\varpi_B = 1_{\check{B}}\lambda$. It follows that $a = 1/V(\check{B}) = 1/V(B)$, since ϖ_B is a probability, and x is uniformly distributed on $\check{B}$. We complete the proof by noting that F in representation (a) is the law of $u(\Gamma)$ and hence is unique.

Now we can return to the Poisson polyhedron Π_0 and interpret the conditional invariance. Let F_0 be the law of $u(\Pi_0)$, so that by (a)

$$Q(K) = \int \frac{V(\check{B} \ominus K)}{V(B)} F_0(dB) \qquad (K \in \mathcal{K}_0)$$

If $K, K' \in \mathcal{K}_0$, we have $Q(K') = Q(K \oplus K')/Q(K)$, by the conditional invariance, and thus

$$Q(K') = \int \frac{V((\check{B} \ominus K) \ominus K')}{V(\check{B} \ominus K)} \frac{V(\check{B} \ominus K)}{Q(K)V(B)} F_0(dB)$$

When the uniqueness of F_0 is taken into account, it follows that the image of the measure $(V(\check{B} \ominus K)/Q(K)V(B))F_0(dB)$ under the mapping $B \to B \ominus \check{K}$ is $F_0(dB)$ itself. In other words, for any measurable function $\varphi \geqslant 0$ invariant under translation on $\mathcal{G}$, we have

$$\int \varphi(B \ominus \check{K}) \frac{V(\check{B} \ominus K)}{V(B)} F_0(dB) = Q(K) \int \varphi(B) F_0(dB) \qquad \text{(6-2-5)}$$

By taking $\varphi(B) = 1_{\{B \neq \varnothing\}}/V(B)$, we get

$$\int_{\{B \ominus \check{K} \neq \varnothing\}} \frac{1}{V(B)} F_0(dB) = Q(K) \int \frac{1}{V(B)} F_0(dB)$$

and this implies $\int (1/V(B)) F_0(dB) < \infty$. Hence it is possible to define a probability F on $\mathcal{G}/\tau$ by putting

$$F(dB)=\left[\int\left(\frac{1}{V(B)}F_0(dB)\right)\right]^{-1}\frac{1}{V(B)}F_0(dB)$$

Clearly, F is the *number law* of the Poisson polyhedron Π. By changing $\varphi(B)$ into $1_{\{B\neq\varnothing\}}\varphi(B)/V(B)$ in (6-2-5), we can rewrite this relationship in terms of the number law F:

$$\int_{\{B\ominus\check{K}\neq\varnothing\}}\varphi(B\ominus\check{K})F(dB)=Q(K)\int\varphi(B)F(dB)$$

In this relationship, $B=u(\Pi)$ is the class in $\mathcal{G}/\tau$ of the Poisson polyhedron Π. By taking $\varphi=1$, we note that the probability for the erosion $\Pi\ominus\check{K}$ not to be empty is equal to $Q(K)$, that is,

$$P(\Pi\ominus\check{K}\neq\varnothing)=Q(K)$$

Hence we can rewrite (6-2-5) in terms of the conditional expectation:

$$E(\varphi(\check{B}\ominus K)|B\ominus\check{K}\neq\varnothing)=E(\varphi(B))$$

In other words, the interpretation of the conditional invariance in terms of the number law is as follows: *conditionally upon* $\Pi\ominus\check{K}\neq\varnothing$, *the erosion* $\Pi\ominus\check{K}$ *is equivalent to* Π *itself.*

If X: $\Pi\rightarrow X(\Pi)$ is a characteristic of the (number law) Poisson polyhedron (i.e., a random variable defined on the space of the convex sets and invariant under translation) such that $X(\varnothing)=0$, (6-2-5) yields

$$E(X(\Pi\ominus\check{K}))=Q(K)E(X(\Pi)) \qquad \text{(6-2-6)}$$

This relationship will be extremely useful in the applications.

6-3. APPLICATIONS

In this section, we consider only Poisson polyhedra associated with a *stationary* Poisson network A, so that the measure λ in (6-2-2) will always be symmetric on the unit sphere S_0. Moreover, the Poisson polyhedra are supposed to be a.s. bounded, that is, as we have seen, the subspace spanned by the support of the measure λ on S_0 is equal to $\mathbf{R}^d$.

Expectation of the Minkowski Functionals $W_k(\Pi)$

In order to evaluate the expectation $E(W_k(\Pi))$ (from the point of view of the number law), let H_j be the hyperplanes of the network A, and K_i the closures of the convex polyhedra they determine. (The network A is locally finite, so that this indexing creates no problem.) If $x_i = \Pi_{K_i} x$ is the projection into K_i of a point $x \in \mathbf{R}^d$ not belonging to K_i, x_i is also the projection of x into a (a.s. unique) linear variety belonging to one of the order $1,2,\dots,d$ networks associated with A. Conversely, if $V = H_{j_1} \cap \cdots \cap H_{j_k}$ is a linear variety in the order-k network, $\Pi_V x$ is the projection of x into a unique Poisson polyhedron K_i. Taking the stationarity into account, it follows that, for any $x \in \mathbf{R}^d$ (except at most for a set negligible with respect to the Lebesgue measure), the sets $\{\Pi_{K_i} x; K_i \not\ni x\}$ and $\{\Pi_{H_j} x\} \cup \{\Pi_{H_{j1} \cap H_{j2}} x\} \cup \cdots$ may be identified with each other. Thus, for any function $\varphi \in \mathbf{C}_{\mathcal{K}}(\mathbf{R}^d)$ and $r \geqslant 0$,

$$\sum_i \int_{K_i \oplus rB} \varphi(\Pi_{K_i} x)\, dx = \sum_j \int_{H_j \oplus rB} \varphi(\Pi_{H_j} x)\, dx + \sum_{j_1 < j_2} \int_{(H_{j_1} \cap H_{j_2}) \oplus rB} \varphi(\Pi_{H_{j1} \cap H_{j2}} x)\, dx + \cdots$$

In other words, for any integer $k = 0,1,\dots,d$, the random Minkowski measures associated with the Poisson polyhedra K_i and the order $1,\dots,d$ networks, respectively, are such that

$$\sum_i W_k{}^{K_i} = \sum_j W_k{}^{H_j} + \sum_{j_1 < j_2} W_k{}^{H_{j1} \cap H_{j2}} + \cdots$$

On the other hand, if V is a $(d-k)$-dimensional linear variety, an easy calculation shows that its Minkowski measures are $W_{k'}{}^V = 0$ if $k' \neq k$ and $W_k{}^V = \left(b_k / \binom{d}{k}\right) \mu_{d-k}^V$. Thus, by taking the expectations in (6-3-1), it follows that

$$E\left(\sum_i W_k{}^{K_i}(dx)\right) = \frac{b_k}{\binom{d}{k}} \nu_k\, dx$$

and, by (6-1-8),

$$E\left(\sum_i W_k{}^{K_i}(dx)\right) = \frac{b_k}{b_{d-k}} W_{d-k}(\Lambda)\, dx$$

In particular, if rB is the ball with radius $r \geqslant 0$, we have

$$E\left(\frac{\sum_i W_k^{K_i}(rB)}{b_d r^d}\right) = \frac{b_k}{b_{d-k}} W_{d-k}(\Lambda)$$

By applying an ergodic theorem (see Miles, 1969), it may be shown that, as $r \uparrow \infty$, the random variables $(1/b_d)\Sigma_i W_k^{K_i}(rB)/(b_d r^d)$ a.s. converge toward $\nu_d = W_0(\Lambda)$ and, in the same way, $\Sigma_i W_k^{K_i}(rB)/(b_d r^d)$ a.s. converge toward $\nu_d E(W_k(\Pi))$, where Π is the Poisson polyhedron considered from the point of view of the number law. Thus we conclude that

$$E(W_k(\Pi)) = \frac{b_k}{b_{d-k}} \frac{W_{d-k}(\Lambda)}{\nu_d} = \frac{b_k}{b_{d-k}} \frac{W_{d-k}(\Lambda)}{W_0(\Lambda)} \tag{6-3-2}$$

In particular, in the isotropic case, the Steiner convex set Λ is the ball with radius a, and

$$E(W_k(\Pi)) = \frac{b_k}{b_{d-k}} a^{k-d} \tag{6-3-2'}$$

The Granulometry with Respect to the Unit Ball

The granulometry of the random set A^c with respect to the unit ball B, as defined in Sections 1-5 and 2-7, is the function $r \to G_B(r)$ such that

$$1 - G_B(r) = P(\{x \in (A^c)_{rB}\}) \qquad (x \in \mathbf{R}^d)$$

where $(A^c)_{rB}$ is the opening of A^c by the ball rB. This function does not depend on the point $x \in \mathbf{R}^d$, because A is stationary. On the other hand, by applying the ergodic theorem, it can be shown that the function G_B and the number law of the Poisson polyhedron Π satisfy the relationship

$$1 - G_B(r) = \frac{E(W_0(\Pi_{rB}))}{E(W_0(\Pi))}$$

By (6-3-2), we already know that the value $E(W_0(\Pi)) = 1/\nu_d$, and we must evaluate $E(W_0(\Pi_{rB}))$ with $\Pi_{rB} = (\Pi \ominus rB) \oplus rB$. But it follows from the conditional invariance that $\Pi \ominus rB$, if nonempty, admits the same law as Π itself. From $P(\Pi \ominus rB \neq \varnothing) = Q(rB) = \exp(-r\psi(B))$, we conclude that

$$1 - G_B(r) = \frac{E(W_0(\Pi \oplus rB))}{E(W_0(\Pi))} \exp(-r\psi(B))$$

By applying the Steiner formula and (6-3-2) and (6-1-5′), it follows that

$$1-G_B(r)$$

$$=\left[\sum_{k=0}^{d}\binom{d}{k}\frac{b_k}{b_{d-k}}W_{d-k}(\Lambda)r^k\right]\exp\left(-2r\left(\frac{d}{b_{d-1}}\right)W_{d-1}(\Lambda)\right) \quad (6\text{-}3\text{-}3)$$

In the isotropic case, this formula reduces to

$$1-G_B(r)=b_d\left(\sum_{k=0}^{d}\binom{d}{k}\frac{b_k}{b_{d-k}}a^k r^k\right)\exp\left(-2\left(\frac{db_d}{b_{d-1}}\right)ar\right) \quad (6\text{-}3\text{-}3')$$

This law is an exponential polynomial the degree of which is the dimension of the euclidean space. If $d=1$, we obtain the gamma law with density $\theta^2 r\exp^{(-\theta r)}$ [$r\geqslant 0$, $\theta=2W_0(\Lambda)$, and $\theta=4a$ in the isotropic case]. Note that the ball rB in $\mathbf{R}^1$ is an interval with length $2r$, so that the linear granulometry admits the density $\beta^2 r\exp^{(-\beta r)}$, $\beta=W_0(\Lambda)$ ($\beta=2a$ in the isotropic case). This law is obviously length weighted, and the corresponding number law is exponential with density $\beta\exp(-\beta r)$.

A Relationship Between the Law of the Volume V and the Law of the Surface Area S

Let $V=V(\Pi)$ be the volume of the Poisson polyhedron Π (considered from the number-law point of view), and $V(r)=V(\Pi\ominus rB)$ the volume of the erosion $\Pi\ominus rB$. We have $Q(rB)=\exp(-\alpha r)$, $\alpha=\psi(B)=(2d/b_{d-1})W_{d-1}(\Lambda)$. Thus it follows from the conditional invariance and (6-2-6) that, for any measurable function φ on $\mathbf{R}_+$,

$$E[\varphi(V(r))]=E(\varphi(V))\exp(-\alpha r)$$

When putting $\varphi(x)=\exp(-\mu x)-1$ ($\mu\geqslant 0$), we see that the Laplace transforms $\Phi(\mu)=E(\exp(-\mu V))$ and $\Phi_r(\mu)=E[\exp(-\mu V(r))]$ are such that

$$\Phi_r(\mu)=1-\exp(-\alpha r)+\Phi(\mu)\exp(-\alpha r)$$

On the other hand, $V(r)$ a.s. admits a derivative $V'(r)$ from the right, and $V'(0)=-S$, because Π is a.s. convex [$S=dW_1(\Pi)$ is the surface area of the polyhedron Π], so that the preceding relationship implies

$$E(S\exp(-\mu V))=\frac{\alpha(1-\Phi(\mu))}{\mu} \quad (6\text{-}3\text{-}4)$$

Then let $h(V)=E(S|V)$ be the *conditional expectation of* S, *given* V, and $F(dV)$ the law of the random volume $V(\Pi)$. The left-hand side in (6-3-4) is the Laplace transform of the measure $h(V)F(dV)$, and the right-hand side is the Laplace transform of the function $\alpha(1-F(V))$. We conclude that the law $F(dV)$ admits a *density* $f(V)$ such that

$$f(V)h(V)=\alpha(1-F(V)) \tag{6-3-5}$$

By an immediate integration, we obtain the expression of the law F in terms of the conditional expectation h, that is,

$$1-F(V)=\exp\left(-\alpha\int_0^V \frac{dx}{h(x)}\right)$$

$$[h(V)=E(S|V), \qquad \alpha=\psi(B)=(2d/b_{d-1})W_{d-1}(\Lambda)].$$

A Relationship Between $\psi(\Pi_0)$ and the Number of $(d-1)$-Faces

We are not able to derive an explicit expression for the distribution of the number N of the $(d-1)$-faces of the Poisson polyhedron $\overline{\Pi}_0$ (considered from the volume-law point of view), but we shall see that the laws of N and $\psi(\Pi_0)$ are closely related. Note that $\psi(\overline{\Pi}_0)$ is the value taken by the functional ψ at $\overline{\Pi}_0 \in \mathcal{K}$ itself. In the isotropic case, $\psi(\Pi_0)$ is proportional to $W_{d-1}(\Pi_0)$ and, in particular, if $d=2$, to the *perimeter* of the Poisson polygon, as follows from (6-1-4).

Let N be the number of $(d-1)$-faces of $\overline{\Pi}_0$, and $P_n=P(\{N=n\})$ (for the volume law). The pure number P_n is not modified by changing ψ to $(1+\lambda)\psi$ $(\lambda \geqslant 0)$. But the Poisson network associated with the functional $(1+\lambda)\psi$ is equivalent to the union of two independent Poisson networks with functionals ψ and $\lambda\psi$, respectively, and the probability for the polyhedron Π_0 [defined by the $(1+\lambda)\psi$ network] to have exactly n faces, all belonging to the ψ-network, is $P_n/(1+\lambda)^n$. This probability may be evaluated in another way, by writing (1) the polyhedron Π_0 defined by the ψ-network admits n faces (probability P_n), and (2) this polyhedron Π_0 does not hit the $\lambda\psi$-network (probability $E_0\,[\exp(-\lambda\psi(\Pi_0))|n]$). It follows that

$$E_0[\exp(-\lambda\psi(\Pi_0))|N]=\frac{1}{(1+\lambda)^N} \tag{6-3-6}$$

(The subscript 0 in E_0 serves to remind us that it is the expectation with respect to the volume law.) In other words, the conditional distribution of

$\psi(\Pi_0)$, given $N=n$, is *gamma* with density

$$\frac{x^{n-1}}{(n-1)!}\exp(-x)$$

Moreover, from the relationship

$$E_0[\exp(-\lambda\psi(\Pi_0))]=\sum_{n=d+1}^{\infty}\frac{P_n}{(1+\lambda)^n} \tag{6-3-6'}$$

we conclude that the distributions of $\psi(\Pi_0)$ and N may be inferred from each other.

If a characteristic X is such that $X(\lambda\Pi_0)=X(\Pi_0)$, $\lambda \geqslant 0$, the same reasoning yields $E_0[\exp(-\lambda\psi(\Pi_0))|N,X]=1/(1+\lambda)^N$, so that, conditionally upon N being fixed, the random variable $\psi(\Pi_0)$ is independent of any nondimensional characteristic $X(\Pi_0)$.

The First Moments of the Volume V

We now suppose that the Poisson network A is *isotropic*, and denote its parameter by a (i.e., the ball aB is the Steiner convex set Λ associated with A). In order to evaluate the first moments of the volume $V=W_0(\Pi_0)$, the surface area $S=dW_1(\Pi_0)$, and the projection area V' of Π_0 into a given hyperplane, we shall use the function g, defined by

$$g(h)=E\left(\int_{\mathbf{R}^d}1_{\Pi_0}(x)1_{\Pi_0}(x+h)\,dx\right)=\int_{\mathbf{R}^d}Q(\{x,x+h\})\,dx$$

Note that $Q(\{x,x+h\})=\exp(-\psi(C))$, where C is the triangle $(o,x,x+h)$ by (6-2-3), and $\psi(C)=a(|x|+|h|+|x+h|)$ by (6-1-4), that is,

$$Q(\{x,x+h\})=\exp(-a(|x|+|h|+|x+h|)) \tag{6-3-7}$$

and $g(h)=\exp(-a|h|)\int\exp(-a|x|)\exp(-a|x+h|)\,dx$. This integral is the convolution product of the exponential function $x\to\exp(-a|x|)$ by itself and may be evaluated by a Fourier transform, for the Fourier transform of $x\to\exp(-a|x|)$ is

$$u\to 2^d a^{-d}\pi^{(d-1)/2}\Gamma\left(\frac{d+1}{2}\right)\left(1+\frac{4\pi^2|u|^2}{a^2}\right)^{(d+1)/2}$$

so that the convolution product $\exp(-a|.|) * \exp(-a|.|)$ admits the transform

$$\left(\frac{2}{a}\right)^{2d} \frac{\pi^{d-1}(\Gamma(d-1)/2)^2}{(1+4\pi^2|u|^2/a^2)^{d+1}}$$

The converse transform yields an explicit expression of this convolution product, in which the Bessel function $K_{1+d/2}$ occurs. Finally, by putting $r=|h|$ and writing $g(r)$ instead of $g(h)$, we find after some calculation that

$$g(r) = \frac{2^{d/2}\pi^{d/2-1}}{a^d d!}\left(\frac{\Gamma(d+1)}{2}\right)^2 \exp^{(-ar)}(ar)^{1+d/2} K_{1+d/2}(ar) \qquad (6\text{-}3\text{-}8)$$

It remains to apply the relationships $g(0)=E_0(V)$ and $\int_{\mathbf{R}^d} g(h)\,dh = E_0(V^2)$. In order to evaluate the integral

$$I = \int_{\mathbf{R}^d} g(h)\,dh = \int\int \exp(-a|x| - a|h| - a|x+h|)\,dx\,dh$$

we note that I is the value at O of the convolution product $f*f*f$ [where f is the function $x \to f(x) = \exp(-a|x|)$] and may be calculated by a Fourier transform. Finally, we determine that the first two moments of V (with respect to the volume law) are

$$E_0(V) = 2^{-d} b_d d!\,a^{-d}$$

$$E_0(V^2) = 2^{2d}\pi^{d-3/2}\frac{\Gamma(d+3/2)[\Gamma((d+1)/2)]^3}{\Gamma(3(d+1)/2)} a^{-2d} \qquad (6\text{-}3\text{-}9)$$

The subscript 0 in E_0 reminds us that E_0 is the expectation with respect to the volume law. By the ergodic theorem, E_0 is such that $E_0(V^k) = E(V^{k+1})/E(V)$, if E is the expectation with respect to the number law. By (6-3-2′), we already know that $E(V) = E(W_0) = 1/b_d a^d$. Thus formulae (6-3-9) yield the *first three moments* of the number distribution of V, that is, explicitly,

$$E(V) = \frac{1}{b_d a^d}$$

$$E(V^2) = 2^{-d} d!\,a^{-2d}$$

$$E(V^3) = 2^{2d}\pi^{(d-3)/2}\frac{\Gamma(1+d/2)\Gamma(d+3/2)[\Gamma((d+1)/2)]^3}{\Gamma(3(d+1)/2)} a^{-3d}$$

In the euclidean space $\mathbf{R}^2$, we have

$$E(V)=\frac{1}{\pi a^2},\qquad E(V^2)=\frac{1}{2a^4},\qquad E(V^3)=\frac{4\pi}{7a^6}$$

and in $\mathbf{R}^3$

$$E(V)=\frac{3}{4\pi a^3},\qquad E(V^2)=\frac{3}{4a^6},\qquad E(V^3)=\frac{21\pi}{8a^9}$$

The First Moments of the Projection Area V'

Now let $V'=V'(\Pi)$ be the $(d-1)$-volume of the projection area of the Poisson polyhedron Π on a given hyperplane in $\mathbf{R}^d$. (Necessarily, we adopt the number-law point of view.) We shall evaluate the first two moments of V'.

From the point of view of the number law, the function g in (6-3-8) is

$$g(r)=\frac{E(VV(r))}{E(V)}$$

where $V(r)=V(\Pi\cap\Pi_r)$ is the volume of the intersection $\Pi\cap\Pi_r$ of Π and its translate Π_r by a translation with modulus $r\geqslant 0$ and direction $u_0\in S_0$. The polyhedron Π is convex, so that $\Pi\cap\Pi_r$ is also the erosion of Π by the set $\{\lambda u_0, 0\leqslant\lambda\leqslant r\}$. Now let ρ be a real $\geqslant 0$, and $\Pi_{\rho+r}$ the translate of Π by $(\rho+r)u_0$. By expression (6-2-6) for the conditional invariance, we have

$$E(V(r)V(r+\rho))=E(VV(\rho))\exp(-2ar)=E(V)g(\rho)\exp^{(-2ar)}\qquad(6\text{-}3\text{-}10)$$

On the other hand, $V(r)$ admits at $r=0$ a derivative from the right, which is $V'(0)=-V'$. By differentiating (6-3-10) once with respect to r and once with respect to ρ, and taking $r=\rho=0$, we obtain

$$E\big((V')^2\big)+E(VV''(0))=-2aE(VV'(0))\qquad(6\text{-}3\text{-}11)$$

But, by using a limited expansion of the Bessel function $K_{1+d/2}$, it follows from (6-3-8) that

$$E(VV(r))=E(V^2)\left(1-ar+\frac{d-1}{2d}a^2r^2-\cdots\right)$$

$(d>1)$. We have already evaluated $E(V^2)=d!2^{-d}a^{-2d}$. Thus it follows that

$$E(VV'(0))=-aE(V^2)=-2^{-d}a^{1-2d}d!$$

$$E(VV''(0))=\frac{d-1}{d}a^2E(V^2)=(d-1)((d-1)!)2^{-d}a^{2-2d}$$

By substituting in (6-3-11), we obtain

$$E(V'^2)=\frac{d+1}{d}a^2E(V^2)=(d+1)(d-1)!2^{-d}a^{2-2d} \qquad (6\text{-}3\text{-}12)$$

Also, $E(V(r))=E(V)\exp(-2ar)$, by the conditional invariance and (6-2-6), and so

$$E(V')=2aE(V)=\frac{2}{b_d}a^{1-d}$$

Thus we have evaluated the first two moments of the projection area V' (considered from the number-law point of view). If $d=2$,

$$E(V')=\frac{2}{\pi a}, \qquad E(V'^2)=\frac{3}{4a^2}, \qquad E(VV')=\frac{1}{2a^3}$$

and, if $d=3$,

$$E(V')=\frac{3}{2\pi a^2}, \qquad E(V'^2)=\frac{1}{a^4}, \qquad E(VV')=\frac{3}{4a^5}$$

The First Moments of the Surface Area S

Now let S_d be the surface area of the isotropic Poisson polyhedron in $\mathbf{R}^d$, that is, the $(d-1)$-volume of its boundary. From (4-1-6), S_d is related to the projection area $V'(u)$ into the hyperplane $u^{\perp}$, $u\in S_0$, by the formula

$$S_d=\frac{db_d}{b_{d-1}}\int V'(u)\varpi(du) \qquad (6\text{-}3\text{-}13)$$

Thus the expectation is from the number-law point of view,

$$E(S_d)=\frac{db_d}{b_{d-1}}E(V')=\frac{2d}{b_{d-1}}a^{1-d}$$

and, from the volume-law point of view,

$$E_0(S_d)=\frac{db_d}{b_{d-1}}E_0(V')=\frac{db_d}{b_{d-1}}\frac{E(VV')}{E(V)}$$

$$=d2^{-d}d!\left(\frac{(b_d)^2}{b_{d-1}}\right)a^{1-d}$$

But $E_0(S_d)$ and $E(S_d^2)$ are related, as we will see from the following heuristic proof. Conditionally upon a hyperplane of the network A containing the origin O, the two polyhedra Π_1 and Π_2 whose boundaries enclose O admit the same probability law, which is the number law weighted by the random variable S_d. Thus, if S_1 and S_2 are the surface areas of Π_1 and Π_2, we have $E(S_1)=E(S_2)=E(S_d^2)/E(S_d)$. On the other hand, if the hyperplane enclosing O is removed, the polyhedron $\Pi_0 = \Pi_1 \cup \Pi_2$ remains, the law of which is now the volume law. The surface area S_0 of Π_0 is $S_0 = S_1 + S_2 - 2F$, F denoting the $(d-1)$-volume of $\Pi_1 \cap \Pi_2$. But F is an arbitrary polyhedron for the network induced in $\mathbf{R}^{d-1}$ (considered from the volume-law point of view), and thus $E(F) = E_0(V_{d-1})$. Finally,

$$E_0(S_d)=E(S_0)=E(S_1)+E(S_2)-2E(F)=\frac{2E(S_d^2)}{E(S_d)}-2E_0(V_{d-1})$$

Taking the preceding results into account, we conclude that

$$E(S_d)=\frac{2d}{b_{d-1}}a^{1-d}$$

$$E(S_d^2)=2^{2-d}d!\left(1+\left(\frac{db_d}{2b_{d-1}}\right)^2\right)a^{2-2d} \tag{6-3-14}$$

In particular, if $d=3$, the surface area of the Poisson polyhedron admits the first two moments

$$E(S)=\frac{6}{\pi a^2}, \qquad E(S^2)=\frac{15}{a^4}$$

and, if $d=2$, the first moments of the perimeter of the Poisson polygon are

$$E(S)=\frac{2}{a}, \qquad E(S^2)=\frac{\pi^2/2+2}{a^2}$$

The Isotropic Poisson Polygons

In the case $d=2$, it is possible to derive the distribution law of the projection area, that is, the *breadth* of an isotropic Poisson polygon (considered from the number-law point of view). We recall that the breadth is the distance between two supporting parallel straight lines. This law is absolutely continuous, its density is

$$f(x)=\frac{4a}{\pi}(2ax)K_{-1}(2ax) \qquad (x \geqslant 0) \tag{6-3-15}$$

where K_{-1} is the (second-order modified) Bessel function with index -1, and its order-n moment ($n \geqslant 0$, integer or not) is

$$m_n=\frac{\Gamma(n+2)}{\sqrt{\pi}}\,\frac{\Gamma((1+n)/2)}{\Gamma(1+n/2)}(2a)^{-n}$$

Proof. First we shall give an explicit procedure that will enable us to construct a (number) Poisson polygon Π. Let A be an isotropic Poisson line network, and suppose that O is a vertex of A (i.e., the intersection of two lines in A). If the euclidean plane has two coordinate axes Ox, Oy, then one (and a.s. only one) of the four polygons containing O as a vertex is included in the half plane $x \geqslant 0$. By the ergodic theorem, it is possible to show that this polygon is equivalent to the number-law polygon Π. From a heuristic point of view, the reason for this is that the Poisson network admits, on the average, the same number of vertices and polygons. Let β_0 and $\beta_1(-\pi/2 \leqslant \beta_0 \leqslant \beta_1 \leqslant \pi/2)$ be the polar angles of the two edges of Π enclosing O. It is easy to see that the distribution of these two random variables admit the density

$$g(\beta_0,\beta_1)=\frac{1}{\pi}\sin(\beta_1-\beta_0) \qquad \left(-\frac{\pi}{2} \leqslant \beta_0 \leqslant \beta_1 \leqslant \frac{\pi}{2}\right)$$

Let now M be the (a.s. unique) vertex of Π the coordinates (x_0,y_0) of which are such that Π is contained in the half plane $x \leqslant x_0$. Clearly, x_0 is the breadth of Π (with respect to the direction of the y-axis). In order to obtain the distribution of x_0, consider the event $\{x_0 \geqslant r\}$, equivalent to {the line $x=r$ hits Π}. First, suppose that β_0 and β_1 are fixed. Then the event $\{x_0 \geqslant r\}$ occurs if one of the two following (disjoint) events is realized:

1. The point $M_1=(r,r\tan\beta_1)$ belongs to Π [probability $\exp(-2ar/\cos\beta_1)$].

2. There exists on the line $x=r$ a point $M=(r,y)$, $r\tan\beta_0 \leqslant y < r\tan\beta_1$, $M \in \Pi$, such that $(r,y+\epsilon) \not\in \Pi$ for any $\epsilon>0$. The corresponding probability is

$$ar\int_{\beta_0}^{\beta_1}\exp\left(\frac{-2ar}{\cos\alpha}\right)(1+\sin\alpha)\,\frac{d\alpha}{\cos^2\alpha}$$

In fact, $P((r,y)\in\Pi$ and $(r,y+\epsilon)\not\in\Pi)=P((r,y)\in\Pi)-P((r,y)\in\Pi$ and $(r,y+\epsilon)\in\Pi)=\exp(-2ar)-\exp[-a(\sqrt{r^2+y^2}+\sqrt{r^2+(y+\epsilon)^2}+\epsilon)]$, and the desired expression follows after an easy calculation. Hence we have

$$P(x_0\geqslant r)=\frac{1}{\pi}\int_{-\pi/2}^{\pi/2}\exp\left(\frac{-2ar}{\cos\beta_1}\right)d\beta_1\int_{-\pi/2}^{\pi/2}\sin(\beta_1-\beta_0)\,d\beta_0$$

$$+\frac{1}{\pi}\int_{-\pi/2}^{\pi/2}d\beta_1\int_{-\pi/2}^{\beta_1}\sin(\beta_1-\beta_0)\,d\beta_1\int_{\beta_0}^{\beta_1}\exp\left(\frac{-2ar}{\cos\alpha}\right)(1+\sin\alpha)\frac{d\alpha}{\cos^2\alpha}$$

After an elementary calculation, we obtain

$$P(x_0\geqslant r)=\frac{4ar}{\pi}\int_0^{\pi/2}\exp\left(\frac{-2ar}{\cos\alpha}\right)\frac{d\alpha}{\cos\alpha}$$

$$+\frac{2}{\pi}\int_0^{\pi/2}\exp\left(\frac{-2ar}{\cos\alpha}\right)d\alpha$$

By differentiating with respect to r, we obtain the density $f(x)$ of the breadth as

$$f(x)=\frac{8a^2x}{\pi}\int_0^{\pi/2}\exp\left(\frac{-ax}{\cos\alpha}\right)\frac{d\alpha}{\cos^2\alpha}$$

and this expression is identical to (6-3-15).

Let us now examine *the distribution of the number N of the sides of the Poisson polygon* Π_0 (from the point of view of the volume law), and denote by S the *perimeter* of Π_0. From (6-1-4) we have $\psi(\Pi_0)=aS$, and it follows from (6-3-6′) that $aE_0(S)=\Sigma nP_n$. But we have calculated $E_0(S)=\pi^2/2a$, and thus

$$E_0(N)=\frac{\pi^2}{2}$$

Now, to work with the number law, we put $p_n=P(N(\Pi)=n)$ and $Z=V/S^2$. Since Z is a nondimensional characteristic, $Z(\Pi_0)$ and $S(\Pi_0)$ are independent conditionally upon N being fixed. Let us denote by $f_n(z)$ and $g_n(z,S)$ the densities of $Z(\Pi_0)$ and $(Z(\Pi),S(\Pi))$ conditionally upon

$N=n$. From the relationship $F_0(d\Pi)=[(V(\Pi)/E(V(\Pi))]\times F(d\Pi)$, it follows that

$$r_n = P_n \frac{E(V)}{E(V|n)} = P_n E(V) E_0\left(\frac{1}{V}\middle| n\right)$$

$$g_n(z,S) = E(V|n)\frac{1}{zS^2} f_n(z) a^n S^{n-1} \frac{\exp(-aS)}{n!}$$

Thus, conditionally upon $N=n$, z and S are independent for the number law, and the law of S is gamma with the density $(a(aS)^{n-3}/(n-3)!)\exp(-aS)$. Thus

$$E(S(\Pi)) = \frac{1}{a}\sum (n-2)p_n$$

$$E\left[(S(\Pi))^2\right] = \frac{1}{a^2}\sum (n-1)(n-2)p_n$$

But we know that $E(S)=2/a$ and $E(S^2)=(\pi^2/2+2)/a^2$. Hence the first two moments of the number N of the sides of the Poisson polygon Π are (for the number law)

$$E(N)=4, \qquad E(N^2)=\frac{\pi^2}{2}+12$$

CHAPTER 7

The Granulometries

From a mathematical point of view, the usual concepts of size distribution are generally ill defined and do not always correspond to precise geometrical properties. In this chapter, we give the more general notion of a granulometric mapping or, simply, a granulometry, that is, a set-valued mapping which satisfies axioms chosen in such a way that they exactly generalize the physical meaning of a size distribution. As a particular case, we will use the granulometry $A \to A_{\lambda K}$ of A with respect to a compact convex set K, as defined in Chapters 1 and 2. The definition and the purely algebraic properties of the granulometries do not involve particular difficulties per se; however, we can apply this notion to a RACS and to its complementary set only if measurability has been proved, and from this standpoint, the most tractable granulometries turn out to be u.s.c. on $\mathcal{F}$ and l.s.c. on $\mathcal{G}$. On the other hand, in practical applications, a closed set can be investigated only from its local properties, a fact which legitimates Definitions 7-1-1 (compact openings) and 7-2-3 (u.s.c. and compact granulometries). But these compact mappings are not so easily characterized, and for this reason we examine first the simplest case, which is, practically, the only useful one, that is, openings compatible with translations, and euclidean granulometries (Sections 7-1 and 7-2). The probabilistic version is given in Section 3. The last two sections examine the topological properties of openings and granulometries on a general LCS space, and may be omitted by the reader interested only in the euclidean case.

7-1. ALGEBRAIC OPENINGS AND CLOSINGS

Let E be a LCS space, $\mathcal{P} = \mathcal{P}(E)$ the class of all subsets of E, $\mathcal{Q} \subset \mathcal{P}$, and ψ a mapping from $\mathcal{Q}$ into $\mathcal{P}$. The mapping ψ is said to be *increasing* if A, $B \in \mathcal{Q}$ and $A \subset B$ imply $\psi(A) \subset \psi(B)$; ψ is said to be *extensive* (resp., antiextensive) if $A \subset \psi(A)$ [resp. $\psi(A) \subset A$] for any $A \in \mathcal{Q}$; ψ is said to be *idempotent* if the range $\psi(\mathcal{Q})$ is enclosed in $\mathcal{Q}$ and $\psi = \psi \circ \psi$. We say that ψ

is a *closing* (resp. an *opening*) if it is increasing, extensive (resp. antiextensive), and idempotent. The notion of granulometry will be linked to these classical algebraic definitions.

Now let $\mathcal{A}^* = \{A\colon A^c \in \mathcal{A}\}$ be the class of the complementary sets of the sets $A \in \mathcal{A}$. The *dual mapping* $\psi^*\colon \mathcal{A}^* \to \mathcal{P}$ is defined by $\psi^*(A) = \complement\psi(A^c)$, $A \in \mathcal{A}^*$, that is, $\psi^* = \complement\circ\psi\circ\complement$. Then ψ^* is increasing if and only if ψ is increasing. It is a closing if and only if ψ is an opening, and conversely. Clearly, $\psi^{**} = \psi$.

Extensions of an Increasing Mapping

If $\mathcal{A} \subset \mathcal{P}$, any increasing mapping $\psi\colon \mathcal{A} \to \mathcal{P}$ admits a *smallest extension* $\underset{\sim}{\psi}$ and a largest extension $\tilde{\psi}$ on $\mathcal{P}$, that is,

$$\left.\begin{aligned} \underset{\sim}{\psi}(B) &= \cup\{\psi(A);\quad A \in \mathcal{A}, A \subset B\} \\ \tilde{\psi}(B) &= \cap\{\psi(A);\quad A \in \mathcal{A}, A \supset B\} \end{aligned}\right\} \quad (B \in \mathcal{P}) \qquad \text{(7-1-1)}$$

The union (resp. the intersection) of the empty family is $\varnothing$ (resp. E), so that $\underset{\sim}{\psi}(B) = \varnothing$ if there exists no $A \in \mathcal{A}$ such that $A \subset B$, and $\tilde{\psi}(B) = E$ if there exists no $A \in \mathcal{A}$ such that $A \supset B$. Obviously, the dual mappings satisfy

$$(\underset{\sim}{\psi})^* = (\tilde{\psi^*}), \qquad (\tilde{\psi})^* = (\underset{\sim}{\psi^*})$$

If ψ is an *opening* on $\mathcal{A}$, its smallest extension $\underset{\sim}{\psi}$ on $\mathcal{P}$ is also an opening.

Proof. Clearly, $\underset{\sim}{\psi}$ is increasing and antiextensive so that we must prove that $\underset{\sim}{\psi}$ is idempotent. For any $B \in \mathcal{P}$, we have

$$\underset{\sim}{\psi}\underset{\sim}{\psi}(B) = \underset{\sim}{\psi}(\cup\{\psi(A); A \in \mathcal{A}, A \subset B\}),$$

and thus

$$\underset{\sim}{\psi}\underset{\sim}{\psi}(B) \supset \cup\{\underset{\sim}{\psi}\psi(A), A \in \mathcal{A}, A \subset B\} = \cup\{\psi(A); A \in \mathcal{A}, A \subset B\} = \underset{\sim}{\psi}(B),$$

because $\underset{\sim}{\psi}$ is increasing and is an extension of the idempotent mapping ψ. But $\underset{\sim}{\psi}$ is an antiextensive, and thus $\underset{\sim}{\psi}\underset{\sim}{\psi} = (B) = \underset{\sim}{\psi}(B)$.

Let us denote by $\mathcal{B}$ the class of the sets $B \in \mathcal{A}$ *invariant* under the opening ψ, that is, $\mathcal{B} = \{B\colon B \in \mathcal{A}, \psi(B) = B\}$. Then the restriction of ψ to $\mathcal{B}$ is the identity function on $\mathcal{B}$. Conversely, ψ *is the smallest extension on* $\mathcal{A}$ *of the identity function on* $\mathcal{B}$, for, $A \in \mathcal{A}$, $B \in \mathcal{B}$, and $B \subset A$ imply $B = \psi(B) \subset \psi(A)$, and so $\psi(A) \supset \cup\{B; B \in \mathcal{B}, B \subset A\}$. On the other hand,

$\psi(A) \in \mathcal{B}$, because ψ is idempotent, and $\psi(A) \subset A$, so that $\psi(A) = \cup\{B; B \in \mathcal{B}, B \subset A\}$.

It follows that $\underset{\sim}{\psi}$ is the smallest extension on $\mathcal{P}$ of the identity function on $\mathcal{B}$, that is, $\underset{\sim}{\psi}(\tilde{D}) = \cup\{B, B \in \mathcal{B}, B \subset D\}$ for any $D \in \mathcal{P}$. Then the class $\underset{\sim}{\mathcal{B}}$ of the sets invariant under $\underset{\sim}{\psi}$ is the class closed under infinite unions generated by $\mathcal{B}$ (this implies $\underset{\sim}{\varnothing} \in \underset{\sim}{\mathcal{B}}$).

By duality, we obtain analogous results if ψ is a closing instead of an opening, and we may state:

Proposition 7-1-1. *Let ψ be an opening (respectively, a closing) on $\mathcal{P}$, and $\mathcal{B} \subset \mathcal{P}$ the class of the sets invariant under ψ. Then $\mathcal{B}$ is closed under infinite unions (resp. intersections), and $\varnothing \in \mathcal{B}$ (resp. $E \in \mathcal{B}$). Further ψ is the smallest (resp. the largest) extension on $\mathcal{P}$ of the identity on $\mathcal{B}$. Conversely, if $\mathcal{B}_0 \subset \mathcal{P}$ is a class of subsets of E, the smallest (resp. the largest) extension on $\mathcal{P}$ of the identity on $\mathcal{B}_0$ is an opening (resp. a closing) ψ, and the class $\mathcal{B}$ of the sets invariant under ψ is the class closed under $\cup$ (resp., under $\cap$) generated by $\mathcal{B}_0$. In particular, $\varnothing \in \mathcal{B}$ (resp. $E \in \mathcal{B}$).*

If ψ is an opening (or a closing) on $\mathcal{P}$, its restriction to a given class $\mathcal{Q} \subset \mathcal{P}$ is still an opening (or a closing) on $\mathcal{Q}$ only if $\mathcal{Q}$ is closed under ψ, that is, $A \in \mathcal{Q}$ implies $\psi(A) \in \mathcal{Q}$. Thus we may ask under what conditions the usual spaces $\mathcal{F}$, $\mathcal{G}$, and $\mathcal{K}$ are closed under ψ.

Proposition 7-1-2. *Let ψ be an opening on $\mathcal{P}$, and $\mathcal{B}$ the class of the ψ-invariant sets. Then:*

a. $\mathcal{F}$ is closed under ψ if and only if $\mathcal{B}$ is closed under the topological closure, that is, $B \in \mathcal{B}$ implies $\overline{B} \in \mathcal{B}$.

b. $\mathcal{K}$ is closed under ψ if and only if $\overline{B} \in \mathcal{B}$ for any relatively compact ψ-invariant set $B \in \mathcal{B}$.

c. $\mathcal{G}$ is closed under ψ if and only if any $B \in \mathcal{B}$ admits a fundamental system of open neighborhoods belonging to $\mathcal{B}$.

Proof. For any $B \in \mathcal{B}$, we have $B = \psi(B) \subset \overline{B}$, and thus $B \subset \psi(\overline{B})$. If $\mathcal{F}$ is closed under ψ, it follows that $\overline{B} \subset \psi(\overline{B})$, since $\psi(\overline{B}) \in \mathcal{F}$. Hence $\overline{B} = \psi(\overline{B}) \in \mathcal{B}$. Conversely, for any $F \in \mathcal{F}$, we have $\psi(F) \subset \overline{\psi(F)} \subset F$. If $\mathcal{B}$ is closed under the topological closure, $\overline{\psi(F)} \in \mathcal{B}$ [since $\psi(F) \in \mathcal{B}$], and thus $\overline{\psi(F)} \subset F$ implies $\overline{\psi(F)} \subset \psi(F)$, that is, $\psi(F) = \overline{\psi(F)}$. Hence statement *a* is proved, and *b* follows.

Let $B \in \mathcal{B}$ and $G \in \mathcal{G}$ be such that $B \subset G$, and thus $B \subset \psi(G) \subset G$. If $\mathcal{G}$ is closed under ψ, $\psi(G) \in \mathcal{G}$ is an open neighborhood of B contained in G. Thus B admits a fundamental system of open neighborhoods, contained in

$\mathcal{B}$. Conversely, if any $B \in \mathcal{B}$ admits a fundamental system of neighborhoods in $\mathcal{B} \cap \mathcal{G}$, for any $G \in \mathcal{G}$, there exists $B \in \mathcal{B} \cap \mathcal{G}$ such that $\psi(G) \subset B \subset G$. It follows that $\psi(G) = B$, and $\psi(G) \in \mathcal{G}$.

Corollary. *Let ψ' be a closing on $\mathcal{P}$, and $\mathcal{B}'$ the class of the sets invariant under ψ'. Then:*

a. $\mathcal{G}$ is closed under ψ' if and only if $\mathcal{B}'$ is closed under the topological opening $B \to \mathring{B}$.

b. $\mathcal{K}$ is closed under ψ' if and only if, for any $B \in \mathcal{B}$, and $K \in \mathcal{K}$ such that $K \subset B$, there exists $B' \in \mathcal{K} \cap \mathcal{B}'$ such that $K \subset B' \subset B$.

c. $\mathcal{F}$ is closed under ψ' if and only if, for any $B \in \mathcal{B}'$ and $F \in \mathcal{F}$ such that $B \supset F$, there exists $B' \in \mathcal{F} \cap \mathcal{B}'$ such that $F \subset B' \subset B$.

Definition 7-1-1. *We say that an opening ψ on $\mathcal{P}$ is compact if the following conditions are fulfilled:*

a. $\mathcal{F}$, $\mathcal{G}$, and $\mathcal{K}$ are closed under ψ.

b. The opening ψ is u.s.c. on $\mathcal{F}$, u.s.c. on $\mathcal{K}$, and l.s.c. on $\mathcal{G}$.

c. The opening ψ is the smallest extension of its restriction to $\mathcal{K}$.

By this definition, $\psi(A) = \cup\{\psi(K), K \subset A, K \in \mathcal{K}\}$ for any $A \in \mathcal{P}$. A complete characterization of compact openings will be given by Proposition 7-4-10. We shall first examine the simplest (and most useful) particular case of the openings compatible with translations on the euclidean space $\mathbf{R}^d$.

τ-Openings and τ-Closings in $\mathbf{R}^d$

If $\mathcal{Q}$ is a class of sets in $\mathbf{R}^d$ closed under translation, a mapping ψ from $\mathcal{Q}$ into $\mathcal{P}$ is said to be compatible with translation or, for brevity, to be a *τ-mapping* if $\psi(A \oplus \{h\}) = \psi(A) \oplus \{h\}$ for any $A \in \mathcal{Q}$ and $h \in \mathbf{R}^d$. In the same way, a closing or an opening on $\mathbf{R}^d$ is called a *τ-closing* or a *τ-opening* if it is compatible with translation. Clearly, a closing or an opening is a τ-mapping if and only if the class $\mathcal{B}$ of its invariant sets is closed under translation.

Let ψ be a τ-opening on $\mathbf{R}^d$, and $\mathcal{B}$ the class of the ψ-invariant sets closed under translation. A subclass $\mathcal{B}_0 \subset \mathcal{B}$ is said to be a *basis of* $\mathcal{B}$ if the class closed under translations and infinite unions generated by $\mathcal{B}_0$ is identical to $\mathcal{B}$. Then, for any

$$A \in \mathcal{P},\ \psi(A) = \cup\{B \oplus \{x\}; x \in \mathbf{R}^d, B \in \mathcal{B}_0, B \oplus \{x\} \subset A\}.$$

But the opening A_B of A by B is

$$A_B=(A\ominus\check{B})\oplus B=\cup\{B\oplus\{x\},x\in\mathbf{R}^d,B\oplus\{x\}\subset A\}.$$

It follows that

$$\psi(A)=\cup\{A_B,B\in\mathcal{B}_0\}$$

In the same way, the closing ψ^*, associated with ψ by duality, admits the representation $\psi^*(A)=\cap\{A^B,B\in\mathcal{B}_0\}$, where $A^B=(A\oplus\check{B})\ominus B$ is the closing of A by B. In other words:

Proposition 7-1-3. *A mapping* ψ: $\mathcal{P}(\mathbf{R}^d)\to\mathcal{P}(\mathbf{R}^d)$ *is a* τ*-opening if and only if it admits the representation* $\psi(A)=\cup\{A_B;B\in\mathcal{B}_0\}$ *for a class* $\mathcal{B}_0\subset\mathcal{P}$. *Then* $\mathcal{B}_0$ *is a basis of the class of* ψ*-invariant sets, and the dual* τ*-closing* ψ^* *admits the representation* $\psi^*(A)=\cap\{A^B,B\in\mathcal{B}_0\}$.

In particular, if $\mathcal{B}_0=\{B\}$ contains only one set B, ψ: $A\to A_B$ is the opening by B, and then ψ is compact, in the sense of Definition 7-1-1, if and only if B is compact. More generally, we will now examine under what conditions a given τ-opening is compact.

Proposition 7-1-4. *Let* ψ *be a* τ*-opening. Then* $\mathcal{F}$ *is closed under* ψ *and* ψ *is u.s.c. on* $\mathcal{F}$ *if and only if there exists a class* $\mathcal{B}_0\subset\mathcal{F}$ *closed in* $\mathcal{F}$, *contained in* $\mathcal{F}_{\{o\}}$ *(i.e.,* $B\in\mathcal{B}_0$ *implies* $O\in B$*), and such that:*

a. $x\in B$ *and* $B\in\mathcal{B}_0$ *imply* $B\oplus\{-x\}\in\mathcal{B}_0$.
b. $\psi(F)=\cup\{F_B,B\in\mathcal{B}_0\}$ *for any* $F\in\mathcal{F}$.

Proof. If $\mathcal{F}$ is closed under ψ, the class $\mathcal{B}$ of the ψ-invariant sets is closed under the topological closure (Proposition 7-1-2). Put $\mathcal{B}'=\mathcal{B}\cap\mathcal{F}$, so that $\psi(F)=\cup\{B',B'\in\mathcal{B}',B'\subset F\}$ for any $F\in\mathcal{F}$. If ψ is u.s.c., $\mathcal{B}'$ is closed in $\mathcal{F}$, for $\{B_n'\}\subset\mathcal{B}'$ and $B=\lim B_n'$ in $\mathcal{F}$ imply $B=\lim\psi(B_n')\subset\psi(B)$, and thus $B=\psi(B)\in\mathcal{B}'$. The desired conditions are satisfied by the class $\mathcal{B}_0=\mathcal{B}'\cap\mathcal{F}_{\{o\}}$. Conversely, let $\mathcal{B}_0$ be a class satisfying these conditions, and let $F\in\mathcal{F}$, and we will prove $\psi(F)\in\mathcal{F}$. If a sequence $\{x_n\}\subset\psi(F)$ converges toward $x\in\mathbf{R}^d$, for any integer n there exist $B_n\in\mathcal{B}_0$ and $y_n\in\mathbf{R}^d$ such that $x_n\in B_n\oplus\{y_n\}\subset F$. But statement a implies $B_n'=B_n\oplus\{y_n-x_n\}\in\mathcal{B}_0$ and thus $O\in B_n'\subset F\oplus\{-x_n\}$. If B is an accumulation point of the sequence $\{B_n'\}$ in $\mathcal{F}$, $B\in\mathcal{B}_0$, for $\mathcal{B}_0$ is closed in $\mathcal{F}$, and $O\in B\subset F\oplus\{-x\}$, that is, $x\in F_B\subset\psi(F)$. Hence $\psi(F)$ is closed. Now let $\{F_n\}\subset\mathcal{F}$ be a sequence such that $\lim F_n=F$ in $\mathcal{F}$, and $\{F_{n_k}\}$ be a subsequence. If a sequence $n_k\to x_{n_k}\in\psi(F_{n_k})$ converges toward x in $\mathbf{R}^d$, for any k there exists $B_{n_k} \in \mathcal{B}_0$ such that $O \in B_{n_k} \subset F_{n_k} \oplus \{-x_{n_k}\}$. If $B \in \mathcal{B}_0'$ is an accumulation point of $\{B_{n_k}\}$, it follows that $O \in B \in F \oplus \{-x\}$. Hence $x \in F_B \subset \psi(F)$, and ψ is u.s.c.

If a τ-opening ψ is u.s.c. on $\mathcal{F}$, its restriction to $\mathcal{K}$ is u.s.c. on $\mathcal{K}$, for $\psi(K) \subset K$, $K \in \mathcal{K}$, implies that $\mathcal{K}$ is closed under ψ, and ψ is u.s.c. for the myope topology on $\mathcal{K}$, because $K_n \downarrow K$ in $\mathcal{K}$ implies that the sets $\psi(K_n) \subset K_n$ are enclosed in a fixed compact set, so that $\psi(K) = \lim \psi(K_n)$ also for the myope topology.

Now let ψ be a τ-opening u.s.c. on $\mathcal{F}$. We may ask under what conditions ψ is the smallest extension on $\mathcal{F}$ of its restriction to $\mathcal{K}$. First we note that the class $\mathcal{B}_0$ closed in $\mathcal{F}$ associated with ψ by Proposition 7-1-4 admits *minimal elements*. More precisely, for any $B \in \mathcal{B}_0$, there exists an $M \in \mathcal{B}_0$ such that $O \in M \subset B$ and $F \notin \mathcal{B}_0$ for any $F \in \mathcal{F}$ containing O and strictly contained in M. (This is an immediate consequence of the Zorn theorem.) The class $\mathcal{M}_0$ of the minimal elements in $\mathcal{B}_0$ is obviously a basis of the class of the ψ-invariant closed sets, so that $\psi(F) = \cup\{F_M, M \in \mathcal{M}_0\}$, for any $F \in \mathcal{F}$, or $\psi(F) = \cup\{M, M \in \mathcal{M}, M \subset F\}$, if $\mathcal{M}$ denotes the class closed under translation of the sets of $\mathcal{M}_0$. It follows that $\psi(F) = \cup\{\psi(K), K \in \mathcal{K}, K \subset F\}$ for any $F \in \mathcal{F}$ if and only if $\mathcal{M} \subset \mathcal{K}$ (or, which is the same, the minimal elements $M \in \mathcal{M}_0$ are compact).

Let ψ' be the smallest extension on $\mathcal{P}$ of the restriction of ψ to $\mathcal{K}$, so that $\psi' = \psi$ on $\mathcal{F}$ if $\mathcal{M} \subset \mathcal{K}$, and ψ' is u.s.c. on $\mathcal{F}$. Then the restriction of ψ' to $\mathcal{G}$ is l.s.c, for $G \in \mathcal{G}$ and $M \in \mathcal{M}$ imply $G_M \in \mathcal{G}$ and

$$\psi'(G) = \cup\{G_M, M \in \mathcal{M}\} \in \mathcal{G}.$$

Moreover, $G_n \uparrow G$ in $\mathcal{G}$ implies

$$\cup \psi'(G_n) = \cup\{(G_n \ominus \check{M}) \oplus M, n > 0, M \in \mathcal{M}\} = \cup\{G_M, M \in \mathcal{M}\} = \psi'(G),$$

for $G \to G_M$ is l.s.c. on $\mathcal{G}$ for any compact set M. Thus ψ' is l.s.c. on $\mathcal{G}$. In other words:

Proposition 7-1-5. *A τ-opening ψ is compact (in the sense of Definition* 7-1-1) *if and only if there exists a class $\mathcal{B}_0$ closed in $\mathcal{F}$, contained in $\mathcal{F}_{\{o\}}$, and such that*:

a. $B \in \mathcal{B}_0$ and $x \in B$ imply $B \oplus \{-x\} \in \mathcal{B}_0$.
b. The class $\mathcal{M}_0$ of the minimal elements in $\mathcal{B}_0$ is $\subset \mathcal{K}$.
c. $\psi(A) = \cup\{A_M, M \in \mathcal{M}_0\}$ for any $A \in \mathcal{P}$.

Example. Let K be a compact set, and $\mathcal{B}_0$ the class of its translates enclosing O. The conditions of Proposition 7-1-5 are satisfied by $\mathcal{B}_0$, and

$\mathcal{M}_0 = \mathcal{B}_0$, so that the opening $A \to A_K$ is compact.

More generally, let $\mathcal{V} \subset \mathcal{K}$ be *compact in* $\mathcal{K}$, and $\mathcal{B}_0$ be the class of the translates K of the elements of $\mathcal{V}$ such that $O \in K$. Clearly, $\mathcal{B}_0$ is compact in $\mathcal{K}$, and thus closed in $\mathcal{F}$ and contained in $\mathcal{F}_{\{o\}}$, and its minimal elements are compact. Moreover, conditions a and c are fulfilled. Thus $A \to \cup \{A_K, K \in \mathcal{V}\}$ is a compact τ-opening.

7-2. THE GRANULOMETRIES

With a view to finding suitable axioms for a granulometric mapping (or, simply, granulometry), let us first analyze what happens in practice when we sieve a material. Sieves are given, whose mesh sizes are characterized by a parameter $\lambda > 0$. By applying the sieve λ to (a material idealized by) a set A, we obtain an *oversize* which is a subset $\psi_\lambda(A) \subset A$. If B is another set containing A, the B-oversize for a given mesh λ is larger than the A-oversize, that is, $A \subset B$ implies $\psi_\lambda(A) \subset \psi_\lambda(B)$. If we compare two different meshes λ and μ such that $\lambda \geqslant \mu$, the μ-oversize is larger than the λ-oversize, that is, $\psi_\lambda(A) \subset \psi_\mu(A)$. Moreover, by applying the largest mesh λ to the μ-oversize, we obtain again the λ-oversize $\psi_\lambda(A)$ itself. In the same way, the μ-oversize of $\psi_\lambda(A)$ is $\psi_\lambda(A)$, so that $\lambda \geqslant \mu$ implies $\psi_\lambda \circ \psi_\mu = \psi_\mu \circ \psi_\lambda = \psi_\lambda$. Hence the following definition:

Definition 7-2-1. *Let E be a set and $\mathcal{Q} \subset \mathcal{P}(E)$. A granulometry on $\mathcal{Q}$ is a one-parameter family ψ_λ, $\lambda > 0$, of mappings from $\mathcal{Q}$ into itself such that*:

1. $\psi_\lambda(A) \subset A$ *for any* $\lambda > 0$ *and* $A \in \mathcal{Q}$.
2. $A, B \in \mathcal{Q}$ *and* $A \subset B$ *imply* $\psi_\lambda(A) \subset \psi_\lambda(B)$ $\quad (\lambda > 0)$.
3. $\lambda \geqslant \mu > 0$ *imply* $\psi_\lambda(A) \subset \psi_\mu(A)$ $\quad (A \in \mathcal{Q})$.
4. $\psi_\lambda \circ \psi_\mu = \psi_\mu \circ \psi_\lambda = \psi_{\mathrm{Sup}(\lambda,\mu)}$ $\quad (\lambda, \mu > 0)$.

We can complete the definition of a granulometry ψ_λ, $\lambda > 0$, by putting $\psi_0(A) = A$, $A \in \mathcal{Q}$, for $\lambda = 0$. Note that condition 3 is superfluous, because it is implied by 1, 2, and 4. On the other hand, for $\lambda = \mu$, condition 4 implies that ψ_λ is idempotent, so that, for any $\lambda > 0$, ψ_λ is an opening on $\mathcal{Q}$. If $\mathcal{B}_\lambda \subset \mathcal{Q}$ is the family of the sets $B \in \mathcal{Q}$ invariant under ψ_λ, $\lambda \geqslant \mu$ implies $\mathcal{B}_\lambda \subset \mathcal{B}_\mu$, because $\psi_\mu \circ \psi_\lambda = \psi_\lambda$ by condition 4. The converse is also true. In other words:

Proposition 7-2-1. *Let E be a set, $\mathcal{Q} \subset \mathcal{P}(E)$, and ψ_λ, $\lambda > 0$, a one-parameter family of mappings from $\mathcal{Q}$ into $\mathcal{P}$. Then ψ_λ, $\lambda > 0$, is a granulometry on $\mathcal{Q}$ if and only if two conditions are fulfilled*:

a. For any $\lambda > 0$, ψ_λ *is an opening on* $\mathcal{Q}$.

b. $\lambda \geqslant \mu > 0$ *implies* $\mathcal{B}_\lambda \subset \mathcal{B}_\mu$ ($\mathcal{B}_\lambda$ *denotes the class of* ψ_λ*-invariant sets in* $\mathcal{Q}$).

Proof. We have just proved the "only if" part. Conversely, suppose that condition *a* and *b* are true. Conditions 1, 2, and 3 of Definition 7-2-1 are obviously satisfied. For any $A \in \mathcal{Q}$ and $\lambda \geqslant \mu$, $\psi_\lambda(A) \in \mathcal{B}_\lambda \subset \mathcal{B}_\mu$ implies $\psi_\mu \psi_\lambda(A) = \psi_\lambda(A)$, that is, $\psi_\mu \circ \psi_\lambda = \psi_\lambda$. It follows that

$$\psi_\lambda = \psi_\lambda \circ \psi_\mu \circ \psi_\lambda \subset \psi_\lambda \circ \psi_\mu \subset \psi_\lambda,$$

because ψ_λ is increasing and idempotent, that is, $\psi_\lambda = \psi_\lambda \varnothing \psi_\mu$, and condition 4 is satisfied.

Corollary. *Any granulometry on* $\mathcal{Q} \subset \mathcal{P}(E)$ *admits an extension to a granumlometry on* $\mathcal{P}(E)$.

Proof. If ψ'_λ is the smallest extension of ψ_λ on $\mathcal{P}$, ψ'_λ is an opening and the family $\mathcal{B}'_\lambda$ of the ψ'_λ-invariant sets is the class closed for the infinite union generated by $\mathcal{B}_\lambda$, so that condition *b* is satisfied by the family $\mathcal{B}'_\lambda$, $\lambda > 0$.

Regularization of a Granulometry

Let ψ_λ, $\lambda > 0$, be a granulometry, which we may always suppose to be defined on $\mathcal{P}(E)$ itself (Proposition 7-2-1, corollary), and, for each $\lambda > 0$, $\mathcal{B}_\lambda$ be the class of the ψ_λ-invariant sets, so that $\mathcal{B}_\lambda \subset \mathcal{B}_\mu$ if $\lambda \geqslant \mu$. For any $\lambda > 0$, put

$$\hat{\mathcal{B}}_\lambda = \bigcup_{\mu > \lambda} \mathcal{B}_\mu; \qquad \check{\mathcal{B}}_\lambda = \bigcap_{\mu < \lambda} \mathcal{B}_\mu \tag{7-2-1}$$

Clearly, $\hat{\mathcal{B}}_\lambda \subset \mathcal{B}_\lambda \subset \check{\mathcal{B}}_\lambda$, and for any $\epsilon > 0$

$$\hat{\mathcal{B}}_\lambda \supset \mathcal{B}_{\lambda+\epsilon}; \qquad \mathcal{B}_\lambda \supset \check{\mathcal{B}}_{\lambda+\epsilon}$$

The mappings $\lambda \to \check{\mathcal{B}}_\lambda$ and $\lambda \to \hat{\mathcal{B}}_\lambda$ are decreasing. If $\check{\psi}_\lambda$ (respectively $\hat{\psi}_\lambda$) denotes the smallest extension on $\mathcal{P}$ of the identity on $\check{\mathcal{B}}_\lambda$ (resp. on $\hat{\mathcal{B}}_\lambda$), $\check{\psi}_\lambda$ (resp. $\hat{\psi}_\lambda$), $\lambda > 0$, is still a granulometry on $\mathcal{P}$ and is called the *regularization from above* (*resp. from below*) *of the granulometry* ψ, $\lambda > 0$.

For any $\epsilon > 0$, it follows from the above considerations that

$$\psi_{\lambda+\epsilon} \subset \hat{\psi}_\lambda \subset \psi_\lambda \subset \check{\psi}_\lambda \subset \psi_{\lambda-\epsilon} \tag{7-2-2}$$

A granulometry $\psi_\lambda, \lambda>0$, is called *regular from above* (*resp. from below*) if $\psi_\lambda=\check{\psi}_\lambda$ (resp. $\psi_\lambda=\hat{\psi}_\lambda$). It is easy to verify that the regularization from above (resp. from below) of a given granulometry is actually regular from above (resp. from below).

If ψ_λ, $\lambda>0$, is a granulometry on $\mathcal{P}(E)$, then, for any $A\subset E$,

$$\hat{\psi}_\lambda(A)=\bigcup_{\mu>\lambda}\cup\{B;B\in\mathcal{B}_\mu,B\subset A\}$$

that is,

$$\hat{\psi}_\lambda(A)=\cup\{\psi_\mu(A),\mu>\lambda\} \tag{7-2-3}$$

On the contrary, the relationship $\check{\psi}_\lambda(A)=\cap\{\psi_\mu(A),\mu<\lambda\}$ does *not* generally hold.

Critical Elements of a Granulometry

Let E be a set and ψ_λ, $\lambda>0$, a granulometry on $\mathcal{P}(E)$. Then a set $M\in\mathcal{P}$ is said to be *critical* at $\lambda=\lambda_0>0$ for the granulometry ψ_λ if

$$M\in\check{\mathcal{B}}_{\lambda_0},\qquad \hat{\psi}_{\lambda_0}(M)=\varnothing,\qquad M\neq\varnothing \tag{7-2-4}$$

or (which is the same) $\psi_\lambda(M)=\varnothing$ if $\lambda>\lambda_0$, and $\psi_\lambda(M)=M$ if $\lambda<\lambda_0$. In the same way, M is said to be *strictly critical* at λ_0 if $M\in\mathcal{B}_{\lambda_0}$, $\hat{\psi}_{\lambda_0}(M)=\varnothing$, and $M\neq\varnothing$. Clearly, a set M strictly critical at λ_0 is critical at λ_0, but the converse is not necessarily true (e.g., a granulometry, regular from below, has no strictly critical elements, but can admit critical elements). Nevertheless, if a granulometry ψ_λ is regular from above, any critical element is strictly critical, for $\psi_\lambda=\check{\psi}_\lambda$.

To see whether critical elements exist, let $A\in\mathcal{P}$ be a nonempty set and put

$$\lambda_A=\text{Inf}\{\lambda:\ \psi_\lambda(A)=\varnothing\}$$

[$\lambda_A=+\infty$ if $\psi_\lambda(A)\neq\varnothing$ for any $\lambda>0$]. In other words, λ_A is defined by $\psi_\lambda(A)=\varnothing$ if $\lambda>\lambda_A$ and by $\psi_\lambda(A)\neq\varnothing$ if $\lambda<\lambda_A$, or (which is the same) by

$$\hat{\psi}_{\lambda_A}(A)=\varnothing,\qquad \psi_\lambda(A)\neq\varnothing\quad \text{for } \lambda<\lambda_A$$

If a strictly critical element M exists in the family $\psi_\lambda(A)$, $\lambda>0$, necessarily $M=\psi_{\lambda_A}(A)$. Conversely, $\psi_{\lambda_A}(A)$ is strictly critical if and only if it is not empty, for, if $\lambda > \lambda_A$, $\psi_\lambda(A)$ is empty and thus is not critical. If $\lambda < \lambda_A$, there exists μ such that $\lambda < \mu < \lambda_A$, and $\psi_\mu(A)$ is not empty. Hence $\psi_\lambda(A)$ is not critical. Finally, if $\psi_{\lambda_A}(A) \neq \varnothing$, it satisfies the condition of the definition and is strictly critical.

In the same way, the only possible element critical in the family $\check{\psi}_\lambda(A)$, $\lambda > 0$, is $\check{\psi}_{\lambda_A}(A)$, and it is actually critical at $\lambda = \lambda_A$ if and only if it is not empty. In other words, *the critical elements of a granulometry* ψ_λ, $\lambda > 0$, are the $\check{\psi}_{\lambda_A}(A) \neq \varnothing$, $A \in \mathcal{P}$.

Example. Let E be the euclidean space $E = \mathbf{R}^d$, B be the unit ball in $\mathbf{R}^d$ (closed in $\mathbf{R}^d$), and $\psi_\lambda(A) = A_{\lambda B}$, $A \in \mathcal{P}$. Then $\mathcal{B}_\lambda = \check{\mathcal{B}}_\lambda = \{A \oplus \lambda B, A \in \mathcal{P}\}$, that is, ψ_λ is regular from above. If A is the open ball with radius λ_0, $\lambda_A = \lambda_0$ but $\psi_{\lambda_0}(A) = \varnothing$, and A is not critical. On the contrary, if A is the closed ball with radius λ_0, $\check{\psi}_{\lambda_0}(A) = A$, and A is critical at $\lambda = \lambda_0$.

Euclidean Granulometries

We suppose now that E is the euclidean space $E = \mathbf{R}^d$. The usual granulometries in $\mathcal{P}(\mathbf{R}^d)$ are compatible with translations, in the sense that the oversize of the translate A_x of $A \in \mathcal{P}$ is the corresponding translate of the oversize of A, that is, $\psi_\lambda(A_x) = \psi_\lambda(A) \oplus \{x\}$, $\lambda > 0$, $x \in \mathbf{R}^d$, $A \in \mathcal{P}$. Moreover, in practice, the family of sieves is homothetic, so that, for a convenient choice of the parameter λ, $\psi_\lambda(\lambda A) = \lambda \psi_1(A)$. Hence we have the following definition:

Definition 7-2-2. *Let $\mathcal{Q} \subset \mathcal{P}(\mathbf{R}^d)$ be closed under translations and positive homothetics and ψ_λ, $\lambda > 0$, be a granulometry on $\mathcal{Q}$. Then ψ_λ is said to be a euclidean granulometry if the following conditions are fulfilled*:

5. *For any $\lambda > 0$, ψ_λ is a τ-opening.*
6. *For any $\lambda > 0$ and $A \in \mathcal{Q}$, $\psi_\lambda(A) = \lambda \psi_1(A/\lambda)$.*

Let, as usual, $\mathcal{B}_\lambda$ denote the class of the ψ_λ-invariant sets in $\mathcal{Q}$. Clearly, condition 5 is equivalent to the following: $\mathcal{B}_\lambda$ is invariant under translations for any $\lambda > 0$, and condition 6 is equivalent to $\mathcal{B}_\lambda = \lambda \mathcal{B}_1$ (i.e., $A \in \mathcal{B}_\lambda$ if and only if $A/\lambda \in \mathcal{B}_1$). The family $\mathcal{B}_1$ cannot be arbitrary, for $\lambda \geqslant \mu$ implies $\lambda \mathcal{B}_1 \subset \mu \mathcal{B}_1$, that is, $\mathcal{B}_1$ is closed under the homothetics with modulus $\geqslant 1$. Conversely, if $\mathcal{B} \subset \mathcal{P}$ is closed under infinite union, translations, and homothetics $\geqslant 1$, then $\mathcal{B}_\lambda = \lambda \mathcal{B}$, $\lambda > 0$, is the invariant family associated with a euclidean granulometry on $\mathcal{P}(\mathbf{R}^d)$.

Let $\mathcal{B} \subset \mathcal{P}$ be a class closed under infinite union, translations, and homothetics $\geqslant 1$. Then a class $\mathcal{B}_0 \subset \mathcal{B}$ is called a *generator* of $\mathcal{B}$ if the class closed under infinite union, translations, and homothetics $\geqslant 1$ generated by $\mathcal{B}_0$ is identical to $\mathcal{B}$. If so, we shall say that $\mathcal{B}_0$ is a generator of the euclidean granulometry ψ_λ: $A \to \cup\{A_{\lambda B}, B \in \mathcal{B}\}$ associated with $\mathcal{B}$. From the relationship $A_{\lambda B} = \cup\{A_{\lambda' B'}, \lambda' \geqslant \lambda, B' \in \mathcal{B}_0\}$, it follows that

$$\psi_\lambda(A) = \cup\{A_{\lambda' B}, \lambda' \geqslant \lambda, B \in \mathcal{B}_0\} \qquad (A \in \mathcal{P})$$

Conversely, for any $\mathcal{B}_0 \subset \mathcal{P}$, this formula defines a euclidean granulometry on $\mathcal{P}$. In other words:

Proposition 7-2-2. *A family ψ_λ, $\lambda > 0$ of mappings from $\mathcal{P}$ ($\mathbf{R}^d$) into itself is a euclidean granulometry if and only if there exists a class $\mathcal{B}_0 \subset \mathcal{P}$ such that*

$$\psi_\lambda(A) = \bigcup_{B \in \mathcal{B}_0} \bigcup_{\mu \geqslant \lambda} A_{\mu B}$$

If so, $\mathcal{B}_0$ is a generator of the euclidean granulometry ψ_λ. Moreover, if $\mathcal{B}_\lambda$ denotes the family of the ψ_λ-invariant sets and $\mathcal{B}$ the class closed under translations, infinite union, and homothetics $\geqslant 1$ generated by $\mathcal{B}_0$, then $\mathcal{B}_\lambda = \lambda\mathcal{B}$ for any $\lambda > 0$.

Example. If $\mathcal{B}_0 = \{B\}$ is reduced to a single element B (not necessarily convex), the mappings ψ_λ defined by

$$\psi_\lambda(A) = \bigcup_{\mu \geqslant \lambda} A_{\mu B} \qquad (\lambda > 0, A \in \mathcal{P})$$

constitute a euclidean granulometry called *the granulometry with respect to the set B*. Moreover, if B is convex, $\mu \geqslant \lambda$ implies $A_{\mu B} \subset A_{\lambda B}$ and $\psi_\lambda(A) = A_{\lambda B}$, so that the present definition agrees with the one given in Chapter 1. If B is compact, the following proposition shows that the converse is true, that is, $\psi_\lambda(A) = A_{\lambda B}$ if and only if the compact set B is *convex*.

Proposition 7-2-3. *Let B be a compact set in $\mathbf{R}^d$. Then the homothetic λB is open with respect to B [i.e., $(\lambda B)_B = \lambda B$] for any $\lambda \geqslant 1$ if and only if B is convex.*

Proof. If $B \in C(\mathcal{K})$, obviously $(\lambda B)_B = \lambda B$ for any $\lambda \geqslant 1$. Conversely, let B be a compact set and, for any $\alpha > 0$, D_α a set such that

$$(1+\alpha)B = B \oplus \alpha D_\alpha \qquad (a)$$

By changing αD_α into $((1+\alpha)B)\ominus\check{B}$, we may suppose that $D_\alpha \in \mathcal{K}$. If C denotes the convex hull, (a) implies

$$(1+\alpha)C(B)=C(B)\oplus\alpha C(D_\alpha) \qquad (a')$$

But $(1+\alpha)C(B)=C(B)\oplus\alpha C(B)$, for these sets are convex, and it follows from Proposition 1-5-3 that

$$C(D_\alpha)=C(B) \qquad (b)$$

and thus

$$D_\alpha \subset C(B) \qquad (b')$$

On the other hand, (a) implies, for any $n>0$,

$$B=((1+\alpha)^{-n}B)\oplus\left(\bigoplus_{k=1}^{n}\frac{\alpha}{(1+\alpha)^k}D_\alpha\right) \qquad (c)$$

and it follows from (b') that

$$\bigoplus_{k=1}^{n}\frac{\alpha}{(1+\alpha)^k}D_\alpha\subset\left(\sum_{k=1}^{n}\frac{\alpha}{(1+\alpha)^k}\right)C(D_\alpha)\subset C(B)$$

so that the Minkowski series $\oplus_{k=1}^{\infty}\alpha(1+\alpha)^{-k}D_\alpha$, obviously convergent in $\mathcal{F}$, is also convergent in $\mathcal{K}$. But the Minkowski sum $\oplus$ is continuous in $\mathcal{K}$, and thus (c) yields

$$B=\frac{\alpha}{1+\alpha}\bigoplus_{k=0}^{\infty}(1+\alpha)^{-k}D_\alpha \qquad (d)$$

Let N be an integer >0, and put

$$B_i(\alpha)=\frac{\alpha}{1+\alpha}\bigoplus_{k=0}^{\infty}(1+\alpha)^{-i-kN}D_\alpha \qquad (i=0,1,\ldots,N-1)$$

so that

$$\begin{aligned}B&=B_0(\alpha)\oplus B_1(\alpha)\oplus\ldots\oplus B_{N-1}(\alpha)\\ B_i(\alpha)&=(1+\alpha)^{-i}B_0(\alpha)\end{aligned} \qquad (e)$$

On the other hand, it follows from (b') that there exists a sequence $\{\alpha_n\}$ such that $\alpha_n\downarrow 0$, $\lim D_{\alpha_n}=D_0$ in $\mathcal{K}$, and $\lim B_0(\alpha_n)=B_0$ in $\mathcal{K}$. Then (e) implies $\lim B_i(\alpha_n)=B_0$, $i=0,1,\ldots,N-1$, and $B=(B_0)^{\oplus N}$. In other words, B

is infinitely divisible for the Minkowski sum, and thus B is convex (Theorem 1-5-1).

Corollary. *Let B be a compact set in $\mathbf{R}^d$. Then the granulometry ψ_λ with respect to B is $\psi_\lambda(A)=A_{\lambda B}$, $A\in\mathcal{P}$, if and only if B is convex.*

Proof. The "if" part is obvious. Conversely, if $\psi_\lambda(A)=A_{\lambda B}$ for any $A\in\mathcal{P}$ and $\lambda>0$, if follows that $\psi_1(\lambda B)=(\lambda B)_B=\lambda B$ if $\lambda\geqslant 1$, and B is convex by the Proposition.

Euclidean Granulometries U.S.C. and Compact

Definition 7-2-3. *A euclidean granulometry ψ_λ, $\lambda>0$, on $\mathcal{P}(\mathbf{R}^d)$ is said to be u.s.c. on $\mathcal{K}$ (resp. on $\mathcal{F}$) if $(\lambda,A)\to\psi_\lambda(A)$ is an u.s.c. mapping from $\mathbf{R}_+\times\mathcal{K}$ (resp. $\mathbf{R}_+\times\mathcal{F}$) into $\mathcal{K}$ (resp. into $\mathcal{F}$). In the same way, ψ_λ is said to be compact if for any $\lambda>0$ the opening ψ_λ is compact (in the sense of Definition 7-1-1).*

Our present goal is to characterize the u.s.c. and compact euclidean granulometries. Clearly, a euclidean granulometry is compact if the opening ψ_{λ_0} is compact for a particular value $\lambda=\lambda_0$, for instance, $\lambda=1$, and the corresponding criterion is given by Proposition 7-1-5. On the other hand, a criterion for the upper semicontinuity is as follows:

Proposition 7-2-4. *Let ψ_λ, $\lambda>0$, be a euclidean granulometry on $\mathcal{P}(\mathbf{R}^d)$, and, for each $\lambda>0$, $\mathcal{B}_\lambda=\lambda\mathcal{B}$ be the family of the ψ_λ-invariant sets. Then the three following conditions are equivalent:*

a. The granulometry ψ_λ is u.s.c. on $\mathcal{F}$ (in the sense of Definition 7-2-3).
b. For any $\lambda>0$, ψ_λ is an u.s.c. opening on $\mathcal{F}$.
c. $\mathcal{B}$ is closed under the topological closure, and $\mathcal{B}\cap\mathcal{F}$ is closed in $\mathcal{F}$.

Proof. Obviously condition a implies b. Condition b and c are equivalent by Propositions 7-1-2 and 7-1-4. If condition b is true, let $\{\lambda_n A_n\}\subset\mathbf{R}_+\times\mathcal{F}$ be a sequence such that $\lim(\lambda_n,A_n)=(\lambda,A)$. From

$$\overline{\lim}\ \psi_{\lambda_n}(A_n)=\overline{\lim}\ \lambda_n\psi_1(A_n/\lambda_n)\subset\lambda\psi_1(A/\lambda)=\psi_\lambda(A),$$

it follows that condition a holds.

Corollary. *Let ψ_λ, $\lambda>0$, be a euclidean granulometry u.s.c. on $\mathcal{F}$ different from the identity mapping. Then a ψ_λ-invariant closed set has no isolated point.*

Proof. Let $B'\in\mathcal{B}$ be a ψ_1-invariant closed set, and $x\in B'$. We may suppose that $x=O$, for $\mathcal{B}$ is invariant under translations. If O is an isolated point of B', there exists an open ball ϵB with center O and radius $\epsilon>0$ disjoint of $B'\backslash\{o\}$. If $\lambda\uparrow\infty$, $\lim\lambda\epsilon B=\mathbf{R}^d$ in $\mathcal{G}$ and so $\lim\lambda(B'\backslash\{o\})=\varnothing$ in $\mathcal{F}$, so that $\lim\lambda B'=\{o\}$ in $\mathcal{F}$. It follows that $\{o\}\in\mathcal{B}$, for $\mathcal{B}$ (by the proposition) is closed. But $\{o\}\in\mathcal{B}$ implies $\mathcal{B}=\mathcal{P}(E)$, since $\mathcal{B}$ is closed under union and translation, and so $\psi_\lambda(A)=A$ for any $A\in\mathcal{P}$.

Example 1. The granulometry with respect to a set $B=\{x_0,y_0\}$ with two distinct elements x_0 and y_0 is a granulometry on $\mathcal{F}$, for it maps $\mathcal{F}$ into itself, but it is not u.s.c. by the corollary.

Example 2. The granulometry with respect to the unit disk C on $\mathcal{P}(\mathbf{R}^2)$ is associated with the class $\mathcal{B}$ closed under translations and infinite union generated by $\{\lambda C,\lambda\geqslant 1\}$. The straight lines belong to the closure $\overline{\mathcal{B}}$ of $\mathcal{B}$ in $\mathcal{F}$, but not to $\mathcal{B}$ itself. Hence this granulometry is not compact.

Example 3. Let $\mathcal{V}$ be a *compact* subset of $C(\mathcal{K})$, that is, $\mathcal{V}$ is closed for the myope topology, and the compact convex sets $B\in\mathcal{V}$ are contained in a fixed compact set. Let $\mathcal{B}$ be the class closed under translation, union, and homothetics $\geqslant 1$ generated by $\mathcal{V}$. The euclidean granulometry ψ_λ associated with $\mathcal{B}$ (which is said to be *the granulometry with respect to the family* $\mathcal{V}$) is *u.s.c. and compact*, and

$$\psi_\lambda(A)=\cup\{A_{\lambda B},B\in\mathcal{V}\}\qquad(A\in\mathcal{P})$$

In particular, if $\mathcal{V}\subset C(\mathcal{K})$ *is finite*, the granulometry with respect to $\mathcal{V}$ is compact and u.s.c. (it is an easy consequence of Proposition 7-1-5 and 7-2-4).

Proposition 7-2-5. *Let ψ_λ be an u.s.c. granulometry on $\mathcal{K}$ (or $\mathcal{F}$) and, for any $K\in\mathcal{K}$, put $\lambda_K=\mathrm{Inf}\{\lambda,\psi_\lambda(K)=\varnothing\}$. Then the critical compact elements are the $\psi_{\lambda_K}(K)$, $K\in\mathcal{K}$, $\lambda_K\neq 0$, $\lambda_K\neq\infty$.*

Proof. For any $\lambda>0$, the upper semicontinuity implies $\psi_\mu(K)\downarrow\psi_\lambda(K)$ if $\mu\uparrow\lambda$, and thus $\psi_\lambda(K)=\cap\{\psi_\mu(K),\mu<\lambda\}$. If $\lambda_K>0$ it follows that $\psi_{\lambda_K}(K)=\cap\{\psi_\lambda(K),\lambda<\lambda_K\}$. But $\psi_{\lambda_K}(K)$ is compact, and $\psi_{\lambda_K}(K)=\varnothing$ would imply

$\psi_\lambda(K)=\varnothing$ for a $\lambda<\lambda_K$, and this is impossible by the definition of λ_K. Then $\psi_{\lambda_K}(K)\neq\varnothing$, and $\psi_{\lambda_K}(K)$ is a critical element. Conversely, any critical compact element K satisfies $K=\psi_{\lambda_K}(K)$.

Other Examples

1. Using the same notation, let $\mathcal{B}_\lambda$ be the class of measurable sets in $\mathbf{R}^d$ which have a Lebesgue measure $\geqslant\lambda$. Then any measurable set A with a Lebesgue measure $=\lambda_0$ is critical at $\lambda=\lambda_0$, for $\psi_\lambda(A)=\varnothing$ if $\operatorname{Mes} A<\lambda$ and $\psi_\lambda(A)=A$ otherwise.

2. Let $\mathcal{B}_0$ be the class of *connected* measurable sets which have a Lebesgue measure $\geqslant 1$, and let ψ_λ be the euclidean granulometry generated by $\mathcal{B}_0$, that is, $\psi_\lambda(A)$ is the union of the connected components of A admitting a Lebesgue measure $\geqslant\lambda$. Any connected measurable set is critical.

3. If B is a convex set, and $\mathcal{B}_0$ the family of the connected sets not enclosed in a translate of B, the euclidean granulometry generated by $\mathcal{B}_0$ corresponds to the usual notion of a sieving.

7-3. GRANULOMETRY OF A RACS AND OF ITS COMPLEMENTARY SET

Let E be a LCS space, A a RACS defined by a probability P on $(\mathcal{F},\sigma_f)$, and ψ_λ, $\lambda>0$, a granulometry u.s.c. on $\mathcal{F}$ (Definition 7-2-3). Then, for each $\lambda>0$, the mapping $A\to\psi_\lambda(A)$ from $\mathcal{F}$ into itself is measurable, and $\psi_\lambda(A)$ is a RACS. More precisely, the mapping $(\lambda,A)\to\psi_\lambda(A)$ is u.s.c. on $\mathbf{R}_+\times\mathcal{F}$ and thus measurable. It follows that the mapping k from $E\times\mathbf{R}_+\times\mathcal{F}$ into $\{0,1\}$, defined by $k(x,\lambda,A)=1$ if $x\in\psi_\lambda(A)$ and $=0$ if $x\notin\psi_\lambda(A)$, is measurable, since the set $k^{-1}(1)=\{(x,\lambda,A):\ x\in\psi_\lambda(A)\}$ is closed, as can easily be verified from the upper semicontinuity. By putting

$$\Lambda(x)=\operatorname{Sup}\{\lambda:\ k(x,\lambda,A)=1\}=\operatorname{Sup}\{\lambda:\ x\in\psi_\lambda(A)\}$$

for any $x\in\mathbf{R}^d$, it follows that the family $\Lambda(x)$, $x\in\mathbf{R}^d$ is a measurable random function. Let us denote by $F_x(.)$ the distribution function of the random variable $\Lambda(x)$ for a given $x\in\mathbf{R}^d$. Then

$$F_x(\lambda)=P(\Lambda(x)<\lambda)=P(x\notin\psi_\lambda(A)) \tag{7-3-1}$$

Proof. $\Lambda(x)<\lambda$ implies $x\notin\psi_\lambda(A)$, and, more precisely,

$$\{\Lambda(x)<\lambda\}=\bigcup_{\epsilon>0}\{x\notin\psi_{\lambda-\epsilon}(A)\}\qquad\text{or}\qquad\{\Lambda(x)\geqslant\lambda\}=\left\{x\in\bigcap_{\epsilon>0}\psi_{\lambda-\epsilon}(A)\right\}$$

The latter event is a.s. equal to $\{x \in \psi_\lambda(A)\}$, for the granulometry ψ_λ is u.s.c. Hence we have (7-3-1).

The function $\lambda \to 1 - F_x(\lambda)$ defined on $\mathbf{R}_+$ is said to be the size distribution of the RACS A, at the point x, with respect to the granulometry ψ_λ. Let us summarize these results.

Proposition 7-3-1. *Let E be a LCS space, ψ_λ, $\lambda > 0$, be a granulometry u.s.c. on $\mathcal{F}(E)$, and A be a RACS. Then $\psi_\lambda(A)$ is a RACS measurable on $R^+ \times \mathcal{F}$. Moreover, $\Lambda(x) = \text{Sup}\{\lambda:\ x \in \psi_\lambda(A)\}$, $x \in E$, is a measurable random function, and $\{\Lambda(x) \geqslant \lambda\} = \{x \in \psi_\lambda(A)\}$ for any $x \in E$. At each $x \in E$, the size distribution of A with respect to the granulometry ψ_λ is the function $\lambda \to 1 - F_x(\lambda)$, defined by*

$$1 - F_x(\lambda) = P(\Lambda(x) \geqslant \lambda) = P(x \in \psi_\lambda(A)) \qquad (\lambda > 0)$$

Note. The realizations of the random function $x \to \Lambda(x)$ are u.s.c. on E.

Proof. Let $\{x_n\} \subset E$ be a sequence such that $\lim x_n = x$, λ_0 an adherent point of $\{\Lambda(x_n)\}$ (in the compactification of $\mathbf{R}_+$), and $k \to \{n_k\}$ a sequence such that $\lambda_{n_k} < \Lambda(x_{n_k})$ for any k, $\lim \Lambda(x_{n_k}) = \lambda_0$ and $\lim \lambda_{n_k} = \lambda_0$. For a fixed $A \in \mathcal{F}$, $\lambda \to \psi_\lambda(A)$ is u.s.c. It follows that $x \in \overline{\lim}\ \psi_{\lambda_{nk}}(A) \subset \psi_{\lambda_0}(A)$, that is, $\Lambda(x) \geqslant \lambda_0$. Thus Λ is a.s. u.s.c. on E.

The Size Distribution of the Pores

If a RACS A represents the solid components of a porous medium, the complementary set A^c (which is a random *open* set) obviously represents *the pores* of the medium. After having defined the size distribution of the "grains" A with respect to a given u.s.c. granulometry, we shall now define the size distribution of the pores by applying the *same* granulometry to the random open set A^c, but only in the case of a euclidean granulometry.

Let $E = \mathbf{R}^d$ be a euclidean space, ψ_λ, $\lambda > 0$, be a *euclidean granulometry u.s.c.* on $\mathcal{K}(\mathbf{R}^d)$, and $\mathcal{B} \subset \mathcal{K}$ be the family of the ψ_1-invariant compact sets. For any $E \in \mathcal{G}$ and $\lambda > 0$, put

$$\psi'_\lambda(G) = \cup\{\psi_\lambda(K), K \in \mathcal{K}, K \subset G\}$$

Then (as can easily be verified), for each $\lambda > 0$, ψ'_λ is a l.s.c. τ-opening on $\mathcal{G}$, and ψ'_λ, $\lambda > 0$, is a euclidean granulometry on $\mathcal{G}$. This granulometry is l.s.c. on $\mathcal{G}$, that is, the mapping $(\lambda, G) \to \psi'_\lambda(G)$ is l.s.c. from $\mathbf{R}^+ \times \mathcal{G}$ into $\mathcal{G}$.

Proof. Let λ be a number >0, $K\in\mathcal{K}$ and $G\in\mathcal{G}$ such that $K\subset\psi'_\lambda(G)$. We must prove that there exist $\lambda_0>\lambda$, $K_0\in\mathcal{K}$ such that $K_0\subset G$, and $\mu<\lambda_0$, $G'\supset K_0$, $G'\in\mathcal{G}$ imply $K\subset\psi'_\mu(G)$. From $K\subset\psi'_\lambda(G)\in\mathcal{G}$, it follows that there exist a ψ_λ-invariant set $B\in\lambda\mathcal{B}$ and $\epsilon>0$ such that the closed ball B_ϵ with radius ϵ satisfies

$$K\subset K\oplus B_\epsilon\subset B\subset B\oplus B_\epsilon\subset G$$

But the homothetics are continuous on $\mathcal{K}$, and so we can find $a>1$ such that $(1/\alpha)K\subset K\oplus B_\epsilon$ and $\alpha B\subset B\oplus B_\epsilon$. Thus

$$K\subset\alpha B\subset B\oplus B_\epsilon\subset G$$

The set αB is $\psi_{\alpha\lambda}$-invariant (because ψ_λ is euclidean), and thus $K\subset\alpha B\subset\psi'_{\alpha\lambda}(G')$ for any open set $G'\supset B\oplus B_\epsilon$. In other words, $G'\supset B\oplus B_\epsilon$ and $\mu\leqslant\alpha\lambda$ imply $K\subset\psi'_\mu(G')$. Hence $(\lambda,G)\to\psi'_\lambda(G)$ is l.s.c. on $\mathcal{G}$, and we can state:

Proposition 7-3-2. *Let ψ_λ, $\lambda>0$, be a euclidean granulometry u.s.c. on $\mathcal{K}(\mathbf{R}^d)$ and*

$$\psi'_\lambda(G)=\cup\{\psi_\lambda(K),K\subset G,K\in\mathcal{K}\}\qquad(G\in\mathcal{G},\lambda>0)$$

Then ψ'_λ, $\lambda>0$, is a l.s.c. euclidean granulometry on $\mathcal{G}$ [i.e., the mapping $(\lambda,G)\to\psi'_\lambda(G)$ is l.s.c. on $\mathbf{R}^+\times\mathcal{G}$].

If the granulometry ψ_λ, $\lambda>0$, is *compact*, which implies that, for each λ, ψ_λ is the smallest extension of its restriction to $\mathcal{K}$, we find

$$\psi'_\lambda(G)=\cup\{\psi_\lambda(K),K\subset G,K\in\mathcal{K}\}=\psi_\lambda(G)$$

and $\psi_\lambda=\psi'_\lambda$, $\lambda>0$, is a l.s.c. euclidean granulometry on $\mathcal{G}$. By applying this granulometry to the random open set A^c, the complement of a given RACS A, we shall obtain the desired notion of a pore size distribution.

Proposition 7-3-3. *Let ψ_λ, $\lambda>0$, be an u.s.c. and compact euclidean granulometry, A a RACS, and P its probability. Then, for any $\lambda>0$, $\psi_\lambda(A^c)$ is a measurable random open set, and, for any measure $\mu\geqslant0$ on $\mathbf{R}^d$,*

$$E[\mu(\psi_\lambda(A^c))]=\int\mu(dx)P(x\in\psi_\lambda(A^c))$$

Moreover, $x\to\Lambda'(x)=\text{Sup}\{\lambda\colon\ x\in\psi_\lambda(A^c)\}$ is a measurable random function, and $\{\Lambda'(x)\leqslant\lambda\}=\{x\not\in\psi_\lambda(A^c)\}$. For any $x\in\mathbf{R}^d$, the function $\lambda\to1-$

$F'_x(\lambda)$, *defined by*

$$1 - F'_x(\lambda) = P(\Lambda'(x) > \lambda) = P(x \in \psi_\lambda(A^c))$$

is said to be the pore size distribution of the RACS A at point x in respect to the granulometry ψ_λ.

Proof. We have just seen that the mapping $(\lambda, G) \to \psi_\lambda(G)$ is l.s.c. from $\mathbf{R}_+ \times \mathcal{G}$ into $\mathcal{G}$, so that $\psi_\lambda(A^c)$ is a measurable random open set for a given $\lambda > 0$. From $\psi_\lambda(G) = \cup\{\psi_\mu(G),\ \mu > \lambda\}$, $G \in \mathcal{G}$, it follows that $\psi_\lambda(G) = \lim \psi_\mu(G)$ in $\mathcal{G}$ if $\mu \downarrow \lambda$. In particular, $\mu \downarrow \Lambda'(x)$ implies $\psi_\mu(A^c) \uparrow \psi_{\Lambda'(x)}(A^c)$, and so $x \notin \psi_{\Lambda'(x)}(A^c)$, since $x \notin \psi_\mu(A^c)$ for any $\mu > \Lambda'(x)$. It follows that $\Lambda'(x) \leqslant \lambda$ if and only if $x \notin \lambda_\lambda(A^c)$. Hence $x \to \Lambda'(x)$ is a random function, and its measurability can be verified in the same way.

Note. The realizations of the random functions $\Lambda'(x)$ are l.s.c. on $\mathbf{R}^d$ [for proof, see the similar statement concerning the upper semicontinuity of $\Lambda(x)$].

7-4. OPENINGS AND GRANULOMETRIES (GENERAL CASE)

In the preceding sections, the τ-openings and the euclidean granulometries on the euclidean space $\mathbf{R}^d$ were thoroughly investigated, because they are the only ones used in practical applications. We now return to the general case of openings and granulometries defined on an arbitrary LCS space E.

We begin with openings and closings on E. In Proposition 7-1-2, we gave the conditions that must be satisfied by the family $\mathcal{B}$ of the ψ-invariant sets in order that $\mathcal{F}$, $\mathcal{G}$, or $\mathcal{K}$ be closed under ψ. More generally, if a class $\mathcal{A} \subset \mathcal{P}$ is closed under ψ, we shall say for brevity that ψ is an opening, or a closing, on $\mathcal{A}$, even if ψ is actually defined on the entire space $\mathcal{P}$. Moreover, if ψ is defined only on $\mathcal{A}$, it is always possible to extend it on $\mathcal{P}$ (see Section 7-1). In what follows, a statement such as "ψ is an u.s.c. opening on $\mathcal{F}$" will always imply that $\mathcal{F}$ is closed under ψ.

Openings and Closings U.S.C. on $\mathcal{F}$ or $\mathcal{K}$

We begin by characterizing the u.s.c. openings.

Proposition 7-4-1. *Let ψ be an opening on $\mathcal{P}$, and $\mathcal{B} \subset \mathcal{P}$ be the family of the ψ-invariant sets. Then ψ is an u.s.c. opening on $\mathcal{K}$ (resp. on $\mathcal{F}$) if and* *only if two conditions are satisfied*:

a. *For any relatively compact* $B \in \mathcal{B}$ (*resp. for any* $B \in \mathcal{B}$) $\bar{B} \in \mathcal{B}$.
b. $\mathcal{B} \cap \mathcal{K}$ *is closed in* $\mathcal{K}$ (*resp.* $\mathcal{B} \cap \mathcal{F}$ *is closed in* $\mathcal{F}$).

Proof. Let us give the proof of the statement concerning the space $\mathcal{K}$. From Proposition 7-1-2, it follows that ψ is an opening on $\mathcal{K}$ if and only if condition a is true, and we must prove that an opening ψ on $\mathcal{K}$ is u.s.c. if and only if $\mathcal{B} \cap \mathcal{K}$ is closed for the myope topology.

Let $\{A_n\} \subset \mathcal{K} \cap \mathcal{B}$ be a sequence such that $\lim A_n = A$ in $\mathcal{K}$. If ψ is u.s.c., it follows that $A = \lim A_n = \lim \psi(A_n) \subset \psi(A)$, and thus $A = \psi(A)$, for ψ is antiextensive. Hence $\mathcal{K} \cap \mathcal{B}$ is closed in $\mathcal{K}$. Conversely, let $\{K_n\} \subset \mathcal{K}$ be a sequence such that $\lim K_n = K$ in $\mathcal{K}$, $\{K_{n_k}\}$ a subsequence, and, for each k, a point $x_{n_k} \in \psi(K_{n_k})$ such that $\lim x_{n_k} = x$ in E. For any k, we have $x_{n_k} \in \psi(K_{n_k}) \subset K_{n_k}$, and $\psi(K_{n_k}) \in \mathcal{B} \cap \mathcal{K}$. The K_n and (a fortiori) the $\psi(K_{n_k})$ are contained in a fixed compact, so that the sequence $\{\psi(K_{n_k})\}$ admits an adherence value $B \in \mathcal{K}$. If $\mathcal{B} \cap \mathcal{K}$ is closed, it follows that $B \in \mathcal{B} \cap \mathcal{K}$ and $x \in B \subset K$. Thus $x \in \psi(K)$, $\overline{\lim}\, \psi(K_n) \subset \psi(K)$, and ψ is u.s.c.

Corollary 1. *Let* $\mathcal{B}_0$ *be a closed subset of* $\mathcal{K}$ (*resp. of* $\mathcal{F}$). *If* $\mathcal{B}_0$ *is closed under finite union, the smallest extension* ψ *of the identity on* $\mathcal{B}_0$ *is an u.s.c. opening on* $\mathcal{K}$ (*resp. on* $\mathcal{F}$).

Proof. Any family $\{B_i, i \in I\} \subset \mathcal{B}_0$ the union of which is relatively compact is such that $\overline{\cup \{B_i, i \in I\}} \in \mathcal{B}_0$, for $\mathcal{B}_0$ is closed in $\mathcal{K}$ and closed under finite union, and conditions a and b are satisfied.

Corollary 2. *A closing* ψ *on* $\mathcal{G}$ *is l.s.c. if and only if the family* $\mathcal{B} \cap \mathcal{G}$ *of the open* ψ*-invariant sets is closed in* $\mathcal{G}$.

Proposition 7-4-2. *Let* ψ *be a closing on* $\mathcal{K}$, *and* $\mathcal{B}_0 = \mathcal{B} \cap \mathcal{K}$ *be the family of the compact* ψ*-invariant sets. Then* ψ *is u.s.c. if and only if any* $B \in \mathcal{B}_0$ *admits a fundamental system of neighborhoods contained in* $\mathcal{B}_0$.

Proof. If $\mathcal{B}_0$ satisfies the stated conditon, then, for any $K \in \mathcal{K}$, $G \in \mathcal{G}$ such that $\psi(K) \subset G$, there exist $G_0 \in \mathcal{G}$, $B_0 \in \mathcal{B}_0$, and $\psi(K) \subset G_0 \subset B_0 \subset G$. Note also that $K \subset G_0$, for ψ is extensive. Then, for any $K' \in \mathcal{K}$, $K' \subset G_0$ implies $K' \subset B_0$ and so $\psi(K') \subset B_0 \subset G$. Hence, by definition, ψ is u.s.c.

Conversely, let ψ be u.s.c. on $\mathcal{K}$, $G \in \mathcal{G}$ and $B \in \mathcal{B}_0$ such that $B \subset G$, that is, $B = \psi(B) \subset G$. Then there exist $G' \in \mathcal{G}$, $G' \supset B$ such that $\psi(K') \subset G$ for any compact set $K' \subset G'$. On the other hand, the compact set B is contained in $G' \in \mathcal{G}$, and we may find $G_0 \in \mathcal{G}$, $K_0 \in \mathcal{K}$ such that

$B \subset G_0 \subset K_0 \subset G'$. But $K_0 \subset G'$ implies $\psi(K_0) \subset G$, and thus $B \subset G_0 \subset \psi(K_0) \subset G$. This achieves the proof, for $\psi(K_0) \in \mathcal{B}_0$.

The closings u.s.c. on $\mathcal{F}$ are characterized by the following:

Proposition 7-4-3. *Let ψ be an opening on $\mathcal{G}$, $\mathcal{B}$ the family of the ψ-invariant sets, and $\mathcal{B}_0 = \mathcal{B} \cap \mathcal{G}$. Then ψ is l.s.c. on $\mathcal{G}$ if and only if, for any $B \in \mathcal{B}_0$ and $K \in \mathcal{K}$ such that $K \subset B$, there exist a relatively compact set $B_0 \in \mathcal{B}_0$ and $K \subset B_0 \subset \overline{B}_0 \subset B$.*

Proof. Let ψ be l.s.c., $K \in \mathcal{K}$, and $K \subset B \in \mathcal{B}_0$. Thus $K \subset \psi(B)$, and there exist $K_0 \in \mathcal{K}$, $K_0 \subset B = \psi(B)$ such that $\psi(G) \supset K$ for any open set $G \supset K_0$. But $B \in \mathcal{G}$, and we can find a relatively compact $G_0 \in \mathcal{G}$ such that $K_0 \subset G_0 \subset \overline{G}_0 \subset B$. It follows that $K \subset \psi(G_0) \subset \overline{G}_0 \subset B$, and the stated condition is satisfied by $\psi(G_0) \in \mathcal{B}_0$.

Conversely, if this condition is satisfied and $G \in \mathcal{G}$, $K \in \mathcal{K}$, $K \subset \psi(G)$, there exists $B_0 \in \mathcal{B}_0$ such that $\overline{B}_0 \in \mathcal{K}$ and $K \subset B_0 \subset \overline{B}_0 \subset \psi(G) \subset G$. Thus $G' \in \mathcal{G}$ and $G' \supset \overline{B}_0$ imply $\psi(G') \supset \psi(B_0) = B_0 \supset K$, and so ψ is l.s.c.

Smallest U.S.C. Upper Bound of an Opening or a Closing on $\mathcal{K}$

If ψ is an opening on $\mathcal{K}$, it follows from Proposition 7-4-1 that there exists a smallest u.s.c. opening ψ' on $\mathcal{K}$ such that $\psi' \subset \psi$, that is, ψ' is the smallest extension on $\mathcal{K}$ of the identity function on the closure $\overline{\mathcal{B} \cap \mathcal{K}}$ of $\mathcal{B} \cap \mathcal{K}$ in $\mathcal{K}$.

Proposition 7-4-4. *Let ψ be an opening on $\mathcal{K}$, and $\mathcal{B}_0$ be the family of the compact ψ-invariant sets. Put $\psi_g(G) = \cup\{\psi(K), K \in \mathcal{K}, K \subset G\}$ for any $G \in \mathcal{G}$, and $\psi_k(K) = \cap\{\psi_g(G), G \in \mathcal{G}, G \supset K\}$ for any $K \in \mathcal{K}$. Then ψ_g is an opening on $\mathcal{G}$ if and only if $\mathcal{B}_0$ encloses a fundamental system of neighborhoods for each $B \in \mathcal{B}_0$. If so, ψ_g is a l.s.c. opening on $\mathcal{G}$, and ψ_k an u.s.c. opening on $\mathcal{K}$. Moreover, ψ_k is the smallest u.s.c. upperbound of ψ on $\mathcal{K}$, and $\psi_g(G) = \cup\{\psi_k(K), K \in \mathcal{K}, K \subset G\}$ for any $G \in \mathcal{G}$.*

Proof. The mapping ψ_g is the restriction to $\mathcal{G}$ of the smallest extension of the restriction of ψ to $\mathcal{K}$, and ψ is an opening on $\mathcal{P}$. From Proposition 7-1-2, condition *c*, it follows that $\tilde{\psi}_g$ is an opening on $\mathcal{G}$ if and only if $\mathcal{B}_0$ satisfies the stated condition. We suppose that this condition is satisfied, and show that ψ_g is l.s.c. by applying the criterion of Proposition 7-4-3. Let $B \in \mathcal{G}$ be a ψ_g-invariant set, that is, $B = \cup\{B_i, i \in I\}$ for a family $B_i, i \in I\} \subset \mathcal{B}_0$. For each $i \in I$, there exist $G_i \in \mathcal{G}$ and $B_i' \in \mathcal{B}_0$ such that

$B_i \subset G_i \subset B_i' \subset B$. If $K \in \mathcal{K}$, $K \subset B$, it follows that

$$K \subset \cup B_i \subset \cup G_i \subset \cup B_i' \subset B.$$

For each $i \in I$, $B_i \subset \psi_g(G_i)$ and thus $K \subset \cup \psi_g(G_i)$. Hence there exists a finite number of indices $i_1, \ldots, i_n \in I$ such that

$$K \subset \bigcup_{k=1}^{n} \psi_g(G_{i_k}) \subset \overline{\cup \psi_g(G_{i_k})} \subset \bigcup_{i \in I} B_i' \subset B$$

The open invariant set $B_0 = \cup_{k=1}^{n} \psi_g(G_{i_k})$ satisfies the criterion of Proposition 7-4-3, and ψ_g is l.s.c.

Let us now show that $\psi_k(K) \in \mathcal{K}$ for any $K \in \mathcal{K}$, for, if $G \supset K$ is open, we can find a relatively compact $G' \in \mathcal{G}$ such that $G \supset \overline{G}' \supset G' \supset K$. It follows that $\psi_g(G) \supset \psi(\overline{G}') \supset \psi_k(K)$, and thus

$$\psi_k(K) = \cap \{ \psi(\overline{G}'), G' \in \mathcal{G}, G' \supset K \}$$

is compact.

Then ψ_k is u.s.c. on $\mathcal{K}$, for, if $\psi_k(K) = \cap \{\psi(\overline{G}'), G' \in \mathcal{G}, \overline{G}' \in \mathcal{K}, G' \supset K\}$ is disjoint of $F \in \mathcal{F}$, the finite intersection property on $\mathcal{K}$ implies that there exists a relatively compact $G' \in \mathcal{G}$, $G' \supset K$, such that $\psi(\overline{G}')$ and (a fortiori) $\psi_g(G')$ is disjoint of F. Then $\psi_k(K') \cap F = \varnothing$ for any compact $K' \subset G'$, and so ψ_k is u.s.c.

Obviously ψ_k is increasing, antiextensive, and $\supset \psi$ on $\mathcal{K}$. Let us show that ψ_k is idempotent on $\mathcal{K}$, so that then we have ψ_k, an u.s.c. opening on $\mathcal{K}$. If $K \in \mathcal{K}$, we have $\psi_k \psi_k(K) = \cap \{\psi_g(G), G \supset \psi_k(K), G \in \mathcal{G}\}$. But $G \supset \psi_k(K)$ implies $\psi_g(G) \supset \psi_k(K)$, for ψ_k is u.s.c. and so we can find an open $G' \supset K$ such that $\psi_k(K') \subset G$ for any compact $K' \subset G'$, and (a fortiori) $\psi_g(G') \subset G$ (because $\psi_k \supset \psi$ on $\mathcal{K}$). It follows that $\psi_g(G') \subset \psi_g(G)$, because ψ_g is an opening. But $G' \supset K$ implies (by definition) $\psi_k(K) \subset \psi_g(G')$, and thus $\psi_k(K) \subset \psi_g(G)$. Then we may write:

$$\psi_k \psi_k(K) = \cap \{ \psi_g(G), G \in \mathcal{G}, \psi_g(G) \supset \psi_k(K) \} \supset \psi_k(K)$$

and so $\psi_k \psi_k(K) = \psi_k(K)$, because ψ_k is antiextensive, that is, ψ_k is idempotent.

Clearly, $\psi_k \supset \psi$ on $\mathcal{K}$. If ψ is u.s.c., this inclusion becomes an equality, for take $K \in \mathcal{K}$, $G \in \mathcal{G}$ such that $\psi(K) \subset G$. If ψ is u.s.c., there exists an open set $G' \supset K$ such that $\psi(K') \subset G$ for any compact $K' \subset G'$, and thus $\psi_g(G') \subset G$ and $\psi_k(K) \subset G$. It follows that $\psi_k(K) = \cap \{G, G \in \mathcal{G}, G \supset \psi(K)\}$

$\subset\psi(K)$, and we have the desired equality. If ψ is not u.s.c. on $\mathcal{K}$, any u.s.c. upper bound $\psi'\supset\psi$ on $\mathcal{K}$ is such that $\psi_k\subset\psi'_k=\psi'$, and so ψ_k is the smallest u.s.c. upper bound of ψ on $\mathcal{K}$. Finally, $\psi(K)\subset\psi_k(K)\subset\psi_g(G)$ for any $G\in\mathcal{G}$, $G\supset K$ implies the last statement.

Proposition 7-4-5. *Let ψ be an opening on $\mathcal{G}$, and $\mathcal{B}_0$ the family of the open ψ-invariant sets. Put $\psi_k(K)=\cap\{\psi(G), G\in\mathcal{G}, G\supset K\}$ for any $K\in\mathcal{K}$, and $\psi_g(G)=\cup\{\psi_k(K), K\in\mathcal{K}, K\subset G\}$ for any $G\in\mathcal{G}$. Then $\mathcal{K}$ is closed under ψ_k if and only if $\mathcal{B}_0$ encloses a fundamental system of neighborhoods of $\overline{B}$ for any relatively compact $B\in\mathcal{B}_0$. If so, ψ_k is an u.s.c. opening on $\mathcal{K}$, and ψ_g is a l.s.c. opening on $\mathcal{G}$. Moreover, ψ_g is the largest l.s.c. lower bound of ψ on $\mathcal{G}$, and $\psi=\psi_g$ if and only if ψ is l.s.c. on $\mathcal{G}$.*

Proof. Suppose that $\mathcal{K}$ is closed under ψ_k, and let $B\in\mathcal{B}_0$ be relatively compact, $G\in\mathcal{G}$, and $\overline{B}\subset G$. $B\subset\overline{B}$ implies $B\subset\psi_k(\overline{B})\subset\overline{B}$, and thus $\overline{B}=\psi_k(\overline{B})$, for $\psi_k(\overline{B})$ is compact. From $\overline{B}\subset G$, it follows that $\overline{B}=\psi_k(\overline{B})\subset\psi(G)\subset G$, and $\psi(G)\in\mathcal{B}_0$. Hence $\mathcal{B}_0$ encloses a fundamental system of neighborhoods of $\overline{B}$.

Conversely, suppose that this condition is satisfied by $\mathcal{B}_0$, and let K be a compact set. If a point $x\notin\psi_k(K)$, there exists an open set $G\supset K$ such that $x\notin\psi(G)$. Then, for any relatively compact open set G' such that $K\subset G'\subset\overline{G}'\subset G$, we have $x\notin\overline{\psi(G')}$, for, if $x\in\overline{\psi(G')}$, there exists $B\in\mathcal{B}_0$ such that $x\in\overline{\psi(G')}\subset B\subset G$ (by the property of $\mathcal{B}_0$), and thus $x\in\psi(B)\subset\psi(G)$, but this is impossible. We conclude that

$$x\notin\cap\{\overline{\psi(G')}, G'\supset K, G'\in\mathcal{G}\},$$

and thus $\psi_k(K)=\cap\{\overline{\psi(G')}, G'\supset K, G'G\}$ is compact.

Clearly, ψ_k is increasing and antiextensive on $\mathcal{K}$. For any $K\in\mathcal{K}$, $\psi_k\psi_k(K)=\cap\{\psi(G), G\in\mathcal{G}, G\supset\psi_k(K)\}$. But $G\supset\psi_k(K)$ implies $\psi(G)\supset\psi_k(K)$, for, from the preceding part of the proof, $\psi_k(K)=\cap\{\overline{\psi(G')}, G'\in\mathcal{G}, G'\supset K\}$, and thus there exists $G'\in\mathcal{G}$ such that $\overline{G}'\in\mathcal{K}$, $G'\supset K$, and $\overline{\psi(G')}\subset G$. It follows that $\psi(G')\subset\psi(G)$ and $\psi_k(K)\subset\psi(G)$. We conclude that $\psi_k\psi_k(K)\supset\psi_k(K)$, and ψ_k is idempotent, that is, is a l.s.c. opening of $\mathcal{K}$. The proof can be achieved in the same way as for Proposition 7-1-6.

For the closings, we shall give only the statements of the proposition, as the corresponding proofs are very similar to the previous ones.

Proposition 7-4-6. *Let ψ be a closing on $\mathcal{K}$, and $\mathcal{B}_0$ the family of the compact ψ-invariant sets. Then $\mathcal{G}$ is closed under ψ_g if and only if, for any $K \in \mathcal{K}$, $G \in \mathcal{G}$, and $B \in \mathcal{B}_0$ such that $K \subset G \subset B$, there exist $B' \in \mathcal{B}_0$, $G' \in \mathcal{G}$, and $K \subset B' \subset G' \subset B$. If so, ψ_g is a l.s.c. closing on $\mathcal{G}$, and ψ_k is an u.s.c. closing on $\mathcal{K}$. Moreover, ψ_k is the smallest u.s.c. upper bound of ψ on $\mathcal{K}$, and $\psi = \psi_k$ if and only if ψ is u.s.c. on $\mathcal{K}$.*

Proposition 7-4-7. *Let ψ be a closing on $\mathcal{G}$, and $\mathcal{B}_0$ the family of the open ψ-invariant sets. Then $\mathcal{K}$ is closed under ψ_k if and only if, for any $G \in \mathcal{G}$, $K \in \mathcal{K}$, and $B \in \mathcal{B}_0$ such that $G \subset K \subset B$, there exist $B' \in \mathcal{B}_0$, $K' \in \mathcal{K}$, and $G \subset B' \subset K' \subset B$. If so, ψ_k is an u.s.c. closing on $\mathcal{K}$, and ψ_g a l.s.c. closing on $\mathcal{G}$. Moreover, ψ_g is the largest l.s.c. lower bound of ψ on $\mathcal{G}$, and $\psi = \psi_g$ if and only if ψ is l.s.c. on $\mathcal{G}$.*

In these statements, ψ_g and ψ_k are defined in exactly the same way as in Propositions 7-4-4 and 7-4-5.

Compact Openings

An opening ψ on $\mathcal{P}$ is said to be *compact* if the two following conditions are satisfied:

a. ψ is the smallest extension of its restriction to $\mathcal{K}$.
b. ψ is an opening u.s.c. on $\mathcal{K}$, u.s.c. on $\mathcal{F}$, and l.s.c. on $\mathcal{G}$.

In order to characterize compact openings, a few preliminary results are required.

Proposition 7-4-8. *Let ψ be an opening on $\mathcal{K}$, and $\mathcal{B}_0$ be the family of the ψ-invariant compact sets. Then there exists a smallest openings ψ' on $\mathcal{F}$ such that $\psi' = \psi$ on $\mathcal{K}$, and the family $\mathcal{B}_0'$ of the ψ-invariant closed sets is generated from $\mathcal{B}_0$ by closed (infinite) union.*

Proof. Let $\mathcal{B}$ be the family of the ψ-invariant sets, so that $\mathcal{B}_0 = \mathcal{B} \cap \mathcal{K}$. It follows from Proposition 7-1-2 that $B \in \mathcal{B}$, $\overline{B} \in \mathcal{K}$ imply $\overline{B} \in \mathcal{B}_0$. If ψ' is an opening on $\mathcal{F}$ such that $\psi = \psi'$ on $\mathcal{K}$, the family $\mathcal{B}'$ of the ψ'-invariant sets contains $\mathcal{B}_0$ and thus $\mathcal{B}' \supset \mathcal{B}_0'$, $\mathcal{B}_0'$ denoting the class generated from $\mathcal{B}_0$ by infinite closed unions. Let ψ_0' be the smallest extension of the identity on $\mathcal{B}_0'$. By Proposition 7-1-2, ψ_0' is an opening on $\mathcal{F}$, and $\psi_0' \supset \psi$ on $\mathcal{K}$. Actually, $\psi_0' = \psi$ on $\mathcal{K}$, for, if $K \in \mathcal{K}$, $\psi_0'(K) = \overline{\cup \{B, B \in \mathcal{B}_0, B \subset K\}} = \overline{\psi(K)} = \psi(K)$.

Note that this smallest extension of ψ on $\mathcal{F}$ is generally not u.s.c.

Proposition 7-4-9. *Let ψ be an opening on $\mathcal{K}$, and $\mathcal{B}_0$ be the family of the compact ψ-invariant sets. Then ψ admits an extension by an u.s.c. opening on $\mathcal{F}$ if and only if $\mathcal{B}_0$ is closed for the relative $\mathcal{F}$-topology on $\mathcal{K}$. If so, ψ is u.s.c. on $\mathcal{K}$, and let ψ' be the smallest extension of ψ by an u.s.c. opening on $\mathcal{F}$; then the family $\mathcal{B}_0'$ of the closed ψ'-invariant sets is the closure $\mathcal{B}_0' = \overline{\mathcal{B}_0}$ of $\mathcal{B}_0 \subset \mathcal{F}$.*

Proof. Let ψ_1 be an extension of ψ by an u.s.c. opening on $\mathcal{F}$, and $\mathcal{B}_1$ be the family of the ψ_1-invariant closed sets. $\mathcal{B}_1$ is closed in $\mathcal{F}$ (Proposition 7-4-2), and $\mathcal{B}_0 = \mathcal{B}_1 \cap \mathcal{K}$, for ψ_1 is an extension of ψ. Thus $\mathcal{B}_0$ is closed for the relative $\mathcal{F}$-topology on $\mathcal{K}$. This implies that $\mathcal{B}_0$ is also closed for the myope topology, and ψ itself is u.s.c. on $\mathcal{K}$.

Conversely, suppose that $\mathcal{B}_0$ is closed for the relative $\mathcal{F}$-topology on $\mathcal{K}$, and put $\mathcal{B}_0' = \overline{\mathcal{B}_0}$ (closure of $\mathcal{B}_0$ in $\mathcal{F}$). Then $\mathcal{B}_0 = \mathcal{B}_0' \cap \mathcal{K}$, and the smallest extension ψ' of the identity on $\mathcal{B}_0'$ is an extension of ψ by an opening u.s.c. on $\mathcal{F}$. Moreover, if $\mathcal{B}_1$ is closed in $\mathcal{F}$ and $\mathcal{B}_0 = \mathcal{B}_1 \cap \mathcal{K}$, clearly $\mathcal{B}_1 \supset \mathcal{B}_0'$ and thus ψ' is the smallest extension of ψ by an opening u.s.c. on $\mathcal{F}$.

Corollary. *Let ψ be an u.s.c. opening on $\mathcal{F}$, and $\mathcal{B}_0'$ the family of the ψ-invariant closed sets. Then ψ is the smallest u.s.c. extension on $\mathcal{F}$ of its restriction to $\mathcal{K}$ if and only if $\mathcal{B}_0' = \overline{\mathcal{B}_0' \cap \mathcal{K}}$, that is, any closed ψ-invariant set is the limit in $\mathcal{F}$ of compact ψ-invariant sets.*

We are now able to characterize the compact openings.

Proposition 7-4-10. *An opening ψ is compact if and only if it admits the representation $\psi(A) = \cup\{B; B \subset A, B \in \mathcal{B}_0\}$, $A \in \mathcal{P}$, for a family $\mathcal{B}_0 \subset \mathcal{K}$ closed under finite union and such that:*

a. $\mathcal{B}_0$ is closed in $\mathcal{K} \cap \mathcal{F}$ for the relative $\mathcal{F}$-topology.
b. $\mathcal{B}_0$ contains a fundamental system of neighborhoods for each $B \in \mathcal{B}_0$.
c. The minimal elements of the closure $\overline{\mathcal{B}_0}$ of $\mathcal{B}$ in $\mathcal{F}$ are compact.

Proof. The "only if" part follows from Propositions 7-4-4 and 7-4-9. Conversely, suppose that conditions *a*, *b*, and *c* are satisfied by $\mathcal{B}_0 \subset \mathcal{K}$, and ψ is the smallest extension of the identity on $\mathcal{B}_0$. Let $\mathcal{B}$ be the family of the ψ-invariant sets, that is, the class closed under $\cup$ generated by $\mathcal{B}_0$. If $B \in \mathcal{B}$, that is, $B = \cup\{B_i, i \in I\}$ for a family $\{B_i, i \in I\} \subset \mathcal{B}_0$ and $\overline{B} \in \mathcal{K}$, it follows that $\overline{B} \in \mathcal{B}_0$ from condition *a*. Thus $\mathcal{K}$ is closed under ψ (Proposition 7-1-2). Let $\mathcal{B}'$ be the family of the closed ψ-invariant sets. Clearly, $\mathcal{B}' \subset \overline{\mathcal{B}_0}$. Conversely, conditions *a* and *c* imply $\overline{\mathcal{B}_0} \subset \mathcal{B}'$ and thus

$\mathcal{B}' = \overline{\mathcal{B}}_0$. If $B \in \mathcal{B}$, it follows that $\overline{B} \in \mathcal{B}_0$, hence $\overline{B} \in \mathcal{B}$, and ψ is u.s.c. on $\mathcal{F}$. Finally, ψ is l.s.c. on $\mathcal{G}$ from Proposition 7-4-4.

Granulometries U.S.C. and Compact

Recall that a granulometry ψ_λ, $\lambda > 0$, is said to be u.s.c. on $\mathcal{F}$ if $(\lambda, F) \to \psi_\lambda(F)$ is an u.s.c. mapping from $\mathbf{R}_+ \times \mathcal{F}$ into $\mathcal{F}$, and to be compact if, for any $\lambda > 0$, the opening ψ_λ is compact. If ψ_λ is u.s.c. and compact, and A a RACS, the RACS $\psi_\lambda(A)$ and the random open set $\psi_\lambda(A^c)$ possess the same properties as in the case of a euclidean u.s.c., compact granulometry (see Section 7-3). The characterization of the compact granulometries is given by Proposition 7-4-10, so we shall now look for the conditions for a granulometry to be u.s.c. Note that, if ψ_λ, $\lambda > 0$, is u.s.c. on $\mathcal{F}$, $F \to \psi_\lambda(F)$ is an u.s.c. opening on $\mathcal{F}$ for each $\lambda > 0$, and thus the family $\mathcal{B}_\lambda$ of the ψ_λ-invariant sets is closed under the topological closure, and $\mathcal{B}_\lambda \cap \mathcal{F}$ is closed in $\mathcal{F}$ (Proposition 7-4-1). In what follows, we suppose that $F \to \psi_\lambda(F)$ is u.s.c. on $\mathcal{F}$ for each $\lambda > 0$, and we write $\mathcal{B}_\lambda$ instead of $\mathcal{B}_\lambda \cap \mathcal{F}$.

Proposition 7-4-11. *Let ψ_λ be a granulometry on $\mathcal{K}$ (resp. on $\mathcal{F}$) such that, for any $\lambda > 0$, ψ_λ is an u.s.c. opening on $\mathcal{K}$ (resp. on $\mathcal{F}$). Then the restriction to $\mathcal{K}$ (resp. to $\mathcal{F}$) of the regularization from above $\check{\psi}_\lambda$ is still u.s.c. on $\mathcal{K}$ (resp. on $\mathcal{F}$). Moreover, for any $A \in \mathcal{K}$ (resp. $A \in \mathcal{F}$) and $\lambda_0 > 0$,*

$$\check{\psi}_{\lambda_0}(A) = \bigcap_{\mu < \lambda_0} \check{\psi}_\mu(A) = \bigcap_{\mu < \lambda_0} \psi_\mu(A)$$

Proof. Let ψ_λ be a granulometry on $\mathcal{K}$, and, for each $\lambda > 0$, $\mathcal{B}_\lambda$ be the family of the ψ_λ-invariant compact sets, which is a closed subset of $\mathcal{K}$. For a given $\lambda_0 > 0$ and $A \in \mathcal{K}$, put

$$A_{\lambda_0} = \bigcap_{\lambda < \lambda_0} \psi_\lambda(A)$$

so that, by (7-2-2), $A_{\lambda_0} \supset \check{\psi}_{\lambda_0}(A)$. With a view to proving the converse inclusion, let us first show that $A_{\lambda_0} \in \mathcal{B}_{\lambda_0}$, that is, $A_{\lambda_0} \in \mathcal{B}_\mu$ for any $\mu < \lambda_0$ (because A_{λ_0} is compact). If $\lambda \uparrow \lambda_0$, we have $\lim \psi_\lambda(A) = A_{\lambda_0}$ in $\mathcal{K}$, and $\overline{\lim}\, \psi_\mu \psi_\lambda(A) \subset \psi_\mu(A_{\lambda_0})$ for a given $\mu < \lambda_0$ (for ψ_μ is u.s.c.). But $\mu < \lambda < \lambda_0$ implies $\psi_\mu \psi_\lambda(A) = \psi_\lambda(A)$, and thus $A_{\lambda_0} = \lim \psi_\lambda(A) \subset \psi_\mu(A_{\lambda_0})$, that is, $A_{\lambda_0} = \psi_\mu(A_{\lambda_0})$, for ψ_μ is an opening. Thus $A_{\lambda_0} \in \mathcal{B}_\mu$ for any $\mu < \lambda_0$, that is, $A_{\lambda_0} \in \check{\mathcal{B}}_{\lambda_0}$, and $A_{\lambda_0} = \check{\psi}_{\lambda_0}(A_{\lambda_0})$.

Then $A_{\lambda_0} \subset A$ yields $A_{\lambda_0} = \psi_{\lambda_0}(A_{\lambda_0}) \subset \psi_{\lambda_0}(A) \subset A_{\lambda_0}$, and thus $A_{\lambda_0} = \check{\psi}_{\lambda_0}(A)$. Hence $\check{\psi}_{\lambda_0}$ is an opening on $\mathcal{K}$. The compact $\check{\psi}_{\lambda_0}$-invariant sets are $A_{\lambda_0} = \check{\psi}_{\lambda_0}(A)$, $A \in \mathcal{K}$, that is, the elements of $\cap \{\mathcal{B}_\lambda, \lambda < \lambda_0\}$. But this family is closed in $\mathcal{K}$, because each $\mathcal{B}_\lambda$ is closed. Then it follows from Proposition 7-4-1 that $\check{\psi}_{\lambda_0}$ is u.s.c.

Proposition 7-4-12. *Let ψ_λ, $\lambda > 0$, be a granulometry on $\mathcal{K}$ (resp. on $\mathcal{F}$) such that, for any $\lambda > 0$, ψ_λ is u.s.c. Then the four following conditions are equivalent:*

1. *The granulometry ψ_λ, $\lambda > 0$, is u.s.c. on $\mathcal{K}$ (resp. on $\mathcal{F}$).*
2. *For any $A \in \mathcal{K}$ (resp. $\in \mathcal{F}$), $\lambda \to \psi_\lambda(A)$ is u.s.c. on $\mathbf{R}_+$.*
3. *For any $\lambda > 0$ and $A \in \mathcal{K}$ (resp., $\in \mathcal{F}$), $\psi_\lambda(A) = \cap_{\mu < \lambda} \psi_\mu(A)$.*
4. *The granulometry ψ_λ, $\lambda > 0$, is regular from above.*

Proof. Condition 1 obviously implies 2. If $\lambda \uparrow \lambda_0$, $\lim \psi_\lambda(A) = A_{\lambda_0} = \cap_{\lambda < \lambda_0} \psi_\lambda(A)$ in $\mathcal{K}$, and thus $A_{\lambda_0} \subset \psi_{\lambda_0}(A)$ if condition 2 is true, that is, $A_{\lambda_0} = \psi_{\lambda_0}(A)$ (because the converse inclusion always holds). Hence condition 2 implies 3. If condition 3 is true, $A_\lambda \supset \check{\psi}_\lambda(A) \supset \psi_\lambda(A)$ implies $A_\lambda = \check{\psi}_\lambda = \psi_\lambda(A)$, and 4 holds.

It follows from Proposition 7-4-11 that condition 4 implies 3. Let $\{\lambda_n\}$ be a sequence such that $\lim \lambda_n \to \lambda_0$, and $\mu < \lambda_0$. For n large enough, we have $\mu < \lambda_n$, and, for any $A \in \mathcal{K}$, $\psi_\mu(A) \supset \psi_{\lambda_n}(A)$. Hence $\overline{\lim}\ \psi_{\lambda_n}(A) \subset \psi_\mu(A)$, and thus

$$\overline{\lim}\ \psi_{\lambda_n}(A) \subset \bigcap_{\mu < \lambda_0} \psi_\mu(A)$$

It follows that condition 3 implies 2. Moreover, if $\{A_n\} \subset \mathcal{K}$ is a sequence such that $\lim A_n = A$, we have $\psi_\mu(\psi_{\lambda_n}(A_n)) = \psi_{\lambda_n}(A_n)$. If condition 2 is true, ψ_μ is u.s.c. For any sequence $k \to n_k$ such that $\lim \psi_{\lambda_{nk}}(A_{n_k}) = A_0$ in $\mathcal{K}$, it follows that $A_0 = \lim \psi_{\lambda_{nk}}(A_{n_k}) \subset \psi_\mu(A_0) \subset A_0$. Thus $A_0 = \psi_\mu(A_0)$ for any $\mu < \lambda_0$. On the other hand, ψ_λ is regular from above (for condition 2 implies 3), and thus $A_0 = \psi_{\lambda_0}(A_0) \in \mathcal{B}_{\lambda_0}$. But $\psi_{\lambda_n}(A_n) \subset A_n$ implies $A_0 \subset A$ and $A_0 = \psi_{\lambda_0}(A_0) \subset \psi_{\lambda_0}(A)$. It follows $\overline{\lim}\ \psi_{\lambda_n}(A_n) \subset \psi_{\lambda_0}(A)$, and condition 2 implies 1.

7-5. OPENINGS AND CLOSINGS L.S.C. ON $\mathcal{K}$ AND $\mathcal{F}$

The l.s.c. mappings from $\mathcal{F}$ or $\mathcal{K}$ into themselves are less useful for the applications than the u.s.c. ones. Nevertheless, it is interesting to characterize at least the continuous closings and continuous openings.

Proposition 7-5-1. *A closing ψ on $\mathcal{F}$ (resp. on $\mathcal{K}$) is l.s.c. if and only if the family $\mathcal{B}$ of the ψ-invariant closed (resp. compact) sets is closed in $\mathcal{F}$ (resp. in $\mathcal{K}$).*

Proof. Let ψ be l.s.c. on $\mathcal{F}$, $\{F_n\}\subset\mathcal{B}$, and $\lim F_n = F$ in $\mathcal{F}$. Then $\psi(F)\subset\underline{\lim}\,\psi(F_n)=\lim F_n = F$, and thus $\psi(F)=F$, as ψ is extensive. Hence $\mathcal{B}$ is closed.

Conversely, if $\mathcal{B}$ is closed in $\mathcal{F}$, let $\{F_n\}\subset\mathcal{F}$ be a sequence such that $\lim F_n = F$ in $\mathcal{F}$, and $k\to n_k$ be a subsequence such that $\lim\psi(F_{n_k})=F_0$ in $\mathcal{F}$. Then $F_{n_k}\subset\psi(F_{n_k})$ implies $F\subset F_0$. But $F_0\in\mathcal{B}$, as $\mathcal{B}$ is closed. It follows that $\psi(F)\subset F_0$ and $\psi(F)\subset \underline{\lim}\,\psi(F_n)$. Thus ψ is l.s.c.

Note. From Propositions 7-5-1 and 7-4-2, a closing ψ on $\mathcal{K}$ is *continuous* if and only if $\mathcal{B}$ is closed in $\mathcal{K}$ and contains a fundamental system of neighborhoods for each $B\in\mathcal{B}$. For instance, if $E=\mathbf{R}^d$ is a euclidean space, the *convex hull* C is a continuous closing on $\mathcal{K}(\mathbf{R}^d)$.

Lemma 7-5-1. *Let $\mathcal{I}$ be the family of finite subsets of a LCS space E. Then any increasing mapping ψ l.s.c. from $\mathcal{K}$ into $\mathcal{F}$ admits a unique l.s.c. extension by a mapping ψ' from $\mathcal{F}$ into itself, and*

$$\psi'(F)=\overline{\cup\{\psi(B), B\in\mathcal{I}, B\subset F\}} \qquad (F\in\mathcal{F})$$

Proof. If there exists such an extension ψ', it admits the indicated representation, for the filtering increasing family $\{B, B\in\mathcal{I}, B\subset F\}$ converges toward F in $\mathcal{F}$, and thus $\psi'(F)\subset\underline{\lim}\,\psi(B)$. But $\psi'(F)\supset\psi(B)$ for any $B\subset F$ implies $\psi'(F)\supset\overline{\lim}\,\psi(B)$, and $\psi'(F)=\lim\psi(B)$.

Conversely, let ψ be l.s.c. from $\mathcal{K}$ into $\mathcal{F}$. The mapping ψ': $\mathcal{F}\to\mathcal{F}$ defined by the formula of the lemma is an extension of ψ, for it follows from the first part of the proof (stated with $K\in\mathcal{K}$ instead of $F\in\mathcal{F}$). Let $\{F_n\}\subset\mathcal{F}$ be a sequence such that $\lim F_n = F$ in $\mathcal{F}$, and

$$x\in\cup\{\psi(B), B\in\mathcal{I}, B\subset F\}.$$

Then there exists a finite set $B\in\mathcal{I}$ such that $B\subset F$, $x\in\psi(B)$, say,

$B=\{x_1,x_2,\ldots,x_k\}$. For each $i=1,2,\ldots,k$, we can find a sequence $n\to y_n{}^i\in F_n$ such that $\lim y_n{}^i=x_i$. Put $B_n=\{x_n{}^i, i=1,\ldots,k\}$ so that $B_n\subset F_n$, and $\lim B_n=B$ in $\mathcal{F}$. It follows that $\psi(B)\subset \underline{\lim}\,\psi(B_n)\subset \underline{\lim}\,\psi'(F_n)$, for ψ is l.s.c., and thus $x\in \underline{\lim}\,\psi'(F_n)$. Hence

$$\cup\{\psi(B),B\in\mathcal{S},B\subset F\}\subset \underline{\lim}\,\psi'(F_n)$$

It follows that $\psi'(F)\subset \underline{\lim}\,\psi'(F_n)$, as $\underline{\lim}\,\psi'(F_n)$ is closed, and so ψ' is l.s.c.

Proposition 7-5-2. *An opening ψ on $\mathcal{K}$ (resp. on $\mathcal{F}$) is l.s.c. if and only if the family $\mathcal{B}$ of the compact (resp. closed) ψ-invariant sets is generated by infinite closed union from a family $\mathcal{B}_0\subset\mathcal{S}$ of finite sets such that, for any $B_0=\{x_1,\ldots,x_k\}\in\mathcal{B}_0$, there exists $G_i\in\mathcal{G}$, $i=1,\ldots,k$, such that $x_i\in G_i$ and $\{y_1,\ldots,y_k\}\in\mathcal{B}_0$ if $y_i\in G_i$, $i=1,\ldots,k$.*

Proof. For any finite set $B\in\mathcal{S}$, we have $\psi(B)\in\mathcal{S}$. Thus, if ψ is l.s.c. on $\mathcal{K}$ (or on $\mathcal{F}$), it follows from Lemma 7-5-1 that $\mathcal{B}$ is generated by $\mathcal{B}_0=\mathcal{B}\cap\mathcal{S}$. Let $B_0=\{x_1,\ldots,x_k\}$ be in $\mathcal{B}_0$, and $G_i'\in\mathcal{G}$, $i=1,\ldots,k$, be disjoint open neighborhoods of the points x_i. Since $B_0=\psi(B_0)$ hits each G_i', there exist a finite number n of open sets G_j, $j=1,\ldots,n$, such that $B_0\cap G_j\neq\varnothing$ and $A\cap G_j\neq\varnothing$ imply $\psi(A)\cap G_i'\neq\varnothing$ ($A\in\mathcal{K}$, $j=1,\ldots,n$, $i=1,\ldots,k$). If so, $\psi(A)$ and (a fortiori) A itself contain at least k distinct points, and thus $n\geqslant k$. On the other hand, B_0 encloses k points and hits each of the G_j (which we may always suppose to be disjoint), so that $n\leqslant k$ and hence $n=k$. The G_j, $j=1,2,\ldots,k$, may be ordered in such a way that $x_i\in G_i$. For any $A=\{y_1,\ldots,y_k\}$ such that $y_i\in G_i$, $i=1,\ldots,k$, we have $\psi(A)\cap G_i'\neq\varnothing$ for each $i=1,\ldots,k$, and $\psi(A)$ contains at least k distinct points. Then $\psi(A)\subset A$ implies $A=\psi(A)\in\mathcal{B}$, and the "only if" part is proved.

Conversely, suppose that the stated condition is satisfied. For any $K\in\mathcal{K}$, we have

$$\psi(K)=\overline{\cup\{B,B\subset K,B\in\mathcal{B}_0\}}$$

Let $\{K_n\}\subset\mathcal{K}$ be a sequence such that $\lim K_n=K$ in $\mathcal{K}$, and $x\in\psi(K)$ such that $x\in B_0$ for a set $B_0=\{x_1,\ldots,x_k\}\in\mathcal{B}_0$. There exist sequences $n\to y_n{}^i\in K_n$ such that $\lim y_n{}^i=x_i$, $i=1,2,\ldots,k$. From the hypothesis, $B_n=\{y_n{}^1,\ldots,y_n{}^k\}\in\mathcal{B}_0$ for n large enough, $B_n\subset K_n$, and there exists $x_n\in B_n\subset K_n$ such that $x=\lim x_n$ (for $x\in B_0$). It follows that $x_n\in\psi(K_n)$, because $B_n\in\mathcal{B}_0$, and $x=\lim x_n$ implies $x\in\underline{\lim}\,\psi(K_n)$. Thus the set $\cup\{B,B\subset K,B\in\mathcal{B}_0\}$ is contained in the closed set $\underline{\lim}\,\psi(K_n)$. It follows that $\psi(K)\subset\underline{\lim}\,\psi(K_n)$, and so ψ is l.s.c.

Corollary. *If the space E is connected, the only openings continuous on $\mathcal{K}$ or on $\mathcal{F}$ are the identity and the trivial mapping $A \to \varnothing$.*

Proof. If ψ is a (nontrivial) opening on $\mathcal{F}$, its restriction to $\mathcal{K}$ is an opening on $\mathcal{K}$ (for ψ is antiextensive) and is continuous for the myope topology if ψ is continuous on $\mathcal{F}$. Thus it is sufficient to prove the statement concerning the space $\mathcal{K}$, for $\mathcal{K}$ is dense in $\mathcal{F}$. The family $\mathcal{B}$ of the compact ψ-invariant sets is closed in $\mathcal{K}$ (Proposition 7-4-1) and generated by a family $\mathcal{B}_0 \subset \mathcal{I}$ satisfying the condition of Proposition 7-5-2. Let $B = \{x_1, \ldots, x_k\}$ be in $\mathcal{B}_0$. From Proposition 7-5-2, there exists $G \in \mathcal{G}$ such that $x_k \in G$ and $\{x_1, \ldots, x_{k-1}, y\} \in \mathcal{B}_0$ for any $y \in G$. Let G_k be the largest open set such that this property holds (i.e., the union of all the suitable $G \in \mathcal{G}$), and $z \in \partial G_k$. Then $\{x_1, \ldots, x_{k-1}, z\} \in \mathcal{B} \cap \mathcal{I} = \mathcal{B}_0$, as $\mathcal{B}$ is closed, and $\{x_1, \ldots, x_{k-1}, z\} = \lim \{x_1, \ldots, x_{k-1}, z_n\}$ for a sequence $\{z_n\} \subset G_k$. In other words, there exists an open neighborhood G' of z such that $\{x_1, \ldots, x_{k-1}, y\} \in \mathcal{B}_0$ for any $y \in G'$. But $G' \subset G_k$, because G_k is maximal, and this contradicts $z \in \partial G_k$. Thus G_k has no boundary points. Hence $G_k = E$, for E is connected. By taking $y = x_{k-1} \in G_k$, we conclude that $\{x_1, \ldots, x_{k-1}\} \in \mathcal{B}_0$. By a recurrence procedure, we have, in the same way, $\{x_1\} \in \mathcal{B}_0$, and the maximal open set G_1 such that $y \in G_1$ implies $\{y\} \in \mathcal{B}_0$ has no boundary point. Hence $G_1 = E$, and $\{x\}$ is ψ-invariant for any $x \in E$. Thus ψ is the identity on $\mathcal{K}$, and the only possible continuous extension on $\mathcal{F}$ is the identity on $\mathcal{F}$.

If $E = \mathbf{R}^d$ is a euclidean space, and $\mathcal{C} \subset \mathcal{P}$ is closed under the *Minkowski sum* $\oplus$, a mapping ψ: $\mathcal{C} \to \mathcal{P}$ is said to be compatible with the Minkowski sum if $\psi(A \oplus A') = \psi(A) \oplus \psi(A')$. If so, we shall say for brevity that ψ is a $\oplus$-mapping.

Proposition 7-5-3. *The only $\oplus$-openings on $\mathcal{K}(\mathbf{R}^d)$ are the identity and the trivial mapping $A \to \varnothing$.*

Proof. Let ψ be an opening on $\mathcal{K}$, and $\psi(K \oplus K') = \psi(K) \oplus \psi(K')$, K, $K' \in \mathcal{K}$. In particular, $\psi(\{o\}) = \psi(\{o\}) \oplus \psi(\{o\})$, and thus $\psi(\{o\}) = \{o\}$ or $\varnothing$, for $\psi(\{o\})$ is compact. If $\psi(\{o\}) = \varnothing$, $\psi(K \oplus \{o\}) = \psi(K) \oplus \varnothing = \varnothing$ for any $K \in \mathcal{K}$. Suppose that $\psi(\{o\}) = \{o\}$. It follows that, for any $x \in \mathbf{R}^d$, $\psi(\{x\}) \oplus \psi(\{-x\}) = \psi(\{x\} \oplus \{-x\}) = \{o\}$. Thus $\psi(\{x\}) \neq \varnothing$ and encloses a single element. Then $\psi(\{x\}) \subset \{x\}$ implies $\psi(\{x\}) = \{x\}$. It follows that ψ is the identity.

Proposition 7-5-4. *Any $\oplus$-closing ψ on $\mathcal{K}(\mathbf{R}^d)$ is continuous and compatible with translations. Its restriction to $C(\mathcal{K})$ is compatible with homothetics $\geqslant 0$, $\psi(A)$ is convex for any $A \in C(\mathcal{K})$, and $\psi(\{x\}) = x$ for any $x \in \mathbf{R}^d$. A closing*

ψ on $\mathcal{K}(\mathbf{R}^d)$ is a $\oplus$-closing if and only if the family of the compact ψ-invariant sets is closed under translations, the Minkowski sum, and infinite intersections.

Proof. Let ψ be a $\oplus$-closing on $\mathcal{K}$.

a. $\psi(\{x\})=\{x\}$ for any $x\in\mathbf{R}^d$, and ψ is a τ-mapping.

For $\{o\}=\{o\}\oplus\{o\}$ yields $\psi(\{o\})=\psi(\{o\})\oplus\psi(\{o\})\in\mathcal{K}$, and thus $\psi(\{o\})=\{o\}$ [because $o\in\psi(\{o\})$]. It follows that $\{o\}=\psi(\{x\})\oplus\psi(\{-x\})$, and $\psi(\{x\})\supset\{x\}$ contains only one element. Thus $\psi(\{x\})=\{x\}$ for any $x\in\mathbf{R}^d$. For any $K\in\mathcal{K}$, it follows that $\psi(K\oplus\{x\})=\{x\}\oplus\psi(K)$, and so ψ is a τ-closing.

b. $\psi(A)$ is convex for any $A\in C(\mathcal{K})$, and $\psi(rA)=r\psi(A)$, r rational $\geqslant 0$.

For $A\in C(\mathcal{K})$ implies $A=(1/n)A^{\oplus n}$ and $\psi(A)=\psi(A/n)^{\oplus n}$. Thus $\psi(A)$ is infinitely divisible for $\oplus$, and $\psi(A)\in C(\mathcal{K})$ (Theorem 1-5-1). Then $\psi(A)=\psi(A/n)^{\oplus n}=n\psi(A/n)$ implies $\psi(rA)=r\psi(A)$ for any r rational $\geqslant 0$.

c. ψ is continuous on $\mathcal{K}$.

Let B be the unit ball. It follows from part b that $\psi((1/n)B)=(1/n)\psi(B)$. Thus, for any $\epsilon>0$, there exists N_ϵ such that $n\geqslant N_\epsilon$ implies $\psi((1/n)B)\subset\epsilon B$. Then let $\{A_n\}\subset\mathcal{K}$ be a sequence converging toward A in $\mathcal{K}$. For n large enough, $A_n\subset A\oplus(1/N_\epsilon)B$ and $A\subset A_n\oplus(1/N_\epsilon)B$ imply

$$\psi(A_n)\subset\psi(A)\oplus\psi\left(\frac{1}{N_\epsilon}B\right)\subset\psi(A)\oplus\epsilon B$$

$$\psi(A)\subset\psi(A_n)\oplus\psi\left(\frac{1}{N_\epsilon}B\right)\subset\psi(A_n)\oplus\epsilon B$$

It follows that $\psi(A)=\lim\psi(A_n)$, and ψ is continuous.

d. For any $\lambda\geqslant 0$ and $A\in C(\mathcal{K})$, $\psi(\lambda A)=\lambda\psi(A)$.

This is an immediate corollary of parts b and c.

e. If ψ is a $\oplus$-closing and thus a τ-closing, the family $\mathcal{B}$ of the compact invariant sets is closed under the translations, the Minkowski sum, and $\cap$. Conversely, let $\mathcal{B}\subset\mathcal{K}$ be closed under $\oplus$, $\cap$, and translations. The largest extension on $\mathcal{K}$ of the identity on $\mathcal{B}$ is the closing on $\mathcal{K}$, defined by

$$\psi(K)=\cap\{B, B\in\mathcal{B}, B\supset K\}\qquad(K\in\mathcal{K})$$

Hence ψ is a τ-closing (for $\mathcal{B}$ is closed under translations). For any K,

$K' \in \mathcal{K}$, it follows that

$$\psi(K \oplus K') = \psi\left(\bigcup_{x \in K'} (K \oplus \{x\}) \right) \supset \bigcup_{x \in K'} \psi(K \oplus \{x\})$$

$$= \bigcup_{x \in K'} \psi(K) \oplus \{x\} = K' \oplus \psi(K)$$

This implies $\psi\psi(K \oplus K') = \psi(K \oplus K') \supset \psi(K' \oplus \psi(K)) \supset \psi(K') \oplus \psi(K)$ for ψ is increasing and idempotent, and thus

$$\psi(K \oplus K') \supset \psi(K) \oplus \psi(K')$$

On the other hand, $\psi(K) \supset K$ and $\psi(K') \supset K'$ imply $\psi(K) \oplus \psi(K') \supset K \oplus K'$. But $\psi(K) \oplus \psi(K') \in \mathcal{B}$, for $\mathcal{B}$ is closed under $\oplus$, and thus $\psi(K) \oplus \psi(K') \supset \psi(K \oplus K')$. Hence $\psi(K \oplus K') = \psi(K) \oplus \psi(K')$.

CHAPTER 8

Increasing Mappings

This chapter, as well as the following one, examines special topics and may be omitted by a reader interested only in applications. It is devoted to increasing mappings, that is, to mappings $\psi: \mathcal{P}(E) \to \mathcal{P}(E')$ such that $A \subset B$ implies $\psi(A) \subset \psi(B)$, and its aim is to generalize some properties encountered when studying openings and closings. Exactly as in Chapter 7, and for the same reason, we examine first the "euclidean" case, that is, the case of increasing mappings, before investigating the more intricate general case in the last two sections.

In this chapter, the word "mapping" will always mean "increasing mapping," unless the contrary is explicitly stated, and, in the same way, we say "τ-mapping" instead of "increasing mapping compatible with translations."

8-1. ALGEBRAIC PROPERTIES OF (INCREASING) τ-MAPPINGS

In this first section, $E = \mathbf{R}^d$ is a *euclidean* space, and we consider only τ-mappings ψ defined on a class $\mathcal{Q} \subset \mathcal{P}$ closed under translations. In other words, if writing A_x instead of $A \oplus \{x\}$,

$$\psi(A_x) = (\psi(A))_x \qquad (A \in \mathcal{Q})$$

In the same way, for any $\mathcal{B} \subset \mathcal{P}$, we denote by $\mathcal{B}_x$ the family of the translates B_x, $B \in \mathcal{B}$, where x is given in $\mathbf{R}^d$.

If $\psi: \mathcal{Q} \to \mathcal{P}$ is a τ-mapping (not necessarily increasing), the family

$$\mathcal{V} = \{A, A \in \mathcal{Q}, O \in \psi(A)\}$$

is said to be the *kernel* of the mapping ψ. Clearly, $x \in \psi(A)$ is equivalent to $A \in \mathcal{V}_x$, for ψ is a τ-mapping. Conversely, if $\mathcal{V} \subset \mathcal{Q}$ is an arbitrary family, the mapping ψ defined by

$$\psi(A) = \{x, x \in \mathbf{R}^d, A \in \mathcal{V}_x\} \qquad (A \in \mathcal{Q})$$

is a τ-mapping, and its kernel is $\mathcal{V}$, so that there is a one-to-one mapping from $\mathcal{P}(\mathcal{Q})$ onto the space of the τ-mappings on $\mathcal{Q}$.

Example 1. For a given $B \in \mathcal{P}$, the *dilatation* $A \to A \oplus \check{B}$ is a τ-mapping on $\mathcal{P}$, and its kernel is

$$\mathcal{V}_B = \{A, A \in \mathcal{P}, A \cap B \neq \varnothing\} \tag{8-1-1}$$

Example 2. In the same way, the *erosion* $A \to A \ominus \check{B}$ has the kernel

$$\mathcal{W}_B = \{A, A \in \mathcal{P}, A \supset B\} \tag{8-1-2}$$

Example 3. The *opening* by B, that is, $A \to A_B = (A \ominus \check{B}) \oplus B$, has the kernel

$$\mathcal{V} = \bigcup_{y \in \check{B}} \mathcal{W}_{B_y} \tag{8-1-3}$$

Example 4. The *closing* by B, that is, $A \to A^B = (A \oplus \check{B}) \ominus B$, has the kernel

$$\mathcal{V} = \bigcap_{y \in \check{B}} \mathcal{V}_{B_y} \tag{8-1-4}$$

Example 5. The *identity* on $\mathcal{P}$ has the kernel

$$\mathcal{P}_0 = \{A, A \in \mathcal{P}, 0 \in A\}$$

In the same way, for any $x \in \mathbf{R}^d$, the kernel of the *translation* $A \to A_x$ is $\mathcal{P}_x = \{A, A \in \mathcal{P}, x \in A\}$.

We recall that a family $\mathcal{V} \subset \mathcal{P}$ is *a filter* if (1), $\varnothing \notin \mathcal{V}$, (2) $\mathcal{V}$ is closed under finite intersection, and (3) $\mathcal{V}$ is $\cup$-hereditary (i.e., $B \in \mathcal{V}$ and $A \supset B$ imply $A \in \mathcal{V}$). A filter $\mathcal{V}$ is said to be *an ultrafilter* if $\mathcal{V}' = \mathcal{V}$ for any other filter $\mathcal{V}'$ such that $\mathcal{V}' \supset \mathcal{V}$. A filter $\mathcal{V}$ is an ultrafilter if and only if, for any $A \in \mathcal{P}$, either $A \in \mathcal{V}$ or $A^c \in \mathcal{V}$ (see, e.g., Bourbaki, 1965a). The kernel $\mathcal{P}_x$ of the translation is an ultrafilter.

Note. Since $\varnothing$ and $\mathbf{R}^d$ are the only elements invariant under translations in $\mathcal{P}(\mathbf{R}^d)$, then, for any τ-mapping ψ, $\psi(\varnothing)$ and $\psi(\mathbf{R}^d)$ are either $\varnothing$ or $\mathbf{R}^d$.

Clearly, $\psi(\varnothing)=\varnothing$ if and only if $\varnothing \notin \mathcal{V}$, and $\psi(\mathbf{R}^d)=\varnothing$ if and only if $\mathbf{R}^d \notin \mathcal{V}$. In what follows, we consider only nontrivial τ-mappings, so that $\psi(\varnothing)=\varnothing$ and $\psi(\mathbf{R}^d)=\mathbf{R}^d$, that is,

$$\varnothing \notin \mathcal{V}, \qquad \mathbf{R}^d \in \mathcal{V}$$

Let us give a few elementary results. The condition for ψ to be extensive or antiextensive is

$$\left.\begin{array}{l}\mathcal{P}_0 \subset \mathcal{V} \Leftrightarrow A \subset \psi(A) \\ \mathcal{P}_0 \supset \mathcal{V} \Leftrightarrow A \supset \psi(A)\end{array}\right\} \qquad (A \in \mathcal{P}) \tag{8-1-5}$$

More generally, if ψ and ψ' are two mappings (not necessarily increasing),

$$\mathcal{V} \subset \mathcal{V}' \Leftrightarrow \psi \subset \psi' \tag{8-1-6}$$

If ψ_i, $i \in I$, is a family of τ-mappings and $\mathcal{V}_i$ is the kernel of ψ_i, then the kernels of $\cap \psi_i$ and $\cup \psi_i$ are $\cap \mathcal{V}_i$ and $\cup \mathcal{V}_i$, respectively.

Also, note that a τ-mapping is *increasing* if and only if $\mathcal{V}$ is $\cup$-hereditary, that is, $B \in \mathcal{V}$ and $A \supset B$ imply $A \in \mathcal{V}$. With the notations of (8-1-2),

$$\psi \text{ increasing} \Leftrightarrow \mathcal{V} = \bigcup_{B \in \mathcal{V}} \mathcal{W}_B \tag{8-1-7}$$

On comparing (8-1-2) and (8-1-7), we conclude that *any increasing τ-mappings admits the representation*

$$\psi(A) = \bigcup_{B \in \mathcal{V}} (A \ominus \check{B})$$

that is, $\psi(A)$ is the union of the erosions of A by the sets $B \in \mathcal{V}$.

Concerning the operations $\cap$ and $\cup$, we find

$$\left.\begin{array}{l}\mathcal{V} \text{ closed under (finite) } \cup \Leftrightarrow \psi(A) \cap \psi(B) \subset \psi(A \cup B) \\ \mathcal{V} \text{ closed under (finite) } \cap \Leftrightarrow \psi(A \cap B) \supset \psi(A) \cap \psi(B)\end{array}\right. \tag{8-1-8}$$

On the other hand, ψ is increasing if and only if $\psi(A \cap B) \subset \psi(A) \cap \psi(B)$. Then it follows from (8-1-7) and (8-1-8) that

$$\psi(A \cap B) = \psi(A) \cap \psi(B) \Leftrightarrow \mathcal{V} \text{ is a filter} \tag{8-1-9}$$

In other words, a nontrivial τ-mapping is increasing and compatible with $\cap$ if and only if its kernel is a filter.

Proposition 8-1-1. *Let ψ be a nontrivial τ-mapping on $\mathcal{P}(\mathbf{R}^d)$. If ψ is extensive and compatible with $\cap$, then ψ is the identity.*

Proof. The kernel $\mathcal{V}$ of ψ is a filter by (8-1-9), and $\mathcal{V} \supset \mathcal{P}_0$ by (8-1-5). It follows that $\mathcal{V} = \mathcal{P}_0$, for $\mathcal{P}_0$ is an ultrafilter.

From the standpoint of duality, if the kernel of ψ is $\mathcal{V}$, then ψ^* has the kernel

$$\mathcal{V}^* = \{A, A \in \mathcal{P}, A^c \notin \mathcal{V}\} \tag{8-1-10}$$

Further ψ^* is increasing if and only if ψ is increasing; $\mathcal{V}^*$ is closed under $\cap$ if and only if $\complement\mathcal{V}$ is closed under $\cup$. Thus

$$\complement\mathcal{V} \text{ closed under } \cup \Leftrightarrow \psi(A \cup B) \supset \psi(A) \cup \psi(B) \tag{8-1-11}$$

and from (8-1-9),

$$\psi(A \cup B) = \psi(A) \cup \psi(B) \Leftrightarrow \mathcal{V} \text{ is an antifilter} \tag{8-1-12}$$

[$\mathcal{V}$ is an antifilter if $\mathcal{V}^*$ is a filter, i.e., (1) $\mathbf{R}^d \in \mathcal{V}$, (2) $\mathcal{V}$ is $\cup$-hereditary, and (3) $\complement\mathcal{V}$ is closed under $\cup$].

Note that $\mathcal{V}$ is an *ultrafilter* if and only if it is at the same time a filter and an antifilter. (A filter is an untrafilter if and only if, for any $A \in \mathcal{P}$, either $A \in \mathcal{V}$ or $A^c \in \mathcal{V}$.) It follows that a τ-mapping ψ on $\mathcal{P}$ is compatible with $\cup$ and $\cap$ if and only if its kernel is an ultrafilter. For any $x \in \mathbf{R}^d$, $\mathcal{P}_x$ is an ultrafilter, and the corresponding translation is compatible with $\cup$ and $\cap$. Although we know from the Zorn theorem that there exist many other ultrafilters, the only ones that we are able to construct effectively are the $\mathcal{P}_x$, $x \in \mathbf{R}^d$. Thus the only τ-mappings compatible with $\cup$ and $\cap$ that can be effectively constructed are the translations (although, theoretically, many others exist).

By duality, it follows from Proposition 8-1-1 that:

Proposition 8-1-2. *If a nontrivial τ-mapping ψ is antiextensive and $\cup$-compatible, ψ is the identity on $\mathcal{P}(\mathbf{R}^d)$.*

Corollary. *The identity is the only nontrivial $\cup$-compatible τ-opening and the only nontrivial $\cap$-compatible τ-closing on $\mathcal{P}(\mathbf{R}^d)$.*

If ψ is an (increasing) τ-mapping, its kernel $\mathcal{V}$ and the kernel $\mathcal{V}^*$ of the dual mapping ψ^* satisfy (8-1-10) and (8-1-7), and thus

$$\mathcal{V}^* = \bigcap_{B \in \mathcal{V}} \mathcal{V}_B \qquad \mathcal{V} = \bigcap_{B \in \mathcal{V}^*} \mathcal{V}_B \tag{8-1-13}$$

Taking (8-1-7), (8-1-13), and (8-1-1) into account, we conclude that

Proposition 8-1-3. *Any (increasing) τ-mapping ψ on $\mathcal{P}(\mathbf{R}^d)$ is a union of erosions, and also an intersection of dilatations. More precisely, if $\mathcal{V}$ and $\mathcal{V}^*$ are the kernels of ψ and ψ^*, respectively, then, for any $A \in \mathcal{P}$,*

$$\psi(A) = \bigcup_{B \in \mathcal{V}} (A \ominus \check{B}) = \bigcap_{B \in \mathcal{V}^*} (A \oplus \check{B})$$

If now the τ-mapping ψ is defined only on a family $\mathcal{A} \subset \mathcal{P}$ (closed under translations), so that its kernel is $\mathcal{V} \subset \mathcal{A}$, let $\underset{\sim}{\psi}$ be its smallest extension and $\tilde{\psi}$ its largest extension on $\mathcal{P}$. Then the kernels $\underset{\sim}{\mathcal{V}}$ and $\tilde{\mathcal{V}}$ of $\underset{\sim}{\psi}$ and $\tilde{\psi}$, respectively, are

$$\underset{\sim}{\mathcal{V}} = \bigcup_{A \in \mathcal{V}} \mathcal{W}_A, \qquad \tilde{\mathcal{V}} = \bigcap_{A \in \mathcal{V}^*} \mathcal{V}_A \tag{8-1-14}$$

and by duality

$$(\underset{\sim}{\mathcal{V}})^* = (\tilde{\mathcal{V}}^*), \qquad (\tilde{\mathcal{V}})^* = (\underset{\sim}{\mathcal{V}}^*)$$

The following result will be very useful in the sequel:

Proposition 8-1-4. *Let $\mathcal{A}$, $\mathcal{B} \subset \mathcal{P}(\mathbf{R}^d)$ be closed under translation, ψ be a (increasing) τ-mapping on $\mathcal{A}$, $\mathcal{V} \subset \mathcal{A}$ its kernel, ψ_b the restriction to $\mathcal{B}$ of the smallest extension $\underset{\sim}{\psi}$ of ψ on $\mathcal{P}$, and ψ_a the restriction to $\mathcal{A}$ of the largest extension $\tilde{\psi}_b$ of ψ_b on $\mathcal{P}$. Then the three following conditions are equivalent:*

1. *$\psi = \psi_a$ on $\mathcal{A}$.*
2. *For any $A \in \mathcal{A}$ and $x \notin \psi(A)$, there exists $B \in \mathcal{B}$ such that $B \supset A$ and $x \notin \psi(A')$ for any $A' \in \mathcal{A}$ contained in B.*
3. *For any $A \in \mathcal{A} \cap \complement \mathcal{V}$, there exists $B \in \mathcal{B}$ such that $B \supset A$ and $A' \notin \mathcal{V}$ for any $A' \in \mathcal{A}$ contained in B.*

Proof. The relationship $\psi_a \supset \psi$ is always true, for $\tilde{\psi}_b \supset \underset{\sim}{\psi}$, because $\tilde{\psi}_b$ is the largest extension of ψ_b, and thus the restrictions ψ_a and ψ of $\tilde{\psi}_b$ and $\underset{\sim}{\psi}$ to

$\mathcal{A}$ satisfy $\psi_a \supset \psi$. Thus condition 1 is equivalent to $\psi_a \subset \psi$. From the definitions, for any $A \in \mathcal{A}$,

$$\psi_a(A) = \cap \{\psi_b(B), B \supset A, B \in \mathcal{B}\}$$

Thus $\psi_a(A) \subset \psi(A)$ is equivalent to $x \in \psi_b(B)$ for any $B \in \mathcal{B}$ such that $B \supset A$ implies $x \in \psi(A)$. This implication is equivalent to the following: for any $x \notin \psi(A)$, there exists $B \in \mathcal{B}$ such that $B \supset A$ and $x \notin \psi_b(B)$. Hence conditions 1 and 2 are equivalent. The third condition is a simple translation of 2 in terms of kernels.

Note. The euclidean structure of $\mathbf{R}^d$ does not intervene in the proof of the equivalence condition 1 $\Leftrightarrow$condition 2, and this equivalence remains valid for E an arbitrary set.

If E is a LCS space, the most interesting case is $\mathcal{A} = \mathcal{K}(E)$, and $\mathcal{B} = \mathcal{G}(E)$. If ψ: $\mathcal{K} \to \mathcal{P}$ is increasing, we find $\psi_g(G) = \cup\{\psi(K), K \subset G, K \in \mathcal{K}\}$ for any $G \in \mathcal{G}$, and $\psi_k(K) = \cap\{\psi_g(G), G \supset K, G \in \mathcal{G}\}$ for any $K \in \mathcal{K}$. Then $\psi = \psi_k$ on $\mathcal{K}$ if and only if, for any $x \notin \psi(K)$, there exists $G \in \mathcal{G}$ such that $G \supset K$ and $x \notin \psi(K')$ for any compact set $K' \subset G$. In particular, *this condition is fulfilled if ψ is an u.s.c. mapping from $\mathcal{K}$ into $\mathcal{K}$ or $\mathcal{F}$.*

8-2. TOPOLOGICAL PROPERTIES OF τ-MAPPINGS

We first examine the (increasing) τ-mappings u.s.c. on $\mathcal{K}$ or $\mathcal{F}$.

Proposition 8-2-1. *A τ-mapping ψ from $\mathcal{K}$ (resp. $\mathcal{F}$) into $\mathcal{P}$ ($\mathbf{R}^d$) is an u.s.c. mapping from $\mathcal{K}$ (resp. $\mathcal{F}$) into $\mathcal{F}$ if and only if its kernel $\mathcal{V}$ is closed in $\mathcal{K}$ (resp. in $\mathcal{F}$). A τ-mapping ψ': $\mathcal{G} \to \mathcal{P}$ is a l.s.c. mapping from $\mathcal{G}$ into itself if and only if its kernel $\mathcal{V}'$ is open in $\mathcal{G}$.*

Proof. If $\mathcal{V}$ is closed, ψ maps $\mathcal{K}$(resp. $\mathcal{F}$) into $\mathcal{F}$, for $x_n \in \psi(A)$ is equivalent to $A_{-x_n} \in \mathcal{V}$, and lim $x_n = x$ implies lim $A_{-x_n} = A_{-x}$ in $\mathcal{K}$ (resp. in $\mathcal{F}$), and thus $A_{-x} \in \mathcal{V}$, that is, $x \in \psi(A)$. Hence $\psi(A) \in \mathcal{F}$. Moreover, for any $K \in \mathcal{K}$,

$$\psi^{-1}(\mathcal{F}_K) = \{A, \psi(A) \cap K \neq \varnothing\} = \bigcup_{x \in K} \mathcal{V}_x$$

In order to prove that $\psi^{-1}(\mathcal{F}_K)$ is closed (i.e., ψ is u.s.c.), let $\{B_n\} \subset \mathcal{V}$, $\{x_n\} \subset K$, and $n \to A_n = B_n \oplus \{x_n\}$ be a sequences such that lim $A_n = A$ in $\mathcal{K}$

(resp. in $\mathcal{F}$). There exists a subsequence $k \to n_k$ such that $\lim x_{n_k} = x \in K$ (as K is compact) and $\lim B_{n_k} = B \in \mathcal{V}$ in $\mathcal{K}$ (resp. in $\mathcal{F}$) (if the basic space is $\mathcal{K}$, the sets $B_n = A_n \oplus \{-x_n\}$ are contained in a fixed compact set, as $\lim A_n = A$ in $\mathcal{K}$, so that the sequence $\{B_n\}$ actually admits accumulation points). Then $A_{n_k} = B_{n_k} \oplus \{x_{n_k}\}$ implies $A = B \oplus \{x\}$, for the Minkowski sum is continuous on $\mathcal{K} \times \mathcal{K}$ (resp. on $\mathcal{F} \times \mathcal{K}$), and $A \in \mathcal{V}_x \subset \psi^{-1}(\mathcal{F}_K)$. Thus ψ is u.s.c.

Conversely, if ψ is an u.s.c. mapping from $\mathcal{K}$ or $\mathcal{F}$ into $\mathcal{F}$, $\psi^{-1}(\mathcal{F}_K)$ is closed for any $K \in \mathcal{K}$. In particular, if $K = \{o\}$, $\psi^{-1}(\mathcal{F}_{\{o\}}) = \mathcal{V}$ is closed. Hence the first statement is proved, and the second one follows by duality.

Now let ψ be a τ-mapping from $\mathcal{K}$ into $\mathcal{P}$, $\mathcal{V} \subset \mathcal{K}$ its kernel, and put

$$\psi_g(G) = \cup \{\psi(K), K \in \mathcal{K}, K \subset G\} \qquad (G \in \mathcal{G})$$

$$\psi_k(K) = \cap \{\psi_g(G), G \in \mathcal{G}, G \supset K\} \qquad (K \in \mathcal{K})$$

Instead of ψ_g, we may consider the dual mapping $\psi_f = \psi_g^*$: $\mathcal{F} \to \mathcal{P}$. If we denote by $\mathcal{V}_g$, $\mathcal{V}_f$, and $\mathcal{V}_k$, respectively, the kernels of the mappings ψ_g, ψ_f, and ψ_k, we find

$$\mathcal{V}_g = \bigcup_{K \in \mathcal{V}} \mathcal{G}_K; \qquad \mathcal{V}_f = \bigcap_{K \in \mathcal{V}} \mathcal{F}_K; \qquad \mathcal{V}_k = \bigcap_{F \in \mathcal{V}_f} \mathcal{K}_F$$

Hence $\mathcal{V}_g$ is open in $\mathcal{G}$, $\mathcal{V}_f$ is closed in $\mathcal{F}$, and $\mathcal{V}_k$ is closed in $\mathcal{K}$. It follows from Proposition 8-2-1 that ψ_g is l.s.c. from $\mathcal{G}$ into itself, ψ_f is u.s.c. from $\mathcal{F}$ into itself, and ψ_k is u.s.c. from $\mathcal{K}$ into $\mathcal{F}$.

Moreover, $\psi_k \supset \psi$ on $\mathcal{K}$, and it follows from Proposition 8-1-4 that $\psi_k = \psi$ if and only if for any $K \notin \mathcal{V}$ there exists $G \in \mathcal{G}$ such that $G \supset K$ and $K' \notin \mathcal{V}$ for any compact set $K' \subset G$, that is, if and only if $\mathcal{V}$ is closed in $\mathcal{K}$. Taking Proposition 8-2-1 into account, we conclude that $\psi = \psi_k$ if and only if ψ is an u.s.c. mapping from $\mathcal{K}$ into $\mathcal{F}$. If ψ' is an u.s.c. mapping from $\mathcal{K}$ into $\mathcal{F}$ such that $\psi' \supset \psi$, it follows that $\psi' = \psi'_k \supset \psi_k$, and thus ψ_k is the smallest u.s.c. mapping $\mathcal{K} \to \mathcal{F}$ such that $\psi_k \supset \psi$. Hence it follows also from Proposition 8-2-1 that $\mathcal{V}_k$ is the closure of $\mathcal{V}$ in $\mathcal{K}$, that is, $\mathcal{V}_k = \overline{\mathcal{V}}$. It is easy to verify that $\cap \{\mathcal{F}_K, K \in \mathcal{V}\} = \cap \{\mathcal{F}_K, K \in \overline{\mathcal{V}}\}$, for each $K \in \overline{\mathcal{V}}$ is the limit in $\mathcal{K}$ of a sequence $\{K_n\} \subset \mathcal{V}$, and thus $\mathcal{V}_f = \cap \{\mathcal{F}_K, K \in \mathcal{V}_k\}$. When starting from a τ-mapping $\mathcal{F} \to \mathcal{P}$, we obtain very similar results, and we may state

Proposition 8-2-2. *Let ψ: $\mathcal{K} \to \mathcal{P}$ be a τ-mapping, $\mathcal{V} \subset \mathcal{K}$ its kernel, ψ_f: $\mathcal{F} \to \mathcal{P}$ and ψ_k: $\mathcal{K} \to \mathcal{P}$ the τ-mappings associated with the kernels $\mathcal{V}_f = \cap \{\mathcal{F}_K, K \in \mathcal{V}\}$ and $\mathcal{V}_k = \cap \{\mathcal{K}_F, F \in \mathcal{V}_f\}$. Then ψ_f and ψ_k are u.s.c. mappings from $\mathcal{F}$ and $\mathcal{K}$ respectively, into $\mathcal{F}$. Moreover, $\mathcal{V}_k = \overline{\mathcal{V}}$ is the* *closure of $\mathcal{V}$ in $\mathcal{K}$, ψ_k is the smallest upper bound of ψ u.s.c. from $\mathcal{K}$ into $\mathcal{F}$, and $\psi=\psi_k$ if and only if ψ is u.s.c. from $\mathcal{K}$ into $\mathcal{F}$. Furthermore:*

$$\mathcal{V}_k=\cap\{\mathcal{K}_F, F\in\mathcal{V}_f\}; \qquad \mathcal{V}_f=\cap\{\mathcal{F}_K, K\in\mathcal{V}_k\} \qquad (8\text{-}2\text{-}1)$$

Now let ψ': $\mathcal{F}\to\mathcal{P}$ be a τ-mapping, $\mathcal{V}'\subset\mathcal{F}$ its kernel, ψ'_k: $\mathcal{K}\to\mathcal{P}$ and ψ'_f: $\mathcal{F}\to\mathcal{P}$ the τ-mappings associated with the kernels $\mathcal{V}'_k=\cap\{\mathcal{K}_F, F\in\mathcal{V}'\}$ and $\mathcal{V}'_f=\cap\{\mathcal{F}_K, K\in\mathcal{V}'_k\}$. Then ψ'_k and ψ'_f are u.s.c. mappings from $\mathcal{K}$ and $\mathcal{F}$ repsectively, into $\mathcal{F}$, $\mathcal{V}'_f$ is the closure of $\mathcal{V}'\subset\mathcal{F}$ and ψ'_f is the smallest upper bound of ψ' u.s.c. from $\mathcal{F}$ into $\mathcal{F}$. Moreover, relationships (8-2-1) *hold for $\mathcal{V}'_k$ and $\mathcal{V}'_f$.*

Corollary. *A τ-mappings ψ: $\mathcal{K}\to\mathcal{P}$ (resp. $\mathcal{F}\to\mathcal{P}$) is an u.s.c. mapping from $\mathcal{K}$ (resp. from $\mathcal{F}$) into $\mathcal{F}$ if and only if it admits the representation $\psi(K)=\cap\{K\oplus\check{F}, F\in\mathcal{B}\}$ for a family $\mathcal{B}$ closed in $\mathcal{F}$ [resp. $\psi(F)=\cap\{F\oplus\check{K}, K\in\mathcal{B}'\}$ for a family $\mathcal{B}'$ closed in $\mathcal{K}$].*

Extension of a Mapping U.S.C. on $\mathcal{K}$

If ψ: $\mathcal{K}\to\mathcal{F}$ is an arbitrary u.s.c. τ-mapping, it is not true in general that it admits an extension to an u.s.c. mapping from $\mathcal{F}$ into itself.

Proposition 8-2-3. *Let ψ: $\mathcal{K}\to\mathcal{F}$ be an u.s.c. τ-mapping, $\mathcal{V}\subset\mathcal{K}$ its kernel, $\mathcal{V}'_f=\overline{\mathcal{V}}$ the closure of $\mathcal{V}$ in $\mathcal{F}$ and ψ'_f: $\mathcal{F}\to\mathcal{F}$ the u.s.c. mapping associated with $\mathcal{V}'_f$. Then ψ admits u.s.c. extensions from $\mathcal{F}$ into itself if and only if one of the two following equivalent conditions is satisfied:*

1. $\mathcal{V}=\mathcal{V}'_f\cap\mathcal{K}$.
2. *ψ is u.s.c. on $\mathcal{K}\cap\mathcal{F}$ for the relative $\mathcal{F}$-topology.*

If one of these equivalent conditions is satisfied, ψ'_f is the smallest extension of ψ by an u.s.c. mapping.

Proof. If ψ': $\mathcal{F}\to\mathcal{F}$ is an u.s.c. extension of ψ, its kernel $\mathcal{V}'$ is closed in $\mathcal{F}$ (Proposition 8-2-1) and $\mathcal{V}=\mathcal{V}'\cap\mathcal{K}$. Thus $\mathcal{V}\subset\mathcal{V}'_f=\overline{\mathcal{V}}\subset\mathcal{V}'$, and (a fortiori) $\mathcal{V}=\mathcal{V}'_f\cap\mathcal{K}$. Hence ψ'_f is the smallest u.s.c. extension of ψ. Conversely, ψ'_f is u.s.c., as $\mathcal{V}'_f$ is closed (Proposition 8-2-1), and is an extension of ψ if and only if $\mathcal{V}=\mathcal{V}'_f\cap\mathcal{K}$. Thus ψ admits u.s.c. extensions if and only if condition 1 is satisfied.

On the other hand, the condition $\mathcal{V}=\overline{\mathcal{V}}\cap\mathcal{K}$ means that, for any sequence $\{K_n\}\subset\mathcal{K}\cap\mathcal{F}$ such that $\lim K_n=K\in\mathcal{K}\cap\mathcal{F}$ in $\mathcal{F}$, $K_n\in\mathcal{V}$, that is, $0\in\psi(K_n)$ implies $0\in\psi(K)$. But ψ is a τ-mapping, and this condition is equivalent to $\overline{\lim}\ \psi(K_n)\subset\psi(K)$, that is, ψ is u.s.c. for the relative $\mathcal{F}$-topology. Thus conditions 1 and 2 are equivalent.

If the conditions of Proposition 8-2-3 are not satisfied, ψ_f' is the smallest u.s.c. mapping from $\mathcal{F}$ into itself such that $\psi_f'\supset\psi$ on $\mathcal{K}$. But ψ_f' might be the trivial mapping $F\to\mathbf{R}^d$, the kernel of which is $\mathcal{V}_f'=\mathcal{F}$. Thus ψ_f' is not the trivial mapping if and only if $\mathcal{V}$ is not dense in $\mathcal{F}$, or (which is the same) if $\varnothing\notin\overline{\mathcal{V}}$. The neighborhoods of $\varnothing$ in $\mathcal{F}$ are the $\mathcal{F}^K$, $K\in\mathcal{K}$, so that $\varnothing\notin\overline{\mathcal{V}}$ if and only if there exists $K_0\in\mathcal{K}$ such that $\mathcal{V}\subset\mathcal{F}_{K_0}$, that is, such that $\psi(K)\subset K\oplus\check{K}_0$ for any $K\in\mathcal{K}$. Hence:

Corollary. *The extension ψ_f' is the smallest u.s.c. mapping from $\mathcal{F}$ into itself such that $\psi_f'\supset\psi$ on $\mathcal{K}$, and ψ_f' is not trivial if and only if ψ is bounded, that is, there exists a fixed compact set K_0 such that $\psi(K)\subset K\oplus K_0$ for any $K\in\mathcal{K}$.*

Formulae (8-2-1) yield a one-to-one correspondence between u.s.c. mappings on $\mathcal{K}$ and $\mathcal{F}$. From this standpoint, note the following results:

Proposition 8-2-4. *Let ψ_k be an u.s.c. mapping from $\mathcal{K}$ into $\mathcal{F}$ such that its kernel satisfies $\mathcal{V}_k=\overline{\mathcal{V}}_k\cap\mathcal{K}$, ψ_f' be its smallest u.s.c. extension on $\mathcal{F}$ associated with the kernel $\mathcal{V}_f'=\overline{\mathcal{V}}_k$, and*

$$\mathcal{V}_k'=\bigcap_{F\in\mathcal{V}_f'}\mathcal{K}_F;\qquad \mathcal{V}_f=\bigcap_{K\in\mathcal{V}_k}\mathcal{F}_K$$

Then $\mathcal{V}_k'=\mathcal{V}_f\cap\mathcal{K}$ and $\mathcal{V}_f=\overline{\mathcal{V}}_k'$, so that ψ_f is the smallest u.s.c. extension of ψ_k' on $\mathcal{F}$. Moreover:

$$\mathcal{V}_k=\bigcap_{F\in\mathcal{V}_f}\mathcal{K}_F;\qquad \mathcal{V}_f'=\bigcap_{K\in\mathcal{V}_k'}\mathcal{F}_K$$

Proof. Let $K\in\mathcal{V}_f\cap\mathcal{K}$ be a compact set, $F\in\mathcal{V}_f'=\overline{\mathcal{V}}_k$ and $\{K_n\}\subset\mathcal{V}_k\subset\mathcal{K}$ be a sequence such that $F=\lim K_n$ in $\mathcal{F}$. Then, for any $n>0$, $K_n\cap K\neq\varnothing$, and this implies $F\cap K\neq\varnothing$, that is, $K\in\mathcal{V}_k'$. Thus $\mathcal{V}_f\cap\mathcal{K}\subset\mathcal{V}_k'$. The converse inclusion is obvious, so that $\mathcal{V}_k'=\mathcal{V}_f\cap\mathcal{K}$. In the same way, $\mathcal{V}_f\supset\mathcal{V}_k'$ implies $\mathcal{V}_f\supset\overline{\mathcal{V}}_k'$. Conversely, let $F\notin\overline{\mathcal{V}}_k'$ be a closed set. There exists an open neighborhood $\mathcal{F}^{K_0}_{G_1,\ldots,G_n}$ $(K_0\in\mathcal{K},\ G_i\in\mathcal{G})$ of F disjoint of $\mathcal{V}_k'$. But $K\in\mathcal{V}_k'$ and $K'\supset K$, $K'\in\mathcal{K}$, imply $K'\in\mathcal{V}_k'$, so that $\mathcal{F}^{K_0}\ni F$ also is disjoint of $\mathcal{V}_k'$. Thus $K_0\cap F=\varnothing$ and $K_0\cap K\neq\varnothing$ for any $K\in\mathcal{V}_k'$. It follows that $K_0\in\cap\{\mathcal{F}_K,\ K\in\mathcal{V}_k'\}=\mathcal{V}_f'$ [formula (8-2-1)], that is, $K_0\in\mathcal{V}_f'\cap\mathcal{K}=\mathcal{V}_k$. Thus $F\notin\mathcal{V}_k'$ implies $F\notin\cap\{\mathcal{F}_{K_0},\ K_0\in\mathcal{V}_k\}$

$= \mathcal{V}_f$, that is, $\mathcal{V}_f \subset \overline{\mathcal{V}}'_k$, and so $\mathcal{V}_f = \overline{\mathcal{V}}'_k$.

More generally:

Proposition 8-2-5. *Let $\psi: \mathcal{K} \to \mathcal{F}$ be a τ-mapping, $\mathcal{V}$ its kernel, ψ_f the τ-mapping defined by the kernel $\mathcal{V}_f = \cap\{\mathcal{F}_K, K \in \mathcal{V}\}$, ψ'_k the restriction of ψ_f to $\mathcal{K}$ associated with the kernel $\mathcal{V}'_k = \mathcal{V}_f \cap \mathcal{K}$, and $\psi'_f: \mathcal{F} \to \mathcal{F}$ associated with $\mathcal{V}'_f = \cap\{\mathcal{F}_K, K \in \mathcal{V}'_k\}$. Then $\mathcal{V}'_f$ is the closure of $\mathcal{V}$ in $\mathcal{F}$, and ψ'_f is the smallest u.s.c. τ-mapping on $\mathcal{F}$ such that $\psi'_f \supset \psi$ on $\mathcal{K}$.*

Proof. If $K \in \mathcal{V}$, K hits any $F \in \mathcal{V}_f$, that is, $K \in \mathcal{V}'_f$ and $\mathcal{V} \subset \mathcal{V}'_f$. Also $\mathcal{V}'_f$ is closed in $\mathcal{F}$ as the intersection of the $\mathcal{F}_K$, $K \in \mathcal{V}'_k$, and thus $\overline{\mathcal{V}} \subset \mathcal{V}'_f$. If $F \not\in \overline{\mathcal{V}}$ is a closed set, there exists $K_0 \in \mathcal{K}$ such that $F \in \mathcal{F}^{K_0}$, that is, $K_0 \in \mathcal{K}^F$, and $\mathcal{V} \cap \mathcal{F}^{K_0} = \varnothing$, that is, $\mathcal{V} \subset \mathcal{F}_{K_0}$. But $\mathcal{V} \subset \mathcal{F}_{K_0}$ is equivalent to $K_0 \in \mathcal{V}_f \cap \mathcal{K}$, and $F \cap K_0 = \varnothing$ implies $F \not\in \mathcal{V}'_f$. Thus $\overline{\mathcal{V}} \supset \mathcal{V}'_f$, and so $\overline{\mathcal{V}} = \mathcal{V}'_f$.

τ-Mappings Compatible with $\cup$ or $\cap$

If $\mathcal{A} \subset \mathcal{P}$ is closed under $\cup$ (resp. under $\cap$), a mapping $\psi: \mathcal{A} \to \mathcal{P}$ is compatible with $\cup$ (resp. with $\cap$), or is a $\cup$-mapping (resp. a $\cap$-mapping) if $\psi(A \cup A') = \psi(A) \cup \psi(A')$ [resp. $\psi(A \cap A') = \psi(A) \cap \psi(A')$] for any A, $A' \in \mathcal{A}$.

Proposition 8-2-6. *Let ψ be an u.s.c. τ-mapping from $\mathcal{K}$ (resp. from $\mathcal{F}$) into $\mathcal{F}$. Then ψ is a $\cap$-mapping if and only if there exists $F_0 \in \mathcal{F}$ such that $\psi(A) = A \ominus \check{F}_0$ for any $A \in \mathcal{K}$ (resp. $A \in \mathcal{F}$).*

Proof. From Proposition 8-2-1 and (8-1-9), a τ-mapping $\psi: \mathcal{K} \to \mathcal{F}$ (resp. $\mathcal{F} \to \mathcal{F}$) is u.s.c. and compatible with $\cap$ if and only if its kernel $\mathcal{V}$ is a closed filter in $\mathcal{K}$ (resp. in $\mathcal{F}$). Thus $F_0 = \cap\{A, A \in \mathcal{V}\}$ belongs to $\mathcal{V}$, for $\mathcal{V}$ is closed in $\mathcal{K}$ (resp. in $\mathcal{F}$) and closed under $\cap$, so that $\mathcal{V} \subset \mathcal{W}_{F_0}$. On the other hand, $A \in \mathcal{V}$ and $B \in \mathcal{W}_A \cap \mathcal{K}$ (resp. $\mathcal{W}_A \cap \mathcal{F}$) imply $B \in \mathcal{V}$, so that $\mathcal{W}_{F_0} \cap \mathcal{K} \subset \mathcal{V}$ ($\mathcal{W}_{F_0} \cap \mathcal{F} \subset \mathcal{V}$). It follows that $\mathcal{V} = \mathcal{W}_{F_0} \cap \mathcal{K}$ (resp. $\mathcal{W}_{F_0} \cap \mathcal{F}$), that is, $\psi(A) = A \ominus \check{F}_0$ for any $A \in \mathcal{K}$ (resp. $A \in \mathcal{F}$). The converse is obvious.

The following lemmas will be used to characterize the $\cup$-mapping.

Lemma 8-2-1. *Let E be an LCS space, G_1, $G_2 \in \mathcal{G}(E)$, $K \in \mathcal{K}(E)$. Then $K \subset G_1 \cup G_2$ if and only if there exist K_1, $K_2 \in \mathcal{K}$ such that $K_1 \subset G_1$, $K_2 \subset G_2$, and $K = K_1 \cup K_2$.*

Proof. The "if" part is obvious. Conversely, let $\{B_n\}$ and $\{B_n'\}$ be two sequences in $\mathcal{G}$ such that, for any n, $\bar{B}_n \in \mathcal{K}$, $\bar{B}_n' \in \mathcal{K}$, $\bar{B}_n \subset B_{n+1}$, $\bar{B}_n' \subset B_{n+1}'$, and $B_n \uparrow G_1$, $B_n' \uparrow G_2$. Then $K \subset G_1 \cup G_2$ implies $K \subset \bar{B}_{n_0} \cup \bar{B}_{n_0}'$ for an integer n_0, and the conditions of the lemma are satisfied by $K_1 = K \cap \bar{B}_{n_0}$ and $K_2 = K \cap \bar{B}_{n_0}'$.

Lemma 8-2-2. *Let E be an LCS space, $K_1, K_2 \in \mathcal{K}(E)$, and $G \in \mathcal{G}(E)$. Then $G \supset K_1 \cap K_2$ if and only if there exist $G_1, G_2 \in \mathcal{G}$ such that $G_1 \supset K_1$, $G_2 \supset K_2$, and $G = G_1 \cap G_2$.*

Proof. The "if" part is obvious. Conversely, let $\{B_n\} \subset \mathcal{G}$, $\{B_n'\} \subset \mathcal{G}$ be two sequences such that $\bar{B}_n \in \mathcal{K}$, $\bar{B}_n' \in \mathcal{K}$, $B_n \supset \bar{B}_{n+1}$, $B_n' \supset \bar{B}_{n+1}'$ for any n and $B_n \downarrow K_1$, $B_n' \downarrow K_2$. It follows that $G^c \cap (\cap_n (\bar{B}_n \cap \bar{B}_n')) = \varnothing$, and thus there exists an integer n_0 such that $G \supset \bar{B}_{n_0} \cap \bar{B}_{n_0}'$. The conditions of the lemma are satisfied by $G_1 = G \cup B_{n_0}$, $G_2 = G \cup B_{n_0}'$.

Recall that $\mathcal{V} \subset \mathcal{K}(E)$ is said to be an *antifilter* in $\mathcal{K}$ if (1) $\mathcal{V}$ is nonempty, (2) $\complement \mathcal{V}$ is closed under $\cup$, and (3) $\mathcal{V}$ is $\cup$*-hereditary* in $\mathcal{K}$, that is, $V \in \mathcal{V}$, $V' \supset V$, $V' \in \mathcal{K}$ imply $V' \in \mathcal{V}$.

Lemma 8-2-3. *Let E be an LCS space and $\mathcal{V} \subset \mathcal{K}(E)$ such that $\varnothing \notin \mathcal{V}$. Then $\mathcal{V}$ is a closed antifilter in $\mathcal{K}$ if and only if $\mathcal{V} = \mathcal{K}_{F_0}$ for a nonempty closed set F_0.*

Proof. The "if" part is obvious. Conversely, let $\mathcal{V}$ be a closed antifilter in $\mathcal{K}$, such that $\varnothing \notin \mathcal{V}$ (i.e., $\mathcal{V} \neq \mathcal{K}$), and $K \in \mathcal{K} \cap \complement \mathcal{V}$. Thus K admits an open neighborhood disjoint of $\mathcal{V}$, and this neighborhood may be assumed to be $\mathcal{K}^F$ for an $F \in \mathcal{F}$ (because $\mathcal{V}$ is $\cup$-hereditary), that is, $K \in \mathcal{K}^F$ and $\mathcal{V} \subset \mathcal{K}_F$. Let $\mathcal{V}_f$ be the family of the closed sets F such that $\mathcal{V} \subset \mathcal{K}_F$, that is, $\mathcal{V}_f = \cap \{\mathcal{F}_K, K \in \mathcal{V}\}$, and thus $\mathcal{V} = \cap \{\mathcal{K}_F, F \in \mathcal{V}_f\}$. The family $\mathcal{V}_f$ is closed in $\mathcal{F}$, as an intersection of closed subsets $\mathcal{F}_K$. In order to prove that $\mathcal{V}_f$ is closed under $\cap$, take $F_1, F_2 \in \mathcal{V}_f$ and $G_1 = F_1^c$, $G_2 = F_2^c$. If $K \in \mathcal{V}$, it follows that $K \not\subset G_1$, $K \not\subset G_2$. Suppose that $K \subset G_1 \cup G_2$. From Lemma 8-2-1, there exist $K_1, K_2 \in \mathcal{K}$ such that $K_1 \subset G_1$, $K_2 \subset G_2$, and $K = K_1 \cup K_2$. One of these two compact sets, say K_1, belongs to $\mathcal{V}$ (for $\complement \mathcal{V}$ is closed under $\cup$, and $K_1 \notin \mathcal{V}$, $K_2 \notin \mathcal{V}$ would imply $K = K_1 \cup K_2 \notin \mathcal{V}$). It follows that $K_1 \in \mathcal{K}_{F_1}$, but this is impossible, for $K_1 \subset G_1 = F_1^c$. We conclude that $K \not\subset G_1 \cup G_2$, and thus $F_1 \cap F_2 \in \mathcal{V}_f$.

Since $\mathcal{V}_f$ is closed under $\cap$ and closed in $\mathcal{F}$, $F_0 = \cap \{F, F \in \mathcal{V}_f\}$ belongs to $\mathcal{V}_f$. If $F_0 = \varnothing$, it follows that $\mathcal{V} = \varnothing$. Thus $F_0 \neq \varnothing$, and $F \in \mathcal{V}_f$ is equivalent to $F \supset F_0$. Hence $\mathcal{V} = \mathcal{K}_{F_0}$.

Lemma 8-2-4. *Let E be an LCS space and $\mathcal{V} \subset \mathcal{F}(E)$ such that $\varnothing \notin \mathcal{V}$. Then $\mathcal{V}$ is a closed antifilter in $\mathcal{F}$ if and only if $\mathcal{V} = \mathcal{F}_{K_0}$ for a nonempty* *compact set K_0. In the same way, a nonempty family $\mathcal{W} \subset \mathcal{G}(E)$ is an open filter in $\mathcal{G}$ if and only if $\mathcal{W} = \mathcal{G}_{K_0}$ for a nonempty compact set K_0.*

Proof. The two statement being equivalent by duality, let us prove, for instance, the first one. The "if" part is obvious. Conversely, let $\mathcal{V}$ be an antifilter closed in $\mathcal{F}$ and $\varnothing \not\in \mathcal{V}$. For any $F \in \mathcal{F}$ such that $F \not\in \mathcal{V}$, there exists $K \in \mathcal{K}$ such that $F \in \mathcal{F}^K$ and $\mathcal{V} \subset \mathcal{F}_K$, because $\mathcal{V}$ is closed and $\cup$-hereditary. Let us denote by $\mathcal{V}_k \subset \mathcal{K}$ the family of the compact sets K such that $\mathcal{V} \subset \mathcal{F}_K$, that is, $\mathcal{V}_k = \cap\{\mathcal{K}_F, F \in \mathcal{V}\}$, and thus $\mathcal{V} = \cap\{\mathcal{F}_K, K \in \mathcal{V}_k\}$. The family $\mathcal{V}_k$ is closed in $\mathcal{K}$. In order to prove that $\mathcal{V}_k$ is closed under $\cap$, take $K_1, K_2 \in \mathcal{V}_k, F \in \mathcal{V}$, and $G = F^C$, so that $K_1 \not\subset G$, $K_2 \not\subset G$. Suppose that $K_1 \cap K_2 \subset G$. Then, from Lemma 8-2-2, there exist $G_1, G_2 \in \mathcal{G}$ such that $K_1 \subset G_1, K_2 \subset G_2, G = G_1 \cap G_2$. Thus $G_1{}^c$, for instance, belongs to $\mathcal{V}$, for $\complement\mathcal{V}$ is closed under $\cup$ and $G^c \in \mathcal{V}$. But $G_1{}^c \in \mathcal{V}$ and $K_1 \cap G_1{}^c = \varnothing$ imply $K_1 \not\in \mathcal{V}_k$. We conclude that $K_1 \cap K_2 \not\subset G$, and $\mathcal{V}_k$ is closed under $\cap$. Being closed under $\cap$ and closed in $\mathcal{F}$, $\mathcal{V}_k$ contains its intersection $K_0 = \cap\{K, K \in \mathcal{V}_k\}$, and so $\mathcal{V} = \mathcal{F}_{K_0}$.

Proposition 8-2-7. *An u.s.c. τ-mapping ψ: $\mathcal{K} \to \mathcal{F}$ (resp. $\mathcal{F} \to \mathcal{F}$) is a $\cup$-mapping if and only if there exists a fixed $F_0 \in \mathcal{F}$ (resp. $K_0 \in \mathcal{K}$) such that $\psi(K) = K \oplus \check{F}_0$ for any $K \in \mathcal{K}$ [resp. $\psi(F) = F \oplus \check{K}_0$ for any $F \in \mathcal{F}$]. If so, ψ is continuous.*

Proof. If follows from (8-1-12) and Proposition 8-2-1 that ψ is an u.s.c. $\cap$-mapping if and only if its kernel $\mathcal{V}$ is a closed antifilter in $\mathcal{K}$ (resp., in $\mathcal{F}$), and the proposition is an immediate corollary of Lemmas 8-2-3 and 8-2-4.

8-3. TOPOLOGICAL COMPLEMENTS

We have seen that there exists a bijective correspondence between the u.s.c. increasing τ-mappings on $\mathcal{K}(\mathbf{R}^d)$ or $\mathcal{F}(\mathbf{R}^d)$ and the $\cup$-hereditary families $\mathcal{V}$ closed in $\mathcal{K}$ or in $\mathcal{F}$. By providing the sets of these families with the relative $\mathcal{F}(\mathcal{K})$ or $\mathcal{F}(\mathcal{F})$ topologies and identifying them with the corresponding spaces of τ-mappings, it would be possible to develop a functional analysis for the u.s.c. mappings. More generally, if E is an arbitrary LCS space, we denote by $\mathcal{F}_u(\mathcal{K}) \subset \mathcal{F}(\mathcal{K})$ *the set of the $\cup$-hereditary families $\mathcal{V} \subset \mathcal{K}(E)$ closed in $\mathcal{K}$*, and by $\mathcal{F}_u(\mathcal{F}) \subset \mathcal{F}(\mathcal{F})$ *the set of the $\cup$-hereditary families $\mathcal{V}' \subset \mathcal{F}(E)$ closed in $\mathcal{F}$*. Note that the empty family $\mathcal{V} = \varnothing$ belongs to $\mathcal{F}_u(\mathcal{K})$ and to $\mathcal{F}_u(\mathcal{F})$.

The mapping $x \to \mathcal{V}(x)$ from E' into $\mathcal{P}(\mathcal{P}(E))$ is said to be the inverse mapping of ψ. For any $x \in E'$, $\mathcal{V}(x)$ is $\cup$-hereditary in $\mathcal{A}$. Conversely, if $\mathcal{V}$ is a mapping from E' into the space $\mathcal{P}_u(\mathcal{A})$ of the $\cup$-hereditary families in $\mathcal{A}$, $\mathcal{V}$ is the inverse mapping of ψ: $A \to \psi(A) = \{x, A \in \mathcal{V}(x)\}$.

In what follows, the space $\mathcal{A} \subset \mathcal{P}(E)$ will always be either $\mathcal{A} = \mathcal{F}(E)$ or $\mathcal{A} = \mathcal{K}(E)$. If $\mathcal{A} = \mathcal{K}$, we shall put $\mathcal{B} = \mathcal{G}$ and $\mathcal{A}' = \mathcal{F}$; if $\mathcal{A} = \mathcal{F}$, $\mathcal{A}' = \mathcal{K}$ and $\mathcal{B} = \{K^c, K \in \mathcal{K}\}$, that is, in both cases $\mathcal{B} = \{A'^c, A' \in \mathcal{A}'\}$. The space $\mathcal{B}$ will always be provided with the topology induced from $\mathcal{A}'$ by the mapping $A' \to A'^c$. We shall also use the notations defined in Proposition 8-1-4, that is, if ψ is an increasing mapping from $\mathcal{A}$ into $\mathcal{P}(E')$, ψ_b is its smallest extension on $\mathcal{B}$ and ψ_a is the largest extension of ψ_b on $\mathcal{A}$. Explicitly:

$$\begin{aligned} \psi_b(B) &= \cup\{\psi(A), A \in \mathcal{A}, A \subset B\} \qquad (B \in \mathcal{B}) \\ \psi_a(A) &= \cap\{\psi(B), B \in \mathcal{B}, B \supset A\} \qquad (A \in \mathcal{A}) \end{aligned} \tag{8-4-2}$$

With these notations, it is possible to state simultaneously analogous results concerning the spaces $\mathcal{F}(E)$ and $\mathcal{K}(E)$.

First, let us investigate under what condition $\psi = \psi_a$ on $\mathcal{A}$. From (8-4-2), $\psi = \psi_a$ if and only if, for any $A \in \mathcal{A}$ and $x \notin \psi(A)$, there exists $B \in \mathcal{B}$ such that $B \supset A$ and $x \notin \psi(A')$ for any $A' \in \mathcal{A}$, $A' \subset B$ (see Proposition 8-1-4). This is equivalent to $\{A \notin \mathcal{V}(x)$ implies that there exists $B \in \mathcal{B}$ such that $B \supset A$ and $\mathcal{V}(x) \subset \mathcal{A}_{B^c}\}$. But $B^c \in \mathcal{A}'$ (by the definition of $\mathcal{B}$), and thus $\mathcal{A}^{B^c}$ is an open neighborhood of A disjoint of $\mathcal{V}(x)$. Hence $\mathcal{V}(x)$ is closed, that is, $\mathcal{V}(x) \in \mathcal{F}_u(\mathcal{A})$. Conversely, if $\mathcal{V}(x)$ is closed and $\cup$-hereditary for any $x \in E'$, $\complement \mathcal{V}(x)$ is $\cap$-hereditary, and for any $A \notin \mathcal{V}(x)$ there exists a $B \in \mathcal{B}$ such that $\mathcal{A}^{B^c}$ is an open neighborhood of A disjoint of $\mathcal{V}(x)$. Hence $\psi = \psi_a$. More precisely:

Proposition 8-4-1. *Let ψ be an increasing mapping from $\mathcal{A}$ into $\mathcal{P}(E')$ and $\mathcal{V}$: $E' \to \mathcal{P}(\mathcal{A})$ defined by $\mathcal{V}(x) = \{A, x \in \psi(A)\}$. Then $\psi = \psi_a$ if and only if $\mathcal{V}$ maps E' into $\mathcal{F}_u(\mathcal{A})$. Moreover, the four following conditions are equivalent:*

1. *ψ is a u.s.c. mapping from $\mathcal{A}$ into $\mathcal{F}(E')$.*
2. *$\mathcal{V}$ is a u.s.c. mapping from E' into $\mathcal{F}_u(\mathcal{A})$.*
3. *For any $A \in \mathcal{A}$ and $x \in E'$, $\psi(A)$ and $\mathcal{V}(x)$ are closed.*
4. *ψ maps $\mathcal{A}$ into $\mathcal{F}(E')$ and $\psi = \psi_a$.*

Proof. We have just proved $\psi = \psi_a$ if and only if $\mathcal{V}(E') \subset \mathcal{F}_u(\mathcal{A})$. If so, $\psi(A)$ is closed for any $A \in \mathcal{A}$ if and only if $A \in \mathcal{V}(x_n)$ and $\lim x_n = x$ in E' implies $A \in \mathcal{V}(x)$, that is, $\mathcal{V}(x_n) \not\subset \mathcal{A}_{A^c}$ and $x = \lim x_n$ imply $\mathcal{V}(x) \not\subset \mathcal{A}_{A^c}$. It follows from Proposition 8-3-2 that this condition is satisfied if and only if

Proposition 8-3-1. *There exists a one-to-one mapping $\mathcal{V} \to \mathcal{V}'$ from $\mathcal{F}_u(\mathcal{K})$ onto $\mathcal{F}_u(\mathcal{F})$, defined by the reciprocal formulae*

$$\mathcal{V} = \bigcap_{F \in \mathcal{V}'} \mathcal{K}_F; \qquad \mathcal{V}' = \bigcap_{K \in \mathcal{V}} \mathcal{F}_K$$

In particular, $\mathcal{V} = \varnothing$ if and only if $\mathcal{V}' = \mathcal{F}$, and $\mathcal{V}' = \varnothing$ if and only if $\mathcal{V} = \mathcal{K}$. More generally, for any $\mathcal{V}_0 \subset \mathcal{K}(E)$ [resp. $\mathcal{V}_0' \subset \mathcal{F}(E)$], put $\mathcal{V}' = \cap\{\mathcal{F}_K, K \in \mathcal{V}_0\}$, $\mathcal{V} = \cap\{\mathcal{K}_F, F \in \mathcal{V}'\}$ (resp. $\mathcal{V} = \cap\{\mathcal{K}_F, F \in \mathcal{V}_0'\}$, $\mathcal{V}' = \cap\{\mathcal{F}_K, K \in \mathcal{V}\}$). Then $\mathcal{V}$ (resp. $\mathcal{V}'$) is the closure in $\mathcal{K}$ (resp. in $\mathcal{F}$) of the $\cup$-hereditary family generated in $\mathcal{K}$ by $\mathcal{V}_0$ (resp. in $\mathcal{F}$ by $\mathcal{V}_0'$).

Proof. Let us prove only the statement concerning $\mathcal{V}_0 \subset \mathcal{K}(E)$. The family $\mathcal{V}$ is closed in $\mathcal{F}$ and $\cup$-hereditary. If $\mathcal{V}_1$ is the $\cup$-hereditary family generated by $\mathcal{V}_0$ in $\mathcal{K}$, $\mathcal{V}_1 \subset \mathcal{V}$, for, if $K \in \mathcal{V}_0$, K hits any $F \in \mathcal{V}'$, and thus $K \in \mathcal{V}$. Hence $\mathcal{V}_0 \subset \mathcal{V}$ and $\mathcal{V}_1 \subset \mathcal{V}$, as $\mathcal{V}$ is $\cup$-hereditary. It follows that $\overline{\mathcal{V}}_1 \subset \mathcal{V}$. Conversely, if $K \in \mathcal{K}$ and $K \notin \overline{\mathcal{V}}_1$, there exists a closed set $F \in \mathcal{F}$ such that $\mathcal{K}^F$ is a neighborhood of K disjoint of the $\cup$-hereditary family $\mathcal{V}_1$, that is, $F \cap K = \varnothing$ and $\mathcal{V}_0 \subset \mathcal{V}_1 \subset \mathcal{K}_F$. This implies $F \in \mathcal{V}'$, as F hits any $K' \in \mathcal{V}_0$. Then $K \notin \mathcal{V}$ follows from $K \notin \mathcal{K}_F$. Hence $\overline{\mathcal{V}}_1 \supset \mathcal{V}$, and so $\mathcal{V} = \overline{\mathcal{V}}_1$.

From a topological standpoint, it turns out that the mapping $\mathcal{V} \to \mathcal{V}'$ is actually a *homeomorphism*. In order to see this, some preliminary results must first be proved.

Proposition 8-3-2. *The spaces $\mathcal{F}_u(\mathcal{F})$ and $\mathcal{F}_u(\mathcal{K})$ are compact. Moreover, $\mathcal{K}$ is an isolated point in $\mathcal{F}_u(\mathcal{K})$, and the empty family $\varnothing$ is an isolated point in $\mathcal{F}_u(\mathcal{F})$. On the contrary, $\varnothing$ and $\mathcal{F}$ are isolated points in $\mathcal{F}_u(\mathcal{K})$ and $\mathcal{F}_u(\mathcal{F})$, respectively, if and only if the LCS space E is compact.*

Proof. Let us prove, for instance, that $\mathcal{F}_u(\mathcal{K})$ is a closed subspace of the compact space $\mathcal{F}(\mathcal{K})$. Let $\{\mathcal{V}_n\} \subset \mathcal{F}_u(\mathcal{K})$ be a sequence such that $\lim \mathcal{V}_n = \mathcal{V}$ in $\mathcal{F}(\mathcal{K})$. If $\varnothing \in \mathcal{V}$, there exists a sequence $n \to K_n \in \mathcal{V}_n$ such that $\varnothing = \lim K_n$ for the myope topology, and this implies $K_n = \varnothing$ for n large enough, that is, $\mathcal{V}_n = \mathcal{K}$, as $\mathcal{V}_n$ is $\cup$-hereditary. Hence $\mathcal{V} = \mathcal{K} \in \mathcal{F}_u(\mathcal{K})$, and $\mathcal{K}$ is an isolated point in $\mathcal{F}_u(\mathcal{K})$. We suppose now that $\varnothing \notin \mathcal{V}$, and prove that $\mathcal{V}$ is $\cup$-hereditary. Let $K \in \mathcal{V}$ and $K' \in \mathcal{K}$, $K' \supset K$. There exists a sequence $n \to K_n \in \mathcal{V}_n$ such that $K = \lim K_n$ in $\mathcal{K}$. This implies $\lim K' \cup K_n = K' \cup K = K'$, for $\cup$ is continuous on $\mathcal{K} \times \mathcal{K}$. But $K' \cup K_n \in \mathcal{V}_n$, because $\mathcal{V}_n$ is $\cup$-hereditary, and thus $K' \in \mathcal{V}$. Hence $\mathcal{V} \in \mathcal{F}_u(\mathcal{K})$, and $\mathcal{F}_u(\mathcal{K})$ is compact. From Proposition 1-2-1, $\varnothing$ is an isolated point in

$\mathcal{F}(\mathcal{K})$ if and only if $\mathcal{K}$ is compact, that is, if and only if E itself is compact. If E is not compact, there exists a sequence $\{A_n\} \subset \mathcal{K}$ without accumulation point, so that there exists no sequence $k \to K_{n_k} \supset A_{n_k}$ converging in $\mathcal{K}$. This implies $\lim \mathcal{F}_{A_n} = \varnothing$ in $\mathcal{F}_u(\mathcal{K})$. Hence $\varnothing$ is isolated in $\mathcal{F}_u(\mathcal{K})$ if and only if E is compact. The proof of the statement concerning $\mathcal{F}_u(\mathcal{F})$ is very similar.

In order to construct a basis for the topologies of the spaces $\mathcal{F}_u(\mathcal{F})$ and $\mathcal{F}_u(\mathcal{K})$, we start with the following lemma.

Lemma 8-3-1. *Let E be an LCS space, $\mathcal{B} \subset \mathcal{G}$ a basis for its topology, and $\mathbf{C} \subset \mathcal{K}$ a family closed under (finite) $\cap$ and such that the class closed under finite $\cup$ and infinite $\cap$ generated by $\mathbf{C}$ is identical to $\mathcal{K}(E)$. Then the topology of $\mathcal{F}(E)$ is generated by the two families $\mathcal{F}_B, B \in \mathcal{B}$, and $\mathcal{F}^C, C \in \mathbf{C}$.*

Proof. The $\mathcal{F}$-topology is generated by the l.s.c.-type open sets $\mathcal{F}_G, G \in \mathcal{G}$, and the u.s.c.-type open sets $\mathcal{F}^K, K \in \mathcal{K}$. Any $G \in \mathcal{G}$ is $G = \cup B_i$ for a family $\{B_i, i \in I\} \subset \mathcal{B}$, so that $\mathcal{F}_G = \mathcal{F}_{\cup B_i} = \cup \mathcal{F}_{B_i}$. Hence $\{\mathcal{F}_B, B \in \mathcal{B}\}$ is a basis for the l.s.c. topology.

Let $\mathbf{C}'$ be the class closed under finite $\cup$ and finite $\cap$ generated by $\mathbf{C}$. Any $K \in \mathcal{K}$ is $K = \cap C_i'$ for a family $\{C_i', i \in I\} \subset \mathbf{C}'$. If $F \in \mathcal{F}^K$, that is, $\cap(C_i' \cap F) = \varnothing$, there exist a finite number of indices $i_1, \ldots, i_k \in I$ such that $F \cap C_{i_1}' \cap \ldots \cap C_{i_k}' = \varnothing$. But $C' = C_{i_1}' \cap \cdots \cap C_{i_k}' \in \mathbf{C}'$ and $C' \supset K$, so that $F \in \mathcal{F}^{C'} \subset \mathcal{F}^K$. Then there exist $C_1, \ldots, C_n \in \mathbf{C}$ such that $C' = C_1 \cup \cdots \cup C_n$, as $C' \in \mathbf{C}'$, and $F \in \mathcal{F}^{C'} = \mathcal{F}^{C_1} \cap \cdots \cap \mathcal{F}^{C_n} \subset \mathcal{F}^K$. It follows that the u.s.c. topology is generated by $\{\mathcal{F}^C, C \in \mathbf{C}\}$.

Example. Let us apply Lemma 8-3-1 to the spaces $E = \mathcal{F}$ and $E = \mathcal{K}$. For brevity, we denote by $\mathcal{A}$ either $\mathcal{F}$ or $\mathcal{K}$ and put $\mathcal{A}' = \mathcal{K}$ if $\mathcal{A} = \mathcal{F}$, $\mathcal{A}' = \mathcal{F}$ if $\mathcal{A} = \mathcal{K}$. As usual, $\mathcal{A}^B_{G_1, \ldots, G_k}$ is the family $\{A, A \in \mathcal{A}, A \cap G_i \neq \varnothing, i = 1, 2, \ldots, k, A \cap B = \varnothing\}$. Then the following classes $\mathcal{B}$ and $\mathbf{C}$:

$$\mathcal{B} = \{\mathcal{A}^{A'}_{G_1, \ldots, G_k}, A' \in \mathcal{A}', k \geqslant 0, G_1, \ldots, G_k \in \mathcal{G}\}$$

$$\mathbf{C} = \{\mathcal{A}^{A^c}_{A_1', \ldots, A_k'}, A \in \mathcal{A}, k \geqslant 0, A_1', \ldots, A_k' \in \mathcal{A}'\}$$

satisfy the condition of the lemma for $\mathcal{A} = \mathcal{F}$ or $\mathcal{K}$. Hence the l.s.c. topology on $\mathcal{F}(\mathcal{A})$ is generated by $\{\mathcal{F}_\Gamma, \Gamma \in \mathcal{B}\}$, and the u.s.c. topology is generated by $\{\mathcal{F}^\chi, \chi \in \mathbf{C}\}$. Note that Γ and χ denote families of subsets in E.

Proposition 8-3-3. *If $\mathcal{A} = \mathcal{K}$ (resp. $\mathcal{A} = \mathcal{F}$) and $\mathcal{A}' = \mathcal{F}$ (resp. $\mathcal{A}' = \mathcal{K}$), u.s.c. topology on $\mathcal{F}_u(\mathcal{A})$ is generated by the family $\{\mathcal{V}, \mathcal{V} \subset \mathcal{A}_{A^c}\}, A \in$ and the l.s.c. topology by the family $\{\mathcal{V}, \mathcal{V} \cap \mathcal{A}^{A'} \neq \varnothing\}, A' \in \mathcal{A}'$.*

Proof. With the notations of the above example, $\chi = \mathcal{A}^{A^c}_{A_1', \ldots, A_k'} \in \mathbf{C}$ $\mathcal{V} \in \mathcal{F}^\chi$, that is, $\mathcal{V} \cap \chi = \varnothing$ if and only if, for any $V \in \mathcal{V}$, either $V \cap A^c =$ or $V \cap A_i' = \varnothing$ for one index $i \in [1, 2, \ldots, k]$. If $\mathcal{V}$ is $\cup$-hereditary, thi equivalent to $V \cap A^c \neq \varnothing$ for any $V \in \mathcal{V}$, that is, $\mathcal{V} \cap \mathcal{A}^{A^c} = \varnothing$, $\mathcal{V} \subset \mathcal{A}_{A^c}$. If $\Gamma = \mathcal{A}^{A'}_{G_1, \ldots, G_k} \in \mathcal{B}$, $\mathcal{V} \in \mathcal{F}_\Gamma$, that is, $\mathcal{V} \cap \mathcal{A}^{A'}_{G_1, \ldots, G_k} \neq \varnothing$ if only if there exists $V \in \mathcal{V}$ such that $V \cap A' = \varnothing$, $V \cap G_i \neq \varnothing$, $i = 1, \ldots, k$ $\mathcal{V}$ is $\cup$-hereditary, this is equivalent to {there exists a $V \in \mathcal{V}$ such $V \cap A' = \varnothing$, that is, $\mathcal{V} \cap \mathcal{A}^{A'} \neq \varnothing$}.

Proposition 8-3-4. *With the same notations, the mapping $\mathcal{V} \to$ $= \cap\{\mathcal{A}_A', A \in \mathcal{V}\}$ is a homeomorphism from $\mathcal{F}_u(\mathcal{A})$ onto $\mathcal{F}_u(\mathcal{A}')$, and reciprocal formula $\mathcal{V} = \cap\{\mathcal{A}_{A'}, A' \in \mathcal{V}'\}$ holds. Moreover, the image u $\mathcal{V} \to \mathcal{V}'$ of any l.s.c. (resp. u.s.c.) open set in $\mathcal{F}_u(\mathcal{A})$ is an u.s.c. (resp. l.s open set in $\mathcal{F}_u(\mathcal{A}')$, and conversely.*

Proof. We know from Proposition 8-3-1 that $\mathcal{V} \to \mathcal{V}'$ is a one-to- mapping from $\mathcal{F}_u(\mathcal{A})$ onto $\mathcal{F}_u(\mathcal{A}')$, and the reciprocal formula holds. the other hand, if $\mathcal{V} \in \mathcal{F}_u(\mathcal{A})$ and $A \in \mathcal{A}$, $\mathcal{V} \not\subset \mathcal{A}_{A^c}$ if and only if t exists $V \in \mathcal{V}$ such that $V \subset A$, that is, if and only if $A \in \mathcal{V}$, as $\cup$-hereditary. But $A \in \mathcal{V} = \cap\{\mathcal{A}_{A'}, A' \in \mathcal{V}'\}$ is equivalent to $\mathcal{V}' \cap \mathcal{A}'^A =$ Thus:

$$\mathcal{V} \subset \mathcal{A}_{A^c} \Leftrightarrow \mathcal{V}' \cap \mathcal{A}'^A \neq \varnothing$$

(8-

$$\mathcal{V}' \subset \mathcal{A}'_{A'^c} \Leftrightarrow \mathcal{V} \cap \mathcal{A}^{A'} \neq \varnothing$$

This achieves the proof when taking Proposition 8-3-2 into account.

8-4. POINT OF VIEW OF THE INVERSE MAPPINGS

In the preceding sections, we investigated the increasing mappings fr $\mathcal{P}(E)$ (or a subset of it) into itself. More generally, now let E and E' two LCS spaces (no longer necessarily identical), $\mathcal{A} \subset \mathcal{P}(E)$ and ψ increasing mapping from $\mathcal{A}$ into $\mathcal{P}(E')$, that is, $A, B \in \mathcal{A}$ and $A \subset B$ i imply $\psi(A) \subset \psi(B)$ in E'. For any $x \in E'$, let us put

$$\mathcal{V}(x) = \psi^{-1}(\mathcal{P}_x(E')) = \{A, A \in \mathcal{A}, x \in \psi(A)\} \qquad (8\text{-}4$$

$\mathcal{V}$ is an u.s.c. mapping from E' into $\mathcal{F}_u(\mathcal{Q})$. Hence conditions 2 and 4 are equivalent.

If ψ is u.s.c. from $\mathcal{Q}$ into $\mathcal{F}(E')$, it follows that $\psi = \psi_a$ (the proof is the same as for Proposition 8-1-4), and thus condition 1 implies 4. Conversely, if condition 2 is true, ψ maps $\mathcal{Q}$ into $\mathcal{F}(E')$ (as condition 2 and 4 are equivalent). Let $K' \in \mathcal{K}(E')$ be a compact set, $\psi^{-1}(\mathcal{F}_{K'}) = \cup\{\mathcal{V}(x), x \in K'\}$, and $n \to A_n \in \psi^{-1}(\mathcal{F}_{K'})$ a sequence such that $\lim A_n = A$ in $\mathcal{Q}$. For any n, there exists $x_n \in K'$ such that $A_n \in \mathcal{V}(x_n)$, and, if $x_0 \in K'$ is an accumulation point of $\{x_n\}$, this implies $A \in \mathcal{V}(x_0)$, for $\mathcal{V}$ is u.s.c. It follows that $A \in \psi^{-1}(\mathcal{F}_{K'})$, and $\psi^{-1}(\mathcal{F}_{K'})$ is closed. Hence ψ is u.s.c., and condition 1 is equivalent to 2. The equivalence of conditions 3 and 4 is obvious.

Proposition 8-4-2. *The mapping $\mathcal{V}$ is l.s.c. from E' into $\mathcal{F}_u(\mathcal{Q})$ if and only if $\psi = \psi_a$ and ψ_b maps $\mathcal{B}$ into $\mathcal{G}(E')$.*

Proof. From Proposition 8-4-1, it follows that $\psi = \psi_a$ is equivalent to $\mathcal{V}(E') \subset \mathcal{F}_u(\mathcal{Q})$, and we suppose this condition to be satisfied. Then $\mathcal{V}$ is l.s.c. if and only if, for any $A' \in \mathcal{Q}'$, the set $\{x\colon \mathcal{V}(x) \cap \mathcal{Q}^{A'} \neq \varnothing\} = \cup\{\psi(A), A \in \mathcal{Q}, A \cap A' = \varnothing\}$ is open in E' (Proposition 8-3-2). But $A \cap A' = \varnothing$ is equivalent to $A \subset A'^c$, and $A'^c \in \mathcal{B}$. Thus ψ is l.s.c. if and only if, for any $B \in \mathcal{B}$, the set $\cup\{\psi(A), A \in \mathcal{Q}, A \subset B\} = \psi_b(B)$ is open in E'.

Corollary. *The mapping ψ is u.s.c. from $\mathcal{Q}$ into $\mathcal{F}(E')$, and ψ_b maps $\mathcal{B}$ into $\mathcal{G}(E')$ if and only if $\mathcal{V}$ is continuous.*

Example. If $E = \mathbf{R}^d$ and ψ is a τ-mapping from $\mathcal{Q}$ into $\mathcal{P}(E)$, $\mathcal{V}(x) = \mathcal{V}_x$ is the translate of its kernel. If the kernel is closed in $\mathcal{Q}$, the mapping $\mathcal{V}$ is continuous, so that ψ is u.s.c. and ψ_b maps $\mathcal{B}$ into $\mathcal{G}(E')$.

Concerning the homeomorphism $\mathcal{V} \to \mathcal{V}'$ (Proposition 8-3-3), we obtain the following results:

Proposition 8-4-3. *Let ψ be a mapping from $\mathcal{Q}$ into $\mathcal{P}(E')$ such that $\psi = \psi_a$, $\mathcal{V}\colon E' \to \mathcal{F}_u(\mathcal{Q})$ be its inverse mapping, $\mathcal{V}'$ be the mapping $E' \to \mathcal{F}_u(\mathcal{Q})$, defined by $\mathcal{V}'(x) = \cap\{\mathcal{Q}'_A, A \in \mathcal{V}(x)\}$, $x \in E'$, and $\psi'\colon \mathcal{Q}' \to \mathcal{P}(E')$ be such that $\psi'(A') = \{x, A' \in \mathcal{V}'(x)\}$, $A' \in \mathcal{Q}'$. Then $\psi' = \psi'_{a'}$ on $\mathcal{Q}'$. Moreover, the four following conditions are equivalent:*

1. *$\mathcal{V}$ is an u.s.c. (resp. l.s.c.) mapping from E' into $\mathcal{F}_u(\mathcal{Q})$.*
2. *$\mathcal{V}'$ is a l.s.c. (resp. u.s.c.) mapping from E' into $\mathcal{F}_u(\mathcal{Q}')$.*
3. *ψ (resp. ψ') is an u.s.c. mapping from $\mathcal{Q}$ (resp. $\mathcal{Q}'$) into $\mathcal{F}(E')$.*
4. *$\psi'_{b'}$ (resp. ψ_b) maps $\mathcal{B}'$ (resp. $\mathcal{B}$) into $\mathcal{G}(E')$.*

Furthermore, the five following conditions are equivalent:

a. ψ *and* ψ' *are u.s.c.*
b. $\mathcal{V}$ *is continuous.*
c. $\mathcal{V}'$ *is continuous.*
d. ψ *is u.s.c., and* ψ_b *maps* $\mathcal{B}$ *into* $\mathcal{G}(E')$.
e. ψ' *is u.s.c., and* ψ'_b *maps* $\mathcal{B}'$ *into* $\mathcal{G}(E')$.

Proof. For any $x \in E'$, $\mathcal{V}'(x)$ is closed in $\mathcal{C}$ (Proposition 8-3-1) and thus $\psi' = \psi'_{a'}$ (Proposition 8-4-1). It follows from Proposition 8-3-3 that $\mathcal{V}$ is u.s.c. (resp. l.s.c.) if and only if $\mathcal{V}'$ is l.s.c. (resp. u.s.c), so that conditions 1 and 2 are equivalent. Then it follows from Proposition 8-4-2 that ψ_b maps $\mathcal{B}$ into $\mathcal{G}(E')$ if and only if ψ' is u.s.c., and ψ'_b maps $\mathcal{B}'$ into $\mathcal{G}(E')$ if and only if ψ is u.s.c. Hence conditions 1, 2, 3, and 4 are equivalent. In the same way, ψ and ψ' are both u.s.c. if and only if ψ is u.s.c. and ψ_b maps $\mathcal{B}$ into $\mathcal{G}(E')$, or if and only if $\mathcal{V}$ is continuous. Hence conditions *a*, *b*, *c*, *d*, and *e* are equivalent.

CHAPTER 9

Integrals and Measures Valued in $\mathcal{K}_o$

We saw in Chapter 1 that, for a given $K \in C(\mathcal{K}(\mathbf{R}^d))$, the family $\{\lambda K, \lambda \geqslant 0\}$ of its homothetics satisfies the semigroup relationship $\lambda K \oplus \mu K = (\lambda + \mu)K$, and that $\lambda \geqslant \mu \geqslant 0$ implies $\lambda K \succcurlyeq \mu K$, where $\succcurlyeq$ is the preordering on $\mathcal{K}$ defined by $A \succcurlyeq B$ if A is open with respect to B.

More generally, it would be interesting to characterize the one-parameter families $\{K(\lambda)\} \subset \mathcal{K}$ such that $K(\lambda) \succcurlyeq K(\mu)$ if $\lambda \geqslant \mu$. The Riemann-Minkowski integral (Section 9-1) provides a first simple example of such a family and, moreover (curiously enough), necessarily takes its values in $C(\mathcal{K})$. On the other hand, the Stieltjes-Minkowski integral affords another example, but no longer necessarily takes its value in $C(\mathcal{K})$. Sections 9-2 and 9-3 constitute a simple translation of the classical integration theory, with the usual duality between the set and functional standpoints. From the set point of view, it is natural to associate with a $\mathcal{K}_o$-valued measure an increasing functional, which turns out to be semiadditive, that is, $I(f+f') = I(f) \oplus I(f')$ if the sets $\{f>0\}$ and $\{f'>0\}$ are disjoint, and (strictly) additive if and only if the measure is valued in $C_0(\mathcal{K})$. Conversely, from the functional point of view, any $\mathcal{K}_0$-valued, semiadditive, convex, and increasing functional can be associated with a $\mathcal{K}_0$-valued measure, which is valued in $C_o(\mathcal{K})$ if and only if the functional is strictly additive. The measures valued in $C(\mathcal{K}_0)$ have been thoroughly investigated, for they can be represented in terms of ordinary integrals of supporting functions (see, e.g., Rockafellar, 1968, 1970, and Valadier, 1970), and we shall not examine them.

9-1. THE RIEMANN-MINKOWSKI INTEGRAL

Let $\{A(\lambda), \lambda \in \mathbf{R}\}$ be a one-parameter family of nonempty compact sets in the euclidean space $\mathbf{R}^d$. Our aim is to construct the *Riemann-Minkowski* (or, for brevity, *R-M*) *integral* $\int_a^b A(\lambda)\,d\lambda$ in $\mathcal{K}'(\mathbf{R}^d)$ which is provided with the myope topology and the Minkowski sum $\oplus$. By changing the parameter λ, it is equivalent to define the R-M integral

$$I=\int_0^1 A(\lambda)\,d\lambda \in \mathcal{K}'(\mathbf{R}^d)$$

We essentially suppose that the mapping $\lambda \to A(\lambda)$ is *continuous*. Thus the image of $[0,1]$ under the mapping A is compact in $\mathcal{K}'$, so that, for any $\lambda \in [0,1]$, $A(\lambda)$ is contained in a fixed compact set $K_0 \in \mathcal{K}'$. Moreover, the mapping A is uniformly continuous on $[0,1]$, so that for any $\epsilon>0$ there exists $\eta(\epsilon)$ such that

$$|\lambda-\lambda'| \leqslant \eta(\epsilon), \qquad \lambda,\lambda' \epsilon [0,1] \Rightarrow \rho(A(\lambda),A(\lambda')) \leqslant \epsilon \tag{9-1-1}$$

ρ denoting the Hausdorff metric on $\mathcal{K}'$.

Let $s=\{x_0,x_1,\ldots,x_n\}$, $x_0=0 \leqslant x_1 \leqslant \cdots \leqslant x_n=1$, be a *subdivision* of the interval $[0,1]$, and put $|s|=\operatorname{Sup}\{|x_i-x_{i-1}|,\ i=1,2,\ldots,n\}$. For any choice $X=(\xi_1,\ldots,\xi_n)$ of points $\xi_i \in [x_{i-1},x_i]$, $i=1,2,\ldots,n$, let us associate with X the compact set

$$I_s(X)=\bigoplus_{i=1}^{n}(x_i-x_{i-1})A(\xi_i) \tag{9-1-2}$$

and denote by $\mathcal{G}_s$ the family $\{I_s(X)\}$ for all the possible choices X.

a. $\mathcal{G}_s$ is compact in $\mathcal{K}'$. For any X, $I_s(X)$ is contained in the fixed compact set $C(K_0)$, and so we have only to prove that $\mathcal{G}_s$ is closed in $\mathcal{K}'$. Let $k \to X_k=(\xi_1(k),\ldots,\xi_n))$ be a sequence of possible choices of the vector X such that the sequence $k \to B_k = I_s(X_k)$ converges toward B in $\mathcal{K}'$. Each $\xi_i(k)$ belongs to a compact interval, so that it is possible to find a subsequence $j \to k_j$ such that $\lim X_{k_j} = X_0 = (\xi_1^{\,0},\ldots,\xi_n^{\,0})$ and X_0 is still a possible choice of X. It follows from the continuity of the Minkowski sum $\oplus$ that $\lim B_{k_j} = I_s(X_0)$. Thus $B=I_s(X_0)\in \mathcal{G}_s$, and hence $\mathcal{G}_s$ is compact in $\mathcal{K}'$.

b. If $|s| \leqslant \eta(\epsilon)$, the diameter of $\mathcal{G}_s$ is $\leqslant \epsilon$. If $|s| \leqslant \eta(\epsilon)$ and X_1, X_2 are two possible choices of X, it follows from (9-1-1) that, for each $i=1,2,\ldots,n$, $A(\xi_i^{\,1}) \subset A(\xi_i^{\,2}) \oplus \epsilon B$, $A(\xi_i^{\,2}) \subset A(\xi_i^{\,1}) \oplus \epsilon B$, B denoting the unit ball. It follows that

$$I_s(X_1)=\bigoplus_{i=1}^{n}(x_i-x_{i-1})A(\xi_i^{\,1}) \subset \epsilon B \oplus \left(\bigoplus_{i=1}^{n}(x_i-x_{i-1})A(\xi_i^{\,2})\right)=I_s(X_2)\oplus \epsilon B$$

and

$$I_s(X_2) \subset I_s(X_1) \oplus \epsilon B$$

so that $\rho(I_s(X_1), I_s(X_2)) \leqslant \epsilon$.

Let us now denote by S the set of the subdivisions s of $[0,1]$, which is filtering for the relationship $\vdash$ defined by $s_1 \vdash s_2$ if the subdivision s_1 is finer than s_2 (i.e., for any $s, s' \in S$, there exists a $s'' \in S$ such that $s'' \vdash s$ and $s'' \vdash s'$).

c. The filtering family $\{\mathcal{I}_s, s \in S\}$ *is converging in* $\mathcal{K}(\mathcal{K}')$. For any $s \in S$ and any possible choice X, $I_s(X) \subset C(K_0)$, that is, the family $\{\mathcal{I}_s, s \in S\}$, is enclosed in the fixed compact set $\mathcal{K}^{(C(K_0))^c}$ in $\mathcal{K}(\mathcal{K})$; hence we need only show that the Cauchy criterion is satisfied by this family with respect to the Hausdorff metric δ in $\mathcal{K}(\mathcal{K}')$. Let $s, s' \in S$ be such that $s \vdash s'$, and X, X' be two possible choices of the vector X associated with s and s', respectively. If $|s'| \leqslant \eta(\epsilon)$ [and thus (a fortiori) $|s| \leqslant \eta(\epsilon)$], we find, as in part *b*,

$$I_s(X) \subset I_{s'}(X') \oplus \epsilon B, \qquad I_{s'}(X') \subset I_s(X) \oplus \epsilon B$$

that is, $\rho(I_s(X), I_{s'}(X')) \leqslant \epsilon$ in $\mathcal{K}'$, and thus $\delta(\mathcal{I}_s, \mathcal{I}_{s'}) \leqslant \epsilon$ in $\mathcal{K}(\mathcal{K}')$.

Now let $s_0 \epsilon S$ be such that $|s_0| \leqslant \epsilon$, and s, s' be the two arbitrary subdivisions finer than s_0. It follows that $\delta(\mathcal{I}_s, \mathcal{I}_{s'}) \leqslant \delta(\mathcal{I}_{s_0}, \mathcal{I}_{s'}) \leqslant 2\epsilon$, and the Cauchy criterion is verified. Thus there exists $\mathcal{I} \in \mathcal{K}(\mathcal{K}')$ such that $\lim\{\mathcal{I}_s, s \in S\} = \mathcal{I}$.

d. $\mathcal{I} = \{I\}$ *for a unique* $I \in \mathcal{K}'$, which will be denoted by $I = \int_0^1 A(\lambda)\, d\lambda$. This is an immediate consequence of part *b*, because the diameter is continuous for the myope topology: the diameter of $\mathcal{I} = 0$, and $\mathcal{I} = \{I\}$ for a unique $I \in \mathcal{K}'$.

e. If we choose a vector X_s *for any* $s \in S$, *the filtering family* $\{I_s(X_s), s \in S\}$ *converges toward* I *in* $\mathcal{K}'$.

For this family is enclosed into the fixed compact set $C(K_0)$ and thus admits an accumulation point $J \in \mathcal{K}'$. But $I_s(X_s) \in \mathcal{I}_s$ and $\lim \mathcal{I}_s = \{I\}$ in $\mathcal{K}(\mathcal{K}')$ imply $J \in \mathcal{I}$, that is, $J = I$. Thus $\lim I_s(X_s) = I$ in $\mathcal{K}'$.

In particular, it follows from part *e* that we may write

$$I = \int_0^1 A(\lambda)\, d\lambda = \lim_{n \to \infty} \bigoplus_{k=1}^{n} \frac{1}{n} A(k/n) \tag{9-1-3}$$

and, more generally, the R-M integral on an arbitrary bounded interval $[a,b]$ is

$$\int_a^b A(\lambda)\, d\lambda = \lim_{|s| \to 0} \bigoplus_i (x_i - x_{i-1}) A(\xi_i)$$

where s is a subdivision $a = x_0 \leqslant x_1 \leqslant \ldots \leqslant x_n = b$ and $\xi_i \in (x_{i-1}, x_i)$. In particular:

$$\left(\int_a^b A(\lambda)\,d\lambda\right) \oplus \left(\int_b^c A(\lambda)\,d\lambda\right) = \int_a^c A(\lambda)\,d\lambda \qquad (a \leqslant b \leqslant c)$$

and, if $A(\lambda)$ and $B(\lambda)$ are two continuous families,

$$\int_a^b (A(\lambda) \oplus B(\lambda))\,d\lambda = \left(\int_a^b A(\lambda)\,d\lambda\right) \oplus \left(\int_a^b B(\lambda)\,d\lambda\right) \qquad (9\text{-}1\text{-}4)$$

Convexity of the Riemann-Minkowski Integral

In order to prove that the R-M integral is always a convex compact set, we shall use the following:

Lemma 9-1-1. *If $A(\lambda) = A \in \mathcal{K}'$ does not depend on λ, $\int_a^b A(\lambda)\,d\lambda = (b-a)$ $C(A)$, $a \leqslant b$. In particular, for any $A \in \mathcal{K}'$ and any continuous family $\{B(\lambda)\} \subset \mathcal{K}'$,*

$$\int_a^b (A \oplus B(\lambda))\,d\lambda = (b-a)C(A) \oplus \int_a^b B(\lambda)\,d\lambda \qquad (a \leqslant b) \quad (9\text{-}1\text{-}5)$$

Proof. The first statement follows from (9-1-3) and Proposition 1-5-5, and then (9-1-5) follows from (9-1-4).

Theorem 9-1-1. *The R-M integral is valued in $C(\mathcal{K}')$. More precisely, for any continuous family $\{A(\lambda)\} \subset \mathcal{K}'$,*

$$\int_a^b A(\lambda)\,d\lambda = \int_a^b C(A(\lambda))\,d\lambda \in C(\mathcal{K}') \qquad (a \leqslant b)$$

Proof. When changing the parameter λ, we see that we have to prove that the R-M integral $I = \int_0^1 (A(\lambda)\,d\lambda$ is $I = \int_0^1 C(A(\lambda))\,d\lambda$, for, if so, $I \in C(\mathcal{K}')$ follows immediately, because $C(\mathcal{K}')$ is closed in $\mathcal{K}'$.

For n integer $\geqslant 0$ and $0 < k \leqslant n$, let us put

$$J_k(n) = \int_{(k-1)/n}^{k/n} A(\lambda)\,d\lambda$$

so that $I=\oplus_{k=1}^{n} J_k(n)$. Let $\eta(\epsilon)\geqslant 0$ be such that (9-1-1) holds for a given $\epsilon\geqslant 0$. Then $(1/n)\leqslant\eta(\epsilon)$ and $(k-1)/n\leqslant\lambda\leqslant k/n$ imply $A(\lambda)\subset A(k/n)\oplus\epsilon B$ and $A(k/n)\subset A(\lambda)\oplus\epsilon B$. On the other hand, the R-M integral is increasing with respect to $\subset$, and it follows from Lemma 9-1-1 that

$$J_k(n)\subset\frac{1}{n}C\left(A\left(\frac{k}{n}\right)\right)\oplus\frac{1}{n}\epsilon B=\frac{1}{n}\left[\epsilon B\oplus C\left(A\left(\frac{k}{n}\right)\right)\right]$$

$$\frac{1}{n}C\left(A\left(\frac{k}{n}\right)\right)\subset\frac{1}{n}(\epsilon B\oplus J_k(n))$$

By taking the Minkowski sum of these relationships, we find

$$I=\bigoplus_{k=1}^{n} J_k(n)\subset\epsilon B\oplus\left[\bigoplus_{k=1}^{n}\frac{1}{n}C\left(A\left(\frac{k}{n}\right)\right)\right]$$

$$\bigoplus_{k=1}^{n}\frac{1}{n}C\left(A\left(\frac{k}{n}\right)\right)\subset I\oplus\epsilon B$$

Thus, if $n\uparrow\infty$, it follows from (9-1-3) that

$$I\subset\epsilon B\oplus\int_0^1 C(A(\lambda))\,d\lambda$$

$$\int_0^1 C(A(\lambda))\,d\lambda\subset I\oplus\epsilon B$$

Hence

$$I=\int_0^1 C(A(\lambda))\,d\lambda$$

The Stieltjes-Minkowski Integral

Let F be a nondecreasing function on $\mathbf{R}$, and $\{A(\lambda),\lambda\in\mathbf{R}\}\subset\mathcal{K}'$ be a continuous one-parameter family in $\mathcal{K}'$. By changing x_i-x_{i-1} into $F(x_i)-F(x_{i-1})$ in the definition of the R-M integral, it is possible to show that the limit

$$\int_a^b A(\lambda)\,dF(\lambda)=\lim_{|s|\to 0}\bigoplus_{i=1}^{n}\big(F(x_i)-F(x_{i-1})\big)A(\xi_i)$$

still exists and defines the Stieltjes-Minkowski (or S-M) integral. But Lemma 9-1-2 is no longer valid, so that Theorem 9-1-1 is not true: the S-M integral is not necessarily valued in $C(\mathcal{K}')$.

The function $\lambda\to B(\lambda)=\int_{0_+}^{\lambda_+}A(\mu)\,dF(\mu)$ is increasing for the preordering $\succcurlyeq$ on $\mathcal{K}'$ (i.e., $K\succcurlyeq K'$ if K is open with respect to K', or $K=K'\oplus K''$ for a $K''\in\mathcal{K}'$), for

$$B(\lambda)=B(\mu)\oplus\int_{\mu_+}^{\lambda_+}A(x)\,dF(x)\qquad(\lambda\geqslant\mu)$$

In other words, there actually exist families in $\mathcal{K}'$ which are increasing for $\succcurlyeq$ and are not valued in $C(\mathcal{K}')$. Nevertheless, we shall not develop this point of view here, but rather devote the last two sections to a simple translation of the classical integration theory.

9.2. RADON MEASURES VALUED IN $\mathcal{K}_0(\mathbf{R}^d)$

In what follows, E is an arbitrary LCS space, and $\mathcal{K}_0(\mathbf{R}^d)=\mathcal{K}_0$ denotes the space of the compact sets in $\mathbf{R}^d$ containing the origin O. With the aim of defining $\mathcal{K}_0$-valued Radon measures on E, we must first investigate the appropriate functional spaces.

The Spaces Φ_k, Φ_g, and $\mathbf{C}_{\mathcal{K}}^+$

Let $\mathbf{R}_+$ be the set of positive real numbers, and $\overline{\mathbf{R}}_+=[0,\infty]$ the compactified half straight line. A function f: $E\to\mathbf{R}_+$ is characterized by its *largest subgraph*:

$$\check{\Gamma}_f=\{(x,r),x\in E,r\in\mathbf{R}_+,r\leqslant f(x)\}$$

which is a subset of the product space $E\times\mathbf{R}_+$. If f is a function from E into $\mathbf{R}_+$ or $\overline{\mathbf{R}}_+$, we shall also characterize it by its *smallest subgraph*:

$$\hat{\Gamma}_f=\{(x,r),x\in E,r\in\mathbf{R}_+,r<f(x)\}$$

which is always a subset of $E\times\mathbf{R}_+$.

Proposition 9-2-1. *A function* f: $E \to \mathbf{R}_+$ *is u.s.c. (resp. a function f from E into* $\mathbf{R}_+$ *or* $\overline{\mathbf{R}}_+$ *is l.s.c.) if and only if* $\check{\Gamma}_f$ *(resp.,* $\hat{\Gamma}_f$*) is closed (resp., open) in* $E \times \mathbf{R}_+$.

Proof. Let f be an u.s.c. function, and $n \to (x_n, r_n) \in \check{\Gamma}_f$ a sequence such that $\lim (x_n, r_n) = (x, r)$ in $E \times \mathbf{R}_+$. For each n, $(x_n, r_n) \in \check{\Gamma}_f$ implies $r_n \leqslant f(x_n)$, and thus $r \leqslant \overline{\lim} f(x_n) \leqslant f(x)$, that is, $(x,r) \in \check{\Gamma}_f$. Hence $\check{\Gamma}_f$ is closed. Conversely, if $\check{\Gamma}_f$ is closed, let $\{x_n\}$ be a sequence such that $\lim x_n = x$ in E. For each n, $(x_n, f(x_n)) \in \check{\Gamma}_f$, and thus $(x, \overline{\lim} f(x_n)) \in \check{\Gamma}_f$, that is, $\overline{\lim} f(x_n) \leqslant f(x)$. Hence f is u.s.c. The proof of the statement concerning the l.s.c. function is analogous.

We shall denote by Φ_g the space of the l.s.c. functions f: $E \to \overline{\mathbf{R}}_+$. It follows from Proposition 9-2-1 that Φ_g can be identified with a subspace of $\mathcal{G}(E \times \mathbf{R}_+)$, that is, the set $\{\hat{\Gamma}_f, f \in \Phi_g\}$ of the open smallest subgraphs. The space Φ_g will always be provided with the corresponding relative topology. An open set $\Gamma \in \mathcal{G}(E \times \mathbf{R}_+)$ is the smallest subgraph of a function $f \in \Phi_g$ if and only if $(x,r) \in \Gamma$ and $0 \leqslant r' \leqslant r$ imply $(x, r') \in \Gamma$, and then $f(x) = \operatorname{Sup}\{r, (x,r) \in \Gamma\}$ for any $x \in E$ such that $\Gamma \cap (\{x\} \times \mathbf{R}_+)$ is nonempty [and $f(x) = 0$ otherwise]. Moreover, it is easy to verify that the set of the open smallest subgraphs is closed in the compact space $\mathcal{G}(E \times \mathbf{R}_+)$, so that the space Φ_g is *compact*.

If f is an u.s.c. positive function, its largest subgraph $\check{\Gamma}_f \in \mathcal{F}(E \times \mathbf{R}_+)$ always contains the set $E_0 = E \times \{o\}$, and we shall denote by Γ'_f the smallest $F \in \mathcal{F}(E \times \mathbf{R}_+)$ such that $\check{\Gamma}_f = F \cup E_0$. In other words, Γ'_f is the closure in $E \times \mathbf{R}_+$ of the set $\check{\Gamma}_f \backslash E_0$, or still $\Gamma'_f = \Gamma_f \cap (\operatorname{Supp} f \times \mathbf{R}_+)$, where $\operatorname{Supp} f$ is the closure in E of the set $\{x, f(x) > 0\}$. We shall denote by Φ_k *the set of the u.s.c. positive functions with compact support on E*, that is, $f \in \Phi_k$ if and only if $\check{\Gamma}_f \in \mathcal{F}(E \times \mathbf{R}_+)$ and $\Gamma'_f \in \mathcal{K}(E \times \mathbf{R}_+)$. We identify Φ_k with the corresponding subset of $\mathcal{F}(E \times \mathbf{R}_+)$, but we will provide it with a topology finer than the relative $\mathcal{F}$-topology, so that Φ_k is *not* a (topological) subspace of $\mathcal{F}(E \times \mathbf{R}_+)$. This Φ_k-topology is generated by two families: (1) $\{f, \Gamma'_f \subset G\}$, $G \in \mathcal{G}(E \times \mathbf{R}_+)$ (u.s.c.-type open sets in Φ_k), and (2) $\{f, \Gamma_f \cap G \neq \varnothing\}$, $G \in \mathcal{G}(E \times \mathbf{R}_+)$ (l.s.c.-type open sets in Φ_k). With this topology, Φ_k is LCS and in many ways analogous to a space $\mathcal{K}$ provided with the myope topology.

If $\chi \in \Phi_k$ and $\gamma \in \Phi_g$, we write $\chi \subset \gamma$ (or $\gamma \supset \chi$) if $\Gamma'_\chi \subset \hat{\Gamma}_\gamma$, that is, $\chi(x) < \gamma(x)$ (strictly) for any x belonging to the support $\overline{\{\gamma > 0\}}$ of γ. For any functions f, g on E, we write $f \leqslant g$ if $f(x) \leqslant g(x)$ for any $x \in E$. With this notation, it is easy to verify that the Φ_k-topology is generated by

1. The family $\{\chi, \chi \subset \gamma\}$, $\gamma \in \Phi_g$ (u.s.c.-type open sets).
2. The family $\{\chi, \chi \nleqslant \chi_0\}$, $\chi_0 \in \Phi_k$ (l.s.c.-type open sets).

In the same way, the Φ_g-topology is generated by

1. The family $\{\gamma, \gamma \supset \chi\}$, $\chi \in \Phi_k$ (l.s.c.-type open sets).
2. The family $\{\gamma, \gamma \not\geqslant \gamma_0\}$, $\gamma_0 \in \Phi_g$ (u.s.c.-type open sets).

Finally, we denote by $\mathbf{C}_{\mathcal{K}}^+$ *the space of the continuous positive functions with compact support on E*, provided with its usual topology. A function φ is in $\mathbf{C}_{\mathcal{K}}^+$ if and only if $\varphi \in \Phi_k \cap \Phi_g$, but the topologies of $\mathbf{C}_{\mathcal{K}}^+$ is coarser than the one generated by the relative topologies of Φ_k and Φ_g. For any $\varphi \in \mathbf{C}_{\mathcal{K}}^+$, $\chi \in \Phi_k$, and $\gamma \in \Phi_g$ we write $\chi \subset \varphi$ if $\Gamma'_\chi \subset \hat{\Gamma}_\varphi$ and $\varphi \subset \gamma$ if $\Gamma'_\varphi \subset \hat{\Gamma}_\gamma$. Then the Φ_k-topology is also generated by the two following families:

$$\{\chi, \chi \subset \varphi\}, \varphi \in \mathbf{C}_{\mathcal{K}}^+, \quad \text{and} \quad \{\chi, \chi \not\leqslant \varphi\}, \varphi \in \mathbf{C}_{\mathcal{K}}^+$$

In the same way, the Φ_g-topology is generated by the families

$$\{\gamma, \gamma \supset \varphi\}, \varphi \in \mathbf{C}_{\mathcal{K}}^+, \quad \text{and} \quad \{\gamma, \gamma \not\geqslant \varphi\} \varphi \in \mathbf{C}_{\mathcal{K}}^+$$

In regard to the monotone convergence of a sequence $\{f_n\}$, the following results will be useful:

$\{f_n\} \subset \Phi_g$, $f_n \uparrow f$ imply $f \in \Phi_g$ and $f = \lim f_n$ in Φ_g.
$\{f_n\} \subset \Phi_g$, $f_n \downarrow f$, and $f \in \Phi_g$ imply $f = \lim f_n$ in Φ_g.
$\{f_n\} \subset \Phi_k$, $f_n \uparrow f$, and $f \in \Phi_k$ imply $f = \lim f_n$ in Φ_k.
$\{f_n\} \subset \Phi_k$, $f_n \downarrow f$ imply $f \in \Phi_k$ and $f = \lim f_n$ in Φ_k.

The *addition* operation $(f, f') \to f + f'$ is u.s.c. from $\Phi_k \times \Phi_k$ into Φ_k, l.s.c. from $\Phi_g \times \Phi_g$ into Φ_g, and continuous from $\mathbf{C}_{\mathcal{K}}^+ \times \mathbf{C}_{\mathcal{K}}^+$ into $\mathbf{C}_{\mathcal{K}}^+$.

Pseudo-Integrals Valued in $\mathcal{K}_0(\mathbf{R}^d)$

Let I be a mapping from $\mathbf{C}_{\mathcal{K}}^+(E)$ into $\mathcal{K}_0(\mathbf{R}^d)$. We shall say that I is *positively homogeneous* if $I(\lambda\varphi) = \lambda I(\varphi)$, $\lambda \geqslant 0$, $\varphi \in \mathbf{C}_{\mathbf{K}}^+$. Also I is *increasing* if $\varphi \leqslant \varphi'$ in $\mathbf{C}_{\mathcal{K}}^+$ implies $I(\varphi) \subset I(\varphi')$; I is *subadditive* if $I(\varphi + \varphi') \subset I(\varphi) \oplus I(\varphi')$, *superadditive* if $I(\varphi + \varphi') \supset I(\varphi) \oplus I(\varphi')$, and *additive* if $I(\varphi + \varphi') = I(\varphi) \oplus I(\varphi')$ $(\varphi, \varphi' \in \mathbf{C}_{\mathcal{K}}^+)$.

In the same way, I is said to be *convex* (resp. *concave*) if it is positively homogeneous and subadditive (resp., superadditive). Finally, I is *positively linear* or is a *C-integral* if it is at the same time convex and concave, that is, positively homogeneous and additive.

With each mapping I: $\mathbf{C}_{\mathcal{K}}^+ \to \mathcal{K}_0(\mathbf{R}^d)$ we associate the mapping CI: $\mathbf{C}_{\mathcal{K}}^+ \to C(\mathcal{K}_0)$ defined by $CI(\varphi) = C(I(\varphi))$, that is, $CI(\varphi)$ is the *convex hull* of $I(\varphi)$. Then I is said to be a *pseudo-integral* if (1) I is *convex* and *increasing*, and (2) *its convex hull CI is a C-integral*. Obviously, any C-integral is a pseudo-integral.

Proposition 9-2-2. *If I is increasing and convex (and, in particular, if it is a pseudo-integral), I is continuous. If I is concave (and, in particular, if it is a C-integral), I is valued in $C(\mathcal{K}_0)$.*

Proof. Let $\{\varphi_n\}\subset \mathbf{C}_{\mathcal{K}}^+$ be a sequence and $\lim \varphi_n=\varphi$ in $\mathbf{C}_{\mathcal{K}}^+$. There exists $K_0\in\mathcal{K}(E)$ such that $\operatorname{Supp}\varphi_n\subset K_0$ for any n, and $\varphi_0\in\mathbf{C}_{\mathcal{K}}^+$ such that $\varphi_0(x)=1$ for any $x\in K_0$. It follows that, for $\epsilon>0$ and n large enough, $\varphi\leqslant\varphi_n+\epsilon\varphi_0$, $\varphi_n\leqslant\varphi+\epsilon\varphi_0$. If I is increasing and convex, this implies $I(\varphi)\subset I(\varphi_n)\oplus\epsilon I(\varphi_0)$ and $I(\varphi_n)\subset I(\varphi)\oplus\epsilon I(\varphi_0)$. Hence $\lim I(\varphi_n)=I(\varphi)$ in $\mathcal{K}_0$, and I is continuous.

For any $\varphi\in\mathbf{C}_{\mathcal{K}}^+$, we may write $\varphi=(1/n)\Sigma_1^n\varphi_i$, $\varphi_i=\varphi$, $i=1,2,\ldots,n$. If I is concave, it follows that $I(\varphi)\supset(1/n)\,(I(\varphi))^{\oplus n}$. Hence $I(\varphi)\supset CI(\varphi)$ (Proposition 1-5-5), and thus $I=CI$, that is, I is valued in $C(\mathcal{K}_0)$.

Extension on Φ_g of a Pseudo-Integral

Although some of the following results are also valid if I is only supposed to be increasing and convex, the letter "I will always denote a *pseudo-integral*, from now on. Our first step is to extend it to the space Φ_g. In the sequel, $\overline{\mathcal{K}}_0$ is used to denote the compactification of $\mathcal{K}_0$, that is, $\mathcal{K}_0\cup\{\omega\}$, where ω is *the point at infinity*. Conventionally, $K\cup\omega=\omega$ and $K\oplus\omega=\omega$ for any $K\in\overline{\mathcal{K}}_0$. Then the mapping I': $\Phi_g\to\overline{\mathcal{K}}_0$, defined by

$$I'(\gamma)=\lim\{I(\varphi),\varphi\in\mathbf{C}_{\mathcal{K}}^+,\varphi\leqslant\gamma\} \tag{9-2-1}$$

is an extension of I. In (9-2-1), "lim" denotes the limit in $\overline{\mathcal{K}}_0$ of the increasing filtering family $I(\varphi)$, $\varphi\leqslant\gamma$. If the closed set

$$F=\overline{\cup\{I(\varphi),\varphi\in\mathbf{C}_{\mathcal{K}}^+,\varphi\leqslant\gamma\}}$$

is compact, $I'(\gamma)=F$, and $I'(\gamma)=\omega$ otherwise. For any $\varphi\in\mathbf{C}_{\mathcal{K}}^+$ and $\epsilon>0$, there exists $\varphi_\epsilon\in\mathbf{C}_{\mathcal{K}}^+$ such that $\varphi_\epsilon\subset\varphi$ and $I(\varphi)\subset I(\varphi_\epsilon)\oplus\epsilon B$ (B is the unit ball in $\mathbf{R}^d$), because I is continuous on $\mathbf{C}_{\mathcal{K}}^+$, and thus the definition of (9-2-1) is equivalent to

$$I'(\gamma)=\lim\{I(\varphi),\varphi\in\mathbf{C}_K^+,\varphi\subset\gamma\} \tag{9-2-1$'$}$$

Proposition 9-2-3. *I' is increasing, convex, and l.s.c. on Φ_g. Moreover, $\gamma_n\uparrow\gamma$ in Φ_g implies $I'(\gamma_n)\uparrow I(\gamma)$ and $I'(\gamma)=\lim I'(\gamma_n)$ in $\overline{\mathcal{K}}_0$.*

Proof. From (9-2-1), I' is increasing. If G is an open set in $\mathbf{R}^d$, $I'(\gamma)\cap G\neq\varnothing$ if and only if $I(\varphi)\neq\varnothing$ for a function $\varphi\in\mathbf{C}_{\mathcal{K}}^+$ such that $\varphi\subset\gamma$, and $\{\gamma:\ \gamma\supset\varphi\}$ is open in Φ_g. Thus $\{\gamma, I'(\gamma)\cap G=\varnothing\}$ is open in Φ_g, and I' is l.s.c. In Φ_g, $\gamma_n\uparrow\gamma$ implies $\gamma=\lim\gamma_n$, and thus $I(\gamma)\subset\underline{\lim}\, I'(\gamma_n)$, for I' is l.s.c. On the other hand, $\overline{\lim}\, I'(\gamma_n)\subset I(\gamma)$ for I' is increasing. Hence $I'(\gamma)=\lim I'(\gamma_n)$.

If γ, $\gamma'\in\Phi_g$, let $\{\varphi_n\}$ and $\{\varphi_n'\}$ be two sequences in $\mathbf{C}_{\mathcal{K}}^+$ such that $\varphi_n\uparrow\gamma$ and $\varphi_n'\uparrow\gamma'$, and λ, $\mu\geqslant 0$. Then $I'(\lambda\gamma+\mu\gamma')=\lim I(\lambda\varphi_n+\mu\varphi_n')\subset\lim(\lambda I(\varphi_n)\oplus\mu I(\varphi_n'))=\lambda I'(\gamma)\oplus\mu I'(\gamma')$, for $(\lambda\varphi_n+\mu\varphi_n)\uparrow(\lambda\gamma+\mu\gamma')$ in Φ_g, and I is convex on $\mathbf{C}_{\mathcal{K}}^+$. Thus I' is convex.

Corollary. *CI' is positively linear on Φ_g. If I is a C-integral, I' is positively linear on Φ_g.*

Extension on Φ_k

For any positive function f on E, let us put

$$I'(f)=\lim\{I'(\gamma),\gamma\in\Phi_g,\gamma\geqslant f\} \qquad (9\text{-}2\text{-}2)$$

That is, $I'(f)$ is the limit in $\overline{\mathcal{K}}_0$ of the decreasing filtering family $I'(\gamma)$, $\gamma\geqslant f$. When using the convention $K\cap\omega=K$ for any $K\in\overline{\mathcal{K}}_0$, (9-2-2) is equivalent to

$$I'(f)=\cap\{I'(\gamma),\gamma\in\Phi_g,\gamma\geqslant f\}$$

Any $\chi\in\Phi_k$ is dominated by a function $\varphi\in\mathbf{C}_{\mathcal{K}}^+$, so that $I'(\chi)\in\mathcal{K}_0$. Moreover,

$$I'(\chi)=\cap\{I'(\gamma),\gamma\in\Phi_g,\gamma\supset\chi\} \qquad (\chi\in\Phi_k) \qquad (9\text{-}2\text{-}2')$$

Proof of (9-2-2'). If $\chi\in\Phi_k$, let us put $J(\chi)=\cap\{I'(\gamma),\gamma\in\Phi_g,\gamma\supset\chi\}$, so that $J(\chi)\supset I'(\chi)$. Conversely, let K be the support of χ, $\epsilon>0$, and $\varphi_0\in\mathbf{C}_{\mathcal{K}}^+$ such that $\varphi_0(x)=1$ for any $x\in K$. Then $\gamma\in\Phi_g$ and $\gamma\geqslant\chi$ imply $\gamma+\epsilon\varphi_0\supset\chi$. Hence $I'(\gamma)\oplus\epsilon I(\varphi_0)\supset I'(\gamma+\epsilon\varphi_0)\supset J(\chi)$. It follows that $I'(\chi)\supset J(\chi)$ and the desired equality.

Proposition 9-2-4. *I' is increasing, convex, and u.s.c. on Φ_k. Moreover, $\chi_n\downarrow\chi$ in Φ_k implies $I'(\chi_n)\downarrow I(\chi)$ and $I(\chi)=\lim I(\chi_n)$ in $\mathcal{K}_0$. Furthermore, for any*

$\gamma \in \Phi_g$, $I'(\gamma) = \lim\{I(\chi), \chi \in \Phi_k, \chi \subset \gamma\}$. *The convex hull* CI' (*and* I' *itself if* I *is a* C-*integral*) *is positively linear on* Φ_k.

Proof. From (9-2-2), I' is increasing. If F is a closed set in $\mathbf{R}^d$, it follows from (9-2-2′) that $I'(\chi) \in \mathcal{K}^F$ if and only if there exists $\gamma \in \Phi_g$ and $\chi \subset \gamma, I'(\gamma) \cap F = \varnothing$. But the sets $\{\chi, \chi \subset \gamma\}$ are open in Φ_k, so that $\{\chi, I'(\chi) \in \mathcal{K}^F\}$ is also open, and hence I' is u.s.c. on Φ_k. Also $\chi_n \downarrow \chi$ implies $\chi = \lim \chi_n$ in Φ_k and thus $I'(\chi_n) \downarrow I'(\chi)$ and $\lim I'(\chi_n) = I'(\chi)$, for I' is increasing and u.s.c. on Φ_k. If χ, $\chi' \in \Phi_k$, let $\{\varphi_n\}$, $\{\varphi_n'\}$ be two sequences in $\mathbf{C}_{\mathcal{K}}^+$ such that $\varphi_n \downarrow \chi$, $\varphi_n \downarrow \chi'$, and λ, $\mu \geqslant 0$. Then $(\lambda\varphi_n + \mu\varphi_n') \downarrow (\lambda\chi + \mu\chi')$, and thus

$$I'(\lambda\chi + \mu\chi') = \lim(\lambda\varphi_n + \mu\varphi_n') \subset \lim(\lambda I(\varphi_n) \oplus \mu I(\varphi_n'))$$

$$= \lambda I'(\chi) \oplus \mu I'(\chi')$$

That is, I' is convex on Φ_k. If I is concave, we obtain the opposite inclusion, that is, I' is concave on Φ_k. If $\gamma \in \Phi_g$, $I'(\gamma) \supset \lim\{I'(\chi), \chi \in \Phi_k, \chi \subset \gamma\}$ by (9-2-2′). Conversely,

$$I'(\gamma) = \lim\{I(\varphi), \varphi \in \mathbf{C}_{\mathcal{K}}^+, \varphi \subset \gamma\},$$

and $\mathbf{C}_{\mathcal{K}}^+ \subset \Phi_k$ implies the opposite inclusion and thus the equality.

The Space Φ of the Pseudo-Integrable Functions

For any positive function f on E, let us now write

$$I'(f) = \lim\{I'(\gamma), \gamma \in \Phi_g, \gamma \geqslant f\}$$
$$I''(f) = \lim\{I'(\chi), \chi \in \Phi_k, \chi \leqslant f\} \tag{9-2-3}$$

and denote by $\overline{\Phi}$ the set of the positive functions f such that $I'(f) = I''(f)$. We shall say that f is *pseudo-integrable* if $I'(f) = I''(f) \in \mathcal{K}_0$ that is, is not ω, and denote by $\Phi \subset \overline{\Phi}$ the set of the pseudo-integrable functions. If $f \in \overline{\Phi}$, we shall write $I(f)$ instead of $I'(f)$ or $I''(f)$. Clearly, $\Phi_g \subset \overline{\Phi}$ and $\Phi_k \subset \Phi$.

For any positive functions f, f', the following inclusions are obvious:

$$I''(f) \subset I'(f)$$

$$I'(f + f') \subset I'(f) \oplus I'(f') \tag{9-2-4}$$

$$CI''(f + f') \supset CI''(f) \oplus CI''(f')$$

Proposition 9-2-5. *A positive function f is pseudo-integrable, that is,* $f\in\Phi$, *if and only if there exist a sequence* $\{\gamma_n\}\subset\Phi_g$ *and a sequence* $\{\chi_n\}\subset\Phi_k$ *such that* $\gamma_n\geqslant f$ *and* $\chi_n\leqslant f$ *for any n, and* $\lim I(\gamma_n-\chi_n)=\{o\}$ *in* $\mathcal{K}_0$. *In the same way,* $f\in\overline{\Phi}$ *and* $I(f)=\omega$ *if and only if there exists an increasing sequence* $n\to\chi_n\leqslant f$ *and* $\lim I(\chi_n)=\omega$ *in* $\overline{\mathcal{K}}_0$.

Proof. The second statement is an immediate consequence of the definition. Let f be a positive function such that $I'(f)\neq\omega$. Then there exists a decreasing sequence $n\to\gamma_n\geqslant f$ in Φ_g such that $I(\gamma_1)\neq\omega$ and $\lim I(\gamma_n)=I'(f)$, and also an increasing sequence $n\to\chi_n\leqslant f$ in Φ_k such that $\lim I(\chi_n)=I''(f)$. The sequence $n\to(\gamma_n-\chi_n)\in\Phi_g$ is decreasing and valued in Φ [as $I(\gamma_1)\neq\omega$]; thus there exists $A=\lim I(\gamma_n-\chi_n)=\cap I(\gamma_n-\chi_n)\in\mathcal{K}_0$. It follows from the second inclusion of (9-2-4) that

$$I(\gamma_n)=I(\gamma_n-\chi_n+\chi_n)\subset I(\gamma_n-\chi_n)\oplus I(\chi_n)$$

Hence $I'(f)\subset A\oplus I''(f)$. If $A=\{o\}$, it follows that $I'(f)=I''(f)$, and so $f\in\Phi$.

Conversely, let us suppose that $f\in\Phi$. The last two inclusions of (9-2-4) imply that the convex hull CI is additive, that is, $CI'(f+f')=CI''(f+f')=CI(f)\oplus CI(f')$ for any $f,f'\in\Phi$. It follows that $CI'(f)=C(A)\oplus CI''(f)$. But $CI'(f)=CI''(f)=CI(f)$, for $f\in\Phi$, and this implies $C(A)=\{o\}$. Hence $A=\{o\}$.

Lemma 9-2-1. *The set* $\overline{\Phi}$ *is closed under* $\vee$, *and* Φ *is closed under* $\wedge$.

Proof. For any positive functions f, f', $f\vee f'=\mathrm{Sup}(f,f')$, and $f\wedge f'=\mathrm{Inf}(f,f')$. If $f,f'\in\Phi$, let $\{\gamma_n\}$, $\{\chi_n\}$ (resp. $\{\gamma_n'\}$, $\{\chi_n'\}$) be two sequences satisfying the conditions of Proposition 9-2-5 with respect to f (resp. to f'). Thus the sequence $n\to I(\gamma_n+\gamma_n'-\chi_n-\chi_n')\subset I(\gamma_n-\chi_n)\oplus I(\gamma_n'-\chi_n')$ converges toward $\{o\}$ in $\mathcal{K}_0$. It follows that $n\to CI(\gamma_n+\gamma_n'-\chi_n-\chi_n')=CI(\gamma_n\vee\gamma_n'-\chi_n\vee\chi_n')\oplus CI(\gamma_n\wedge\gamma_n'-\chi_n\wedge\chi_n')$ also converges toward $\{o\}$, and this implies $\{o\}=\lim CI(\gamma_n\vee\gamma_n'-\chi_n\vee\chi_n')=\lim CI(\gamma_n\wedge\gamma_n'-\chi_n\wedge\chi_n')$. Hence (a fortiori) $\lim I(\gamma_n\vee\gamma_n'-\chi_n\vee\chi_n')=\lim I(\gamma_n\wedge\gamma_n'-\gamma_n\wedge\gamma_n')=\{o\}$. Thus $f\vee f'\in\Phi$ and $f\wedge f'\in\Phi$ (Proposition 9-2-5).

If $f\in\overline{\Phi}$ and $I(f)=\omega$, there exists a sequence $n\to\chi_n\leqslant f$ in Φ_k and $\lim I(\chi_n)=\omega$ in $\mathcal{K}_0$. For any positive function f', $f\vee f'\geqslant\chi_n$ implies $I''(f\vee f')=\omega$. Hence $f\vee f'\in\overline{\Phi}$, by Proposition 9-2-5, and $I(f\vee f')=\omega$.

Lemma 9-2-2. *The space Φ is closed under addition, and I is convex and increasing on Φ. Moreover, $f, f' \in \Phi$ and $f \geqslant f'$ imply $f - f' \in \Phi$. Furthermore, $\bar{\Phi}$ is closed under $+$, and I is convex and increasing on $\bar{\Phi}$.*

Proof. If $f, f' \in \Phi$, let $\{\gamma_n\}$, $\{\chi_n\}$ (resp. $\{\gamma'_n\}$, $\{\chi'_n\}$) be two sequences satisfying the conditions of Proposition 9-2-5 with respect to f (resp. with respect to f'). Then $I(\gamma_n + \gamma'_n - \chi_n - \chi'_n) \subset I(\gamma_n - \chi_n) \oplus I(\gamma'_n - \chi'_n)$ implies that $\lim I(\gamma_n + \gamma'_n - \chi_n - \chi'_n) = \{o\}$, and $f + f \in \Phi$ (Proposition 9-2-5). If $f - f' \geqslant 0$, $\gamma_n - \chi'_n \geqslant f - f' \geqslant (\chi_n - \gamma'_n)_+$ and $\gamma_n - \chi'_n - (\chi_n - \gamma'_n)_+ \leqslant \gamma_n - \chi'_n - (\chi_n - \gamma'_n) = (\gamma_n - \chi_n) + (\gamma'_n - \chi'_n)$. It follows that $I(\gamma_n - \chi'_n - (\chi_n - \gamma'_n)_+) \subset I(\gamma_n - \chi_n) \oplus I(\gamma'_n - \chi'_n)$ and $\lim I(\gamma_n - \chi'_n - (\chi_n - \gamma'_n)_+) = \{o\}$ in $\mathcal{K}_0$. Hence $f - f' \in \Phi$ (Proposition 9-2-5). The statement concerning $\bar{\Phi}$ is obvious.

Proposition 9-2-6. *The set $\bar{\Phi}$ is closed under countable addition. For any sequence $\{f_n\} \subset \Phi$,*

$$I\left(\sum f_n\right) = \lim I\left(\sum_1^N f_n\right) \subset \bigoplus_{n=1}^{\infty} I(f_n)$$

and

$$CI\left(\sum f_n\right) = \bigoplus_{n=1}^{\infty} CI(f_n)$$

Proof. Let $\{f_n\}$ be a sequence in $\bar{\Phi}$. If $I(f_{n_0}) = \omega$ for an integer n_0, $f = \sum f_n$ implies $f \geqslant f_{n_0}$ and thus $I''(f) \supset I(f_{n_0}) = \omega$, that is, $f \in \bar{\Phi}$, $I(f) = \omega$. Let us suppose that $f_n \in \Phi$ for any n, and let ϵ be a number >0 and $\{\epsilon_n\} \subset \mathbf{R}_+$ be a sequence such that $\sum \epsilon_n = \epsilon$. It follows from Proposition 9-2-5 that for each n there exist $\gamma_n \in \Phi_g$, $\gamma_n \geqslant f_n$, and $\chi_n \in \Phi_k$, $\chi_n \leqslant f_n$, such that

$$(a) \qquad I(\gamma_n - \chi_n) \subset \epsilon_n B$$

(B is the unit ball in $\mathbf{R}^d$). On the other hand, $\gamma = \sum \gamma_n \in \Phi_g$, and $\chi = \sum \gamma_n$ is such that $\gamma - \chi \in \Phi_g$, so by Proposition 9-2-3

$$(b) \qquad I(\gamma) = \lim_N I\left(\sum_1^N \gamma_n\right), \qquad I(\gamma - \chi) = \lim_N I\left(\sum_1^N (\gamma_n - \chi_n)\right)$$

Let us first suppose that $I(\gamma) \neq \omega$. Then (a) yields

$$(c) \qquad I\left(\sum_1^N (\gamma_n - \chi_n)\right) \subset \bigoplus_{n=1}^{N} I(\gamma_n - \chi_n) \subset \epsilon B$$

and thus (b) implies $I(\gamma-\chi)\subset\epsilon B$. Hence $f\in\Phi$ (Proposition 9-2-5). On the other hand, $I(\Sigma_1^N\gamma_n)\subset I(\Sigma_1^N f_n)\oplus\epsilon B\subset\lim_N I(\Sigma_1^N f_n)\oplus\epsilon B$ implies $I(\gamma)\subset\lim_N I(\Sigma_1^N f_n)\oplus\epsilon B$ (Proposition 9-2-3). Hence

$$I(f)\subset I(\gamma)\subset\left(\lim_N I\left(\sum_1^N f_n\right)\right)\oplus\epsilon B$$

It follows that

$$I(f)=\lim_N I\left(\sum_1^N f_n\right),\quad\text{and}\quad I(f)\subset\bigoplus_{n=1}^{\infty}I(f_n)$$

(Lemma 9-2-2). In the same way

$$CI(f)=\lim_N\bigoplus_{i=1}^{N}CI(f_n)=\bigoplus_{i=1}^{\infty}CI(f_n)$$

If now $I(\gamma)=\omega$, it follows that

$$CI(\gamma)=\omega\quad\text{and}\quad\lim CI\left(\sum_1^N\gamma_n\right)=\lim\bigoplus_1^N CI(\gamma_n)=\omega$$

But (a) yields $CI(f_n)\subset CI(\gamma_n)\subset CI(f_n)\oplus\epsilon_n B$, and thus

$$\bigoplus_{n=1}^{N}CI(\gamma_n)\subset\epsilon B\oplus\left(\bigoplus_{n=1}^{N}CI(f_n)\right)$$

Hence

$$\lim\bigoplus_{n=1}^{N}CI(f_n)\lim CI\left(\sum_1^N f_n\right)=\omega$$

Thus $f\in\overline{\Phi}$ and $I(f)=\omega$.

Proposition 9-2-7. *The set $\overline{\Phi}$ is closed under monotone convergence $\uparrow$, and $f_n\uparrow f$ in $\overline{\Phi}$ implies $I(f_n)\uparrow I(f)$ and $\lim I(f_n)=I(f)$ in $\mathcal{K}_0$. Also Φ is closed under monotone convergence $\downarrow$, and $f_n\downarrow f$ in Φ implies $I(f_n)\downarrow I(f)$ and $\lim I(f_n)=I(f)$ in $\mathcal{K}_0$.*

Proof. If $\{f_n\}\subset\overline{\Phi}$ and $f_n\uparrow f$, put $h_1=f_1$ and $h_n=f_n-f_{n-1}$, so that $f_N=\Sigma_1^N h_n, f=\Sigma h_n$. Then it follows from Proposition 9-2-6 that $f\in\overline{\Phi}$ and $I(f)=\lim I(f_n)$.

Let now $\{f_n\}\subset\Phi$ be such that $f_n\downarrow 0$. We have $f_1-f_n\in\Phi$ (Lemma 9-2-2), and $(f_1-f_n)\uparrow f_1$. Hence $I(f_1)=\lim I(f_1-f_n)$ in $\mathcal{K}_0$ by the first part of the proof. On the other hand, $\lim I(f_n)=\cap I(f_n)=A$ in $\mathcal{K}_0$, and $\lim CI(f_n) = C(A)$. But CI is additive on Φ (Proposition 9-2-2), and so $CI(f_1) = CI(f_1-f_n)\oplus CI(f_n)$. It follows that, for $n\uparrow\infty$, $CI(f_1)=C(A)\oplus CI(f_1)$, that is, $C(A)=\{o\}$ and $A=\{o\}=I(o)$.

Now let $\{f_n\}\subset\Phi$ be a sequence such that $f_n\downarrow f$. Again we have $f_1-f_n\in\Phi$ (Lemma 9-2-2), $(f_1-f_n)\uparrow(f_1-f)$, and thus $f_1-f\in\Phi$ and $\lim I(f_1-f_n) = I(f_1-f)$ by the first part of the proof. On the other hand, $f=f_1-(f_1-f) \in\Phi$ (Lemma 9-2-2), and thus $f_n-f\in\Phi$. It follows that $\lim I(f_n-f)=\{o\}$, for $(f_n-f)\downarrow 0$. Hence $I(f_n)\subset I(f)\oplus I(f_n-f)$ implies $\overline{\lim}\ I(f_n)\subset I(f)$, and thus $\lim I(f_n)=I(f)$, for $f_n\geqslant f$.

More generally:

Fatou-Lebesgue Lemma. *Let $\{f_n\}\subset\Phi$ be a sequence dominated by a fixed $g\in\Phi$. Then $\overline{\lim}\ I(f_n)\subset I(\lim\mathrm{Sup} f_n)$ and $\lim I(f_n)\supset I(\lim\mathrm{Inf} f_n)$. In particular, if $\{f_n\}$ converges pointwise toward f and is dominated by a fixed $g\in\Phi$, then $f\in\Phi$ and $\lim I(f_n)=I(f)$ in $\mathcal{K}_0$.*

Proof. We have $\mathrm{Sup}\{f_m, m\geqslant n\}\in\overline{\Phi}$ (Lemma 9-2-1 and Proposition 9-2-7), and $\mathrm{Sup}\{f_m, m\geqslant n\}\leqslant g\in\Phi$ implies $\mathrm{Sup}\{f_m, m\geqslant n\}\in\Phi$. Hence $I(\lim\mathrm{Sup} f_n)=\lim I(\mathrm{Sup}_{m\geqslant n} f_n)$. Thus $I(f_n)\subset I(\mathrm{Sup}_{m\geqslant n} f_m)$ implies $\overline{\lim}\ I(f_n) \subset I(\lim\mathrm{Sup} f_n)$. The proof is analogous for $\underline{\lim}\ I(f_n)$.

Lemma 9-2-3. *For any $f\in\Phi$ and $a\geqslant 0$, $f\wedge a\in\Phi$.*

Proof. Let $\{K_n\}\subset\mathcal{K}(E)$ be a sequence such that $K_n\uparrow E$. We have $1_{K_n}\in\Phi_k\subset\Phi$, and thus $f\wedge(a1_{K_n})\in\Phi$ (Lemma 9-2-1). On the other hand, $(f\wedge a1_{K_n})\uparrow(f\wedge a)$, so that $(f\wedge a)\in\overline{\Phi}$ (Proposition 9-2-7). Since $f\wedge a \leqslant f\in\Phi$, it follows that $f\wedge a\in\Phi$.

Lemma 9-2-4. *For any $f\in\Phi$ and $a\geqslant b>0$, the indicators $1_{\{f>b\}}$ and $1_{\{a\geqslant f>b\}}$ are pseudo-integrable.*

Proof. If $\beta>b\geqslant 0$, let us put $r_\beta=((f\wedge\beta)-(f\wedge b))/(\beta-b)$. Then $r_\beta\in\Phi$ (Lemmas 9-2-3 and 9-2-2). If $\beta\downarrow b$, it follows that $r_\beta\uparrow 1_{\{f>b\}}\in\Phi$ (Proposition 9-2-7), and in particular $1_{\{f>b\}}\in\Phi$ if $b>0$ [for $1_{\{f>b\}}\leqslant(1/b)f$]. Finally, $1_{\{a\geqslant f>b\}}=1_{\{f>b\}}-1_{\{f>a\}}\in\Phi$ (Lemma 9-2-2).

As an immediate corollary, we obtain:

Proposition 9-2-8. *For any $f\in\overline{\Phi}$, put*

$$r_n=\sum_{k=1}^{\infty} k2^{-n}1_{\{(k+1)2^{-n}\geqslant f>k2^{-n}\}}$$

Then $r_n \uparrow f$, *in* $\overline{\Phi}$, $\lim I(r_n) = I(f)$ *in* $\overline{\mathcal{K}}_0$, *and*

$$CI(f) = \lim_n \bigoplus_{k=1}^{\infty} (k2^{-n}) CI\left(1_{\{(k+1)2^{-n} \geq f > k2^{-n}\}}\right)$$

The $\mathcal{K}_0$-Integrals

A pseudo-integral I is said to be a $\mathcal{K}_0$-integral if $I(\varphi+\varphi') = I(\varphi) \oplus I(\varphi')$ for any $\varphi, \varphi' \in \mathbf{C}_{\mathcal{K}}^{+}$ with disjoint supports. If $\varphi, \varphi' = 0$, Supp φ and Supp φ' are not necessarily disjoint, but, if $\{\epsilon_n\} \subset \mathbf{R}_+$ and $\epsilon_n \downarrow 0$, the functions $\varphi_n = (\varphi - \epsilon_n)_+$ and $\varphi'_n = (\varphi' - \epsilon_n)_+$ are in $\mathbf{C}_{\mathcal{K}}^{+}$ and their supports are disjoint. Then $I(\varphi_n + \varphi'_n) = I(\varphi_n) \oplus I(\varphi'_n)$, $\varphi_n \uparrow \varphi$, and $\varphi'_n \uparrow \varphi'$ imply $I(\varphi+\varphi') = I(\varphi) \oplus I(\varphi')$. More generally:

Proposition 9-2-9. *If I is a $\mathcal{K}_0$ integral, $f, f' \in \overline{\Phi}$ and $\{f>0\} \cap \{f'>0\} = \varnothing$ imply $I(f+f') = I(f) \oplus I(f')$.*

Proof. The proposition holds if $f, f' \in \mathbf{C}_{\mathcal{K}}^{+}$, and also if $f, f' \in \Phi_g$ or $f, f' \in \Phi_k$, as can be easily verified. If $f \in \overline{\Phi}$ and $I(f) = \omega$, $I(f+f') = \omega = I(f) \oplus I(f')$. Suppose that $f, f' \in \Phi$, and let $n \to \chi_n \leqslant f$, $n \to \chi'_n \leqslant f'$ be two sequences in Φ_k such that $I(f) = \lim I(\chi_n)$ and $I(f') = \lim I(\chi'_n)$ in $\mathcal{K}_0$. If $\{f>0\} \cap \{f'>0\} = \varnothing$, a fortiori $\chi_n \chi'_n = 0$, and $I(\chi_n + \chi'_n) = I(\chi_n) \oplus I(\chi'_n)$. It follows that $I(f) \oplus I(f') = \lim I(\chi_n \oplus \chi'_n) \subset I(f+f')$, and thus $I(f) \oplus I(f') = I(f+f')$, for I is convex on Φ.

Corollary. *If I is a $\mathcal{K}_0$-integral, then, for any $f \in \Phi$,*

$$I(f) = \lim_n \bigoplus_{k=1}^{\infty} (k2^{-n}) I\left(1_{\{(k+1)2^{-n} \geq f > k2^{-n}\}}\right)$$

Proof. Let r_n be defined as in Proposition 9-2-8. It follows from Proposition 9-2-6, 9-2-8, and 9-2-9 that

$$I(r_n) = \bigoplus_{k=1}^{\infty} (k2^{-n}) I\left(1_{\{(k+1)2^{-n} \geq f > k2^{-n}\}}\right)$$

and $I(f) = \lim I(r_n)$.

If I is a $\mathcal{K}_0$-integral on a LCS space E, and $\overline{\Phi}$ the functional space on which I can be extended, $\overline{\Phi}$ contains the borelian functions on E, and, in particular, the indicator 1_B of any set B belonging to the borelian tribe $\mathcal{B}(E)=\mathcal{B}$. By putting $I(B)=I(1_B)$, I becomes a mapping from $\mathcal{B}$ into $\overline{\mathcal{K}}_0$. Also I is increasing on $\mathcal{B}$, $I(\varnothing)=\{o\}$, and $B_n\uparrow B$ in $\mathcal{B}$ implies $I(B)=\lim I(B_n)$ in $\overline{\mathcal{K}}_0$. Moreover, $B_n\downarrow B$ in $\mathcal{B}$ and $I(B_{n_0})\neq\omega$ for an integer n_0 imply $I(B)=\lim I(B_n)$ in $\mathcal{K}_0$. Furthermore, I is additive on $\mathcal{B}$, with respect to the Minkowski sum $\oplus$ (Proposition 9-2-9), and thus also σ-additive. Finally, $I(K)\neq\omega$ if $K\in\mathcal{K}(E)$. Hence the mapping I: $\mathcal{B}\rightarrow\overline{\mathcal{K}}_0$ may be called a Radon measure (valued in $\overline{\mathcal{K}}_0$).

It follows from Proposition 9-2-8 that $I(f)$ may be written as

$$\int_E f(x)I(dx),$$

where the integration symbol has the meaning of a limit of integrals of step functions. This suggests the development of a theory of abstract measures valued in $\mathcal{K}_0$.

9-3. ABSTRACT MEASURES VALUED IN $\mathcal{K}_0$

If E is an arbitrary set provided with a σ-algebra $\mathcal{B}$, a mapping μ: $\mathcal{B}\rightarrow\mathcal{K}_0$ will be called a measure valued in $\mathcal{K}_0$, or a $\mathcal{K}_0$-*measure*, if three conditions are satisfied:

1. $\mu(B)\subset\mu(B')$ if $B\subset B'$ in $\mathcal{B}$.
2. $\mu(B\cup B')=\mu(B)\oplus\mu(B')$ if B, B' are disjoint sets in $\mathcal{B}$.
3. $B_n\downarrow\varnothing$ in $\mathcal{B}$ implies $\lim\mu(B_n)=\{o\}$ in $\mathcal{K}_0$.

Note that $\mu(E)$ is a compact set, so that $\mu(B)\subset\mu(E)$ for any $B\in\mathcal{B}$, and *the range* $\mu(\mathcal{B})$ *is relatively compact in* $\mathcal{K}_0$. Axioms 1, 2, and 3 imply the following relationships:

$I(\varnothing)=\{o\}$.
$B_n\uparrow B$ or $B_n\downarrow B$ in $\mathcal{B}$ implies $\mu(B)=\lim\mu(B_n)$ in $\mathcal{K}_0$.
$\mu(\cup B_n)=\oplus\mu(B_n)$ if $B_1, B_2\ldots$ are disjoints sets in $\mathcal{B}$.

Let us prove, for instance, the sequential monotone continuity. If $B_n\uparrow B$ in $\mathcal{B}$, it follows that $(B\setminus B_n)\downarrow\varnothing$ and $\lim\mu(B\setminus B_n)=\{o\}$ in $\mathcal{K}_0$ (Axiom 3). On the other hand, $\mu(B)=\mu(B_n)\oplus\mu(B\setminus B_n)$ (Axiom 2) and $\lim\mu(B_n)=\overline{\cup\mu(B_n)}\in\mathcal{K}_0$ [for $\{\mu(B_n)\}$ is increasing and bounded in $\mathcal{K}_0$] imply $\lim\mu(B_n)=\mu(B)$ by the continuity of $\oplus$. If $B_n\downarrow B$ in $\mathcal{B}$, $(B_n\setminus B)\downarrow\varnothing$ and $\mu(B_n\setminus B)\downarrow\{o\}$. Hence $\mu(B_n)=\mu(B)\oplus\mu(B_n\setminus B)\downarrow\mu(B)$ in $\mathcal{K}_0$, and so $\lim\mu(B_n)=\mu(B)$.

In the same way, a mapping μ from $\mathcal{B}$ into the compactification $\overline{\mathcal{K}_0}$ is called a *σ-finite* $\mathcal{K}_0$*-measure* if there exists a sequence $\{E_n\} \subset \mathcal{B}$ such that $E_n \uparrow E$, $\mu(E_n) \in \mathcal{K}_0$, $\mu(B \cap E_n) \uparrow \mu(B)$ for any $B \in \mathcal{B}$, and for any n the restriction μ_n of μ to E_n satisfies the three axioms given above.

A σ-finite $\mathcal{K}_0$-measure μ is *increasing and σ-additive*, and $B_n \uparrow B$ in $\mathcal{B}$ implies $\lim \mu(B_n) = \mu(B)$ in $\overline{\mathcal{K}}_0$.

Let us prove only the monotone continuity. If $B_n \uparrow B$ in $\mathcal{B}$, and $\mu(B) \neq \omega$,

$$\mu(B) = \lim_m \mu(B \cap E_m) = \overline{\cup \mu(B \cap E_m)}$$

$$\mu(B_n) = \lim_m \mu(B_n \cap E_m) = \overline{\cup \mu(B_n \cap E_m)}$$

and thus

$$\lim \mu(B_n) = \overline{\cup \mu(B_n)} = \overline{\bigcup_n \overline{\bigcup_m \mu(B_n \cap E_m)}} = \overline{\bigcup_{m,n} \mu(B_n \cap E_m)}$$

$$= \overline{\bigcup_m \mu(B \cap E_m)} = \mu(B)$$

If $\mu(B) = \omega$, for any $K \in \mathcal{K}(\mathbf{R}^d)$, $\mu(B \cap E_m)$ is not contained in K for m large enough. Hence $\mu(B_n \cap E_m) \not\subset K$ for n, m large enough, and $\mu(B_n) \not\subset K$, that is, $\lim \mu(B_n) = \omega$.

In the same way, $B_n \downarrow B$ and $B_{n_0} \subset E_{m_0}$ for two integers n_0, m_0 imply $\lim \mu(B_n) = \mu(B)$ in $\mathcal{K}_0$, that is, $\mu(B) = \cap \mu(B_n)$.

Note. A $\mathcal{K}_0$-measure μ is *increasing* with respect to the preordering $\succcurlyeq$ on $\mathcal{K}_0$ defined by $A \succcurlyeq B$ if $A_B = A$, for $B \subset B'$ implies $\mu(B') = \mu(B) \oplus \mu(B' \cap B^c)$, that is, $\mu(B) \preccurlyeq \mu(B')$.

The Integral Associated with a $\mathcal{K}_0$-Measure

Let μ be a $\mathcal{K}_0$-measure on $(E, \mathcal{B})$, and $\mathcal{E}_+$ be the set of the positive step functions on E (i.e., $f \in \mathcal{E}_+$ if $f = \Sigma x_i 1_{B_i}$, $x_i \geqslant 0$, $x_i \neq x_j$ if $i \neq j$, and $\{B_i\} \subset \mathcal{B}$ is a finite partition of E). For any $f \in \mathcal{E}_+$, we define the integral $\mu(f)$ by

$$\mu(f) = \oplus x_i \mu(B_i)$$

For another $f' \in \mathcal{E}_+$, say $f' = \Sigma x_j' 1_{B_j'}$,

$$f + f' = \sum_{i,j} (x_i + x_j') 1_{B_i \cap B_j'} \quad \text{and} \quad \mu(f+f') = \bigoplus_{i,j} (x_i + x_j') \mu(B_i \cap B_j')$$

so that the relationship $(\lambda + \mu)K \subset \lambda K \oplus \mu K$ $(K \in \mathcal{K}_0)$ yields

$$\mu(f+f') \subset \mu(f) \oplus \mu(f')$$

Moreover, if $ff' = 0$, it follows that $f + f' = \Sigma x_i 1_{B_i} + \Sigma x_j' 1_{B_j'}$, because $B_i \cap B_j' = \varnothing$ if $x_i x_j' \neq 0$, and so $\mu(f+f') = (\oplus x_i \mu(B_i)) \oplus (\oplus x_j' \mu(B_j'))$, that is, $\mu(f+f') = \mu(f) \oplus \mu(f')$. Then, by using the classical arguments of integration theory, it is possible to extend μ on the class $\mathcal{I}_+$ closed under $\uparrow$ generated by $\mathcal{E}_+$ in such a way that

$\mu(1_B) = \mu(B) \qquad (B \in \mathcal{B})$.
$f_n \uparrow f$ or $f_n \downarrow f$ in $\mathcal{I}_+$ implies $\lim \mu(f_n) = \mu(f)$.
μ is increasing and convex on $\mathcal{I}_+$.
$\mu(f+f') = \mu(f) \oplus \mu(f')$ if $f, f' \in \mathcal{I}_+$ and $f, f' = 0$.

Moreover, the Fatou-Lebesgue lemma holds, as well as the dominated convergence theorem.

Note that μ is additive if and only if it is valued in $C(\mathcal{K}_0)$, as $1_B = (1/n) \Sigma_1^n 1_B$ and μ-additive implies that $\mu(B) = (1/n)\mu(B)^{\oplus n}$, that is, $\mu(B)$ is infinitely divisible for $\oplus$ [i.e., $\mu(B) \in C(\mathcal{K}_0)$].

Suppose now that E is a LCS space, $\mathcal{B}$ its borelian tribe, $\{E_n\} \subset \mathcal{K}(E)$ a sequence such that $E_n \uparrow E$ and $E_n \subset \mathring{E}_{n+1}$, and μ a σ-finite $\mathcal{K}_0$-measure on $\mathcal{B}$ such that $\mu(E_n) \in \mathcal{K}_0$ for any n, that is, $\mu(K) \neq \omega$ for any $K \in \mathcal{K}(E)$. (A σ-finite measure with this property is said to be *regular.*) For any $\varphi \in \mathbf{C}_{\mathcal{K}}^+$, supp$\varphi$ is contained in an E_n, and the integral $\mu(\varphi)$ exists and belongs to $\mathcal{K}_0$. The mapping $\varphi \to \mu(\varphi)$ is increasing, convex (and thus continuous: Proposition 9-2-2) on $\mathbf{C}_{\mathcal{K}}^+$, that is, there is a $\mathcal{K}_0$-integral I such that $I(\varphi) = \mu(\varphi)$, $\varphi \in \mathbf{C}_{\mathcal{K}}^+$. The $\mathcal{K}_0$-integral I can be extended on $\overline{\Phi} \supset \mathcal{I}_+$, and we still denote by I this extension (which is valued on $\overline{\mathcal{K}}_0$). Then it is not difficult to verify that $I(f) = \mu(f)$ for any $f \in \mathcal{I}_+$; more precisely:

Proposition 9-3-1. *Let E be a LCS space and $\mathcal{B}$ its borelian tribe. Then, with each regular σ-finite measure μ on $\mathcal{B}$ valued in $\overline{\mathcal{K}}_0$, is associated a unique $\mathcal{K}_0$-integral I such that $\mu(B) = I(1_B)$ for any $B \in \mathcal{B}$, and conversely. Moreover, for any $f \in \Phi$, there exist f_1 and $f_2 \in \mathcal{I}_+$ such that $f_1 \leqslant f \leqslant f_2$ and $\mu(f_1) = \mu(f_2)$. Furthermore, I is a C-integral if and only if μ is valued in $C(\mathcal{K}_0)$.*

Bibliography

Alfsen, E.M., *Compact Convex Sets, and Boundary Integrals*, Springer, Berlin, 1971.

Ahmad, S., Eléments aléatoires dans les espaces vectoriels topologiques, *Ann. Inst. Henri Poincaré*, B,**II** No. 2, 95–135, 1965.

Aumann, R.J., Integrals of set valued functions. *J. Math. An. Appl.*, **12**, p. 1–12, 1965.

Berge, C., *Espaces Topologiques et Fonctions Multivoques*, Dunod, Paris, 1959.

Blaschke, W., *Vorlesungen über integral Geometrie*, Teubner, Leipzig, 1936–37.

Bonnesen, T., and W. Fenchel, *Theorie der konvexen Körper*, Springer, Berlin, 1934.

Bourbaki, N., *Eléments de Mathématiques*, Fasc. II: *Topologie Générale*, Hermann, Paris, 1965a.

Bourbaki, N., *Eléments de Mathématique*, Fasc. XIII: *Intégration*, Hermann, Paris, 1965b.

Brothers, J.E., Integral geometry in homogeneous spaces., *Trans. Am. Math. Soc.*, **124**, 480–517, 1966.

Choquet, G., Theory of capacities, *Ann. Inst. Fourier*, **V**, 131–295, 1953-54.

Choquet, G., Le théorème de représentation intégrale dans les ensembles convexes compacts, *Ann. Inst. Fourier*, **10**, 333–444, 1960.

Datka, R., Measurability properties of set valued mappings in Banach spaces, *SIAM J. Control*, **8**, No. 2, 226–238, 1970.

Debreu, G., Integration of correspondences, in *5th Berkeley Symposium on Mathematical and Statistical Probability*, Vol. II, Part I, pp. 351–372.

Delfiner, P., A Generalization of the Concept of Size, 3rd International Congress for Stereology, Bern, 1970.

Delfiner, P., Etude Morphologique des Milieux Poreux, et Automatisation des Mesures en Lames Minces, Thesis, Nancy, 1971.

Dellacherie, C., Ensembles aléatoires, in *Séminaire de Probabilités*, Vol. III, No. 88, Springer, Berlin, 1969, pp. 97–114.

Dellacherie, C., *Capacités et Processus Stochastiques*, Springer, Berlin, 1972.

Deltheil, R., *Probabilités Géometriques*, Gauthier-Villars, Paris, 1926.

Fara, H.D., and A.E. Scheidegger, *J. Geophys. Res.*, **66**, No. 10, 3279–3284 (October 1961).

Federer, H., *Geometric Measure Theory*, Springer, Berlin, 1969.

Feller, W., *An Introduction to Probability Theory and Its Applications*, John Wiley Sons, New York, 1966.

Guelfand, I.M., and Y. Vilenkin, *Generalized Functions*, V, Moscow, 1961.

Haas, A., G. Matheron, and J. Serra, Morphologie mathématique et granulométries en place, *Ann. Mines*, **XI**, 736–753, and **XII**, 767–782 (1967).

Hadwiger, H., *Vorlesungen Über Inhalt, Oberfläche und Isoperimetrie*, Springer, Berlin, 1957.

Halmos, P., *Measure Theory*, Van Nostrand, Princeton, N.J., 1950.

Hormander, L., Sur la fonction d'appui des ensembles convexes dans un espace localement convexe, *Ark. Math.*, **3**, No. 12, 181–186, 1954.

Hukuhara, M., Sur l'application semicontinue dont la valeur est un convexe compact, *Funkcialoj Ekvacioj*, **10**, 43–66, 1967.

Joly, J.L., Une Famille de Topologies et de Convergences sur l'Ensemble des Fonctionelles Convexes, Thesis, Grenoble, 1970.

Kendall, D.G., Foundations of a theory of random sets, in *Stochastic Geometry*, (E. F. Harding and D. G. Kendall, eds), John Wiley Sons, New York, 1973, pp. 322–376.

Kendall, M.G., and P.A.P. Moran, *Geometrical Probabilities*, Hafner, New York, 1963.

Klein, J.C., and J. Serra, The texture analyzer, *J. Microsc.*, **95**, Part 2, 349–356 (April 1971).

Landkoff, N.S., *Osnovy Sovremennoï Teorii Potentziala*, Izd. Nauka, Moscow, 1966.

Matheron, G., *Eléments pour une Théorie des Milieux Poreux*, Masson, Paris, 1967.

Matheron, G., Théorie des ensembles aléatoires, in *Cahiers du Centre de Morph. Math.*, Fasc. 4, Fontainebleau, 1969.

Matheron, G., Ensembles aléatoires, ensembles semi-markoviens et polyèdres poissoniens, *Adv. Appl. Prob.*, December 1972.

Meyer, P.A., *Probabilités et Potentiel*, Hermann, Paris, 1966.

Michael, E., Topologies on spaces of subsets, *Trans. Am. Math. Soc.*, **71**, 152–182, 1951.

Miles, R.E., Random polygons determined by random lines in a plane, *Proc. Natl. Acad. Sci.* (*U.S.A.*), **52**, 901–907 and 1157–1160, 1964.

Miles, R.E., Poisson flats in euclidean space, *Adv. Appl. Prob.*, **1**, 211–237, 1969.

Miles, R.E., A synopsis of Poisson flats in euclidean spaces, *Izv. Akad. Nauk. Arm. SSR.*, **3**, 263–285, 1970.

Miles, R.E., Poisson flats in euclidean spaces, Part II, *Adv. Appl. Prob.*, **3**, 1–43, 1971.

Miles, R.E., The random division of space, *Suppl. Adv. Appl. Prob.*, 243–266, 1972.

Moore, G.A., L.L. Wyman, and H.M. Joseph, *Cmooents on the Possibilities of Performing Quantitative Metallographic Analysis with a Computer in Quantitative Microscopy*, McGraw-Hill, New York, 1968.

Moran, P.A.P., A note on recent research in geometric probability, *J. Appl. Prob.*, **3**, 453–463, 1966.

Moran, P.A.P., A second note on recent research in geometric probability, *Adv. Appl. Prob.*, **1**, 73–89, 1969.

Morean, J.J., *Fonctionelles Convexes*, College de France, 1966-67.

Mosco, U., Convergence of convex sets and of solutions of variational inequalities, *Adv. Math.*, **3**, No. 4, 510–585.

Nachbin, L., *The Haar Integral*, Van Nostrand, Princeton, N.J., 1965.

Neveu, J., *Bases Mathématiques du Calcul des Probabilités*, Masson, Paris, 1964.

Rockafellar, R.T., Integrals which are convex functionals, *Pacific J. Math.*, **24**, No. 3, 525–539, 1968.

Rockafellar, R.T., *Convex Analysis*, Princeton University Press, Princeton, N.J., 1970.

Santalo, L.A., *Introduction to Integral Geometry*, Hermann, Paris, 1953.

Serra, J., Introduction à la morphologie mathématique, *Cahiers du Centre de Morph. Math.*, Fasc. 3, Fontainebleau, 1969.

Stoka, M.I., *Geometrie Integrala*, Bucharest, 1967.

Valadier, M., Contribution à l'Analyse Convexe, Thesis, Paris, 1970.

Van Cutsem, B., Eléments Aléatoires à Valeurs Convexes Compactes, Thesis, Grenoble, 1971.

Wijsman, R.A., Convergence of sequences of sets, cones and functions, Part I, *Bull. Am. Math. Soc.*, **70**, 186–188, 1964; Part II, *Trans. Am. Math. Soc.*, **123**, 32–45, 1966.

Yosida, K., *Functional Analysis*, Springer, Berlin, 1968.

Index

Published by Carrick Publishing,
28 Miller Road, Ayr, KA7 2AY
0292 266679
First published 1985

Printed in Great Britain by
Antony Rowe Ltd.
Chippenham

THE SCOTTISH COMPANION

First Edition 1985

Carrick Publishing Ayr

THE SCOTTISH COMPANION

Applied Probability and Statistics (Continued)

LEWIS · Stochastic Point Processes
MANN, SCHAFER and SINGPURWALLA · Methods for Statistical Analysis of Reliability and Life Data
MILTON · Rank Order Probabilities: Two-Sample Normal Shift Alternatives
OTNES and ENOCHSON · Digital Time Series Analysis
RAO and MITRA · Generalized Inverse of Matrices and Its Applications
SARD and WEINTRAUB · A Book of Splines
SEAL · Stochastic Theory of a Risk Business
SEARLE · Linear Models
THOMAS · An Introduction to Applied Probability and Random Processes
WHITTLE · Optimization under Constraints
WILLIAMS · Regression Analysis
WONNACOTT and WONNACOTT · Econometrics
YOUDEN · Statistical Methods for Chemists
ZELLNER · An Introduction to Bayesian Inference in Econometrics

Tracts on Probability and Statistics

BILLINGSLEY · Ergodic Theory and Information
BILLINGSLEY · Convergence of Probability Measures
CRAMÉR and LEADBETTER · Stationary and Related Stochastic Processes
JARDINE and SIBSON · Mathematical Taxonomy
KINGMAN · Regenerative Phenomena
RIORDAN · Combinatorial Identities
TAKACS · Combinatorial Methods in the Theory of Stochastic Processes

Proposition 8-3-1. *There exists a one-to-one mapping* $\mathcal{V} \to \mathcal{V}'$ *from* $\mathcal{F}_u(\mathcal{K})$ *onto* $\mathcal{F}_u(\mathcal{F})$, *defined by the reciprocal formulae*

$$\mathcal{V} = \bigcap_{F \in \mathcal{V}'} \mathcal{K}_F; \qquad \mathcal{V}' = \bigcap_{K \in \mathcal{V}} \mathcal{F}_K$$

In particular, $\mathcal{V} = \varnothing$ *if and only if* $\mathcal{V}' = \mathcal{F}$, *and* $\mathcal{V}' = \varnothing$ *if and only if* $\mathcal{V} = \mathcal{K}$. *More generally, for any* $\mathcal{V}_0 \subset \mathcal{K}(E)$ [*resp.* $\mathcal{V}_0' \subset \mathcal{F}(E)$], *put* $\mathcal{V}' = \cap\{\mathcal{F}_K, K \in \mathcal{V}_0\}$, $\mathcal{V} = \cap\{\mathcal{K}_F, F \in \mathcal{V}'\}$ (*resp.* $\mathcal{V} = \cap\{\mathcal{K}_F, F \in \mathcal{V}_0'\}$, $\mathcal{V}' = \cap\{\mathcal{F}_K, K \in \mathcal{V}\}$). *Then* $\mathcal{V}$ (*resp.* $\mathcal{V}'$) *is the closure in* $\mathcal{K}$ (*resp. in* $\mathcal{F}$) *of the* $\cup$*-hereditary family generated in* $\mathcal{K}$ *by* $\mathcal{V}_0$ (*resp. in* $\mathcal{F}$ *by* $\mathcal{V}_0'$).

Proof. Let us prove only the statement concerning $\mathcal{V}_0 \subset \mathcal{K}(E)$. The family $\mathcal{V}$ is closed in $\mathcal{F}$ and $\cup$-hereditary. If $\mathcal{V}_1$ is the $\cup$-hereditary family generated by $\mathcal{V}_0$ in $\mathcal{K}$, $\mathcal{V}_1 \subset \mathcal{V}$, for, if $K \in \mathcal{V}_0$, K hits any $F \in \mathcal{V}'$, and thus $K \in \mathcal{V}$. Hence $\mathcal{V}_0 \subset \mathcal{V}$ and $\mathcal{V}_1 \subset \mathcal{V}$, as $\mathcal{V}$ is $\cup$-hereditary. It follows that $\overline{\mathcal{V}}_1 \subset \mathcal{V}$. Conversely, if $K \in \mathcal{K}$ and $K \notin \overline{\mathcal{V}}_1$, there exists a closed set $F \in \mathcal{F}$ such that $\mathcal{K}^F$ is a neighborhood of K disjoint of the $\cup$-hereditary family $\mathcal{V}_1$, that is, $F \cap K = \varnothing$ and $\mathcal{V}_0 \subset \mathcal{V}_1 \subset \mathcal{K}_F$. This implies $F \in \mathcal{V}'$, as F hits any $K' \in \mathcal{V}_0$. Then $K \notin \mathcal{V}$ follows from $K \notin \mathcal{K}_F$. Hence $\overline{\mathcal{V}}_1 \supset \mathcal{V}$, and so $\mathcal{V} = \overline{\mathcal{V}}_1$.

From a topological standpoint, it turns out that the mapping $\mathcal{V} \to \mathcal{V}'$ is actually a *homeomorphism*. In order to see this, some preliminary results must first be proved.

Proposition 8-3-2. *The spaces* $\mathcal{F}_u(\mathcal{F})$ *and* $\mathcal{F}_u(\mathcal{K})$ *are compact. Moreover,* $\mathcal{K}$ *is an isolated point in* $\mathcal{F}_u(\mathcal{K})$, *and the empty family* $\varnothing$ *is an isolated point in* $\mathcal{F}_u(\mathcal{F})$. *On the contrary,* $\varnothing$ *and* $\mathcal{F}$ *are isolated points in* $\mathcal{F}_u(\mathcal{K})$ *and* $\mathcal{F}_u(\mathcal{F})$, *respectively, if and only if the LCS space E is compact.*

Proof. Let us prove, for instance, that $\mathcal{F}_u(\mathcal{K})$ is a closed subspace of the compact space $\mathcal{F}(\mathcal{K})$. Let $\{\mathcal{V}_n\} \subset \mathcal{F}_u(\mathcal{K})$ be a sequence such that lim $\mathcal{V}_n = \mathcal{V}$ in $\mathcal{F}(\mathcal{K})$. If $\varnothing \in \mathcal{V}$, there exists a sequence $n \to K_n \in \mathcal{V}_n$ such that $\varnothing = \lim K_n$ for the myope topology, and this implies $K_n = \varnothing$ for n large enough, that is, $\mathcal{V}_n = \mathcal{K}$, as $\mathcal{V}_n$ is $\cup$-hereditary. Hence $\mathcal{V} = \mathcal{K} \in \mathcal{F}_u(\mathcal{K})$, and $\mathcal{K}$ is an isolated point in $\mathcal{F}_u(\mathcal{K})$. We suppose now that $\varnothing \notin \mathcal{V}$, and prove that $\mathcal{V}$ is $\cup$-hereditary. Let $K \in \mathcal{V}$ and $K' \in \mathcal{K}$, $K' \supset K$. There exists a sequence $n \to K_n \in \mathcal{V}_n$ such that $K = \lim K_n$ in $\mathcal{K}$. This implies lim $K' \cup K_n = K' \cup K = K'$, for $\cup$ is continuous on $\mathcal{K} \times \mathcal{K}$. But $K' \cup K_n \in \mathcal{V}_n$, because $\mathcal{V}_n$ is $\cup$-hereditary, and thus $K' \in \mathcal{V}$. Hence $\mathcal{V} \in \mathcal{F}_u(\mathcal{K})$, and $\mathcal{F}_u(\mathcal{K})$ is compact. From Proposition 1-2-1, $\varnothing$ is an isolated point in

$\mathcal{F}(\mathcal{K})$ if and only if $\mathcal{K}$ is compact, that is, if and only if E itself is compact. If E is not compact, there exists a sequence $\{A_n\} \subset \mathcal{K}$ without accumulation point, so that there exists no sequence $k \to K_{n_k} \supset A_{n_k}$ converging in $\mathcal{K}$. This implies $\lim \mathcal{F}_{A_n} = \varnothing$ in $\mathcal{F}_u(\mathcal{K})$. Hence $\varnothing$ is isolated in $\mathcal{F}_u(\mathcal{K})$ if and only if E is compact. The proof of the statement concerning $\mathcal{F}_u(\mathcal{F})$ is very similar.

In order to construct a basis for the topologies of the spaces $\mathcal{F}_u(\mathcal{F})$ and $\mathcal{F}_u(\mathcal{K})$, we start with the following lemma.

Lemma 8-3-1. *Let E be an LCS space, $\mathcal{B} \subset \mathcal{G}$ a basis for its topology, and $\mathbf{C} \subset \mathcal{K}$ a family closed under (finite) $\cap$ and such that the class closed under finite $\cup$ and infinite $\cap$ generated by $\mathbf{C}$ is identical to $\mathcal{K}(E)$. Then the topology of $\mathcal{F}(E)$ is generated by the two families $\mathcal{F}_B, B \in \mathcal{B}$, and $\mathcal{F}^C, C \in \mathbf{C}$.*

Proof. The $\mathcal{F}$-topology is generated by the l.s.c.-type open sets $\mathcal{F}_G, G \in \mathcal{G}$, and the u.s.c.-type open sets $\mathcal{F}^K, K \in \mathcal{K}$. Any $G \in \mathcal{G}$ is $G = \cup B_i$ for a family $\{B_i, i \in I\} \subset \mathcal{B}$, so that $\mathcal{F}_G = \mathcal{F}_{\cup B_i} = \cup \mathcal{F}_{B_i}$. Hence $\{\mathcal{F}_B, B \in \mathcal{B}\}$ is a basis for the l.s.c. topology.

Let $\mathbf{C}'$ be the class closed under finite $\cup$ and finite $\cap$ generated by $\mathbf{C}$. Any $K \in \mathcal{K}$ is $K = \cap C_i'$ for a family $\{C_i', i \in I\} \subset \mathbf{C}'$. If $F \in \mathcal{F}^K$, that is, $\cap(C_i' \cap F) = \varnothing$, there exist a finite number of indices $i_1, \ldots, i_k \in I$ such that $F \cap C_{i_1}' \cap \ldots \cap C_{i_k}' = \varnothing$. But $C' = C_{i_1}' \cap \cdots \cap C_{i_k}' \in \mathbf{C}'$ and $C' \supset K$, so that $F \in \mathcal{F}^{C'} \subset \mathcal{F}^K$. Then there exist $C_1, \ldots, C_n \in \mathbf{C}$ such that $C' = C_1 \cup \cdots \cup C_n$, as $C' \in \mathbf{C}'$, and $F \in \mathcal{F}^{C'} = \mathcal{F}^{C_1} \cap \cdots \cap \mathcal{F}^{C_n} \subset \mathcal{F}^K$. It follows that the u.s.c. topology is generated by $\{\mathcal{F}^C, C \in \mathbf{C}\}$.

Example. Let us apply Lemma 8-3-1 to the spaces $E = \mathcal{F}$ and $E = \mathcal{K}$. For brevity, we denote by $\mathcal{Q}$ either $\mathcal{F}$ or $\mathcal{K}$ and put $\mathcal{Q}' = \mathcal{K}$ if $\mathcal{Q} = \mathcal{F}$, $\mathcal{Q}' = \mathcal{F}$ if $\mathcal{Q} = \mathcal{K}$. As usual, $\mathcal{Q}^B_{G_1, \ldots, G_k}$ is the family $\{A, A \in \mathcal{Q}, A \cap G_i \neq \varnothing, i = 1, 2, \ldots, k, A \cap B = \varnothing\}$. Then the following classes $\mathcal{B}$ and $\mathbf{C}$:

$$\mathcal{B} = \left\{ \mathcal{Q}^{A'}_{G_1, \ldots, G_k}, A' \in \mathcal{Q}', k \geqslant 0, G_1, \ldots, G_k \in \mathcal{G} \right\}$$

$$\mathbf{C} = \left\{ \mathcal{Q}^{A^c}_{A_1', \ldots, A_k'}, A \in \mathcal{Q}, k \geqslant 0, A_1', \ldots, A_k' \in \mathcal{Q}' \right\}$$

satisfy the condition of the lemma for $\mathcal{Q} = \mathcal{F}$ or $\mathcal{K}$. Hence the l.s.c. topology on $\mathcal{F}(\mathcal{Q})$ is generated by $\{\mathcal{F}_\Gamma, \Gamma \in \mathcal{B}\}$, and the u.s.c. topology is generated by $\{\mathcal{F}^\chi, \chi \in \mathbf{C}\}$. Note that Γ and χ denote families of subsets in E.

Proposition 8-3-3. *If $\mathcal{A} = \mathcal{K}$ (resp. $\mathcal{A} = \mathcal{F}$) and $\mathcal{A}' = \mathcal{F}$ (resp. $\mathcal{A}' = \mathcal{K}$), the u.s.c. topology on $\mathcal{F}_u(\mathcal{A})$ is generated by the family $\{\mathcal{V}, \mathcal{V} \subset \mathcal{A}_{A^c}\}, A \in \mathcal{A}$, and the l.s.c. topology by the family $\{\mathcal{V}, \mathcal{V} \cap \mathcal{A}^{A'} \neq \varnothing\}$, $A' \in \mathcal{A}'$.*

Proof. With the notations of the above example, $\chi = \mathcal{A}^{A^c}_{A'_1,\ldots,A'_k} \in \mathbf{C}$ and $\mathcal{V} \in \mathcal{F}^{\chi}$, that is, $\mathcal{V} \cap \chi = \varnothing$ if and only if, for any $V \in \mathcal{V}$, either $V \cap A^c \neq \varnothing$ or $V \cap A'_i = \varnothing$ for one index $i \in [1, 2, \ldots, k]$. If $\mathcal{V}$ is $\cup$-hereditary, this is equivalent to $V \cap A^c \neq \varnothing$ for any $V \in \mathcal{V}$, that is, $\mathcal{V} \cap \mathcal{A}^{A^c} = \varnothing$, or $\mathcal{V} \subset \mathcal{A}_{A^c}$. If $\Gamma = \mathcal{A}^{A'}_{G_1,\ldots,G_k} \in \mathcal{B}$, $\mathcal{V} \in \mathcal{F}_{\Gamma}$, that is, $\mathcal{V} \cap \mathcal{A}^{A'}_{G_1,\ldots,G_k} \neq \varnothing$ if and only if there exists $V \in \mathcal{V}$ such that $V \cap A' = \varnothing$, $V \cap G_i \neq \varnothing$, $i = 1, \ldots, k$. If $\mathcal{V}$ is $\cup$-hereditary, this is equivalent to {there exists a $V \in \mathcal{V}$ such that $V \cap A' = \varnothing$, that is, $\mathcal{V} \cap \mathcal{A}^{A'} \neq \varnothing$}.

Proposition 8-3-4. *With the same notations, the mapping $\mathcal{V} \to \mathcal{V}' = \cap\{\mathcal{A}'_A, A \in \mathcal{V}\}$ is a homeomorphism from $\mathcal{F}_u(\mathcal{A})$ onto $\mathcal{F}_u(\mathcal{A}')$, and the reciprocal formula $\mathcal{V} = \cap\{\mathcal{A}_{A'}, A' \in \mathcal{V}'\}$ holds. Moreover, the image under $\mathcal{V} \to \mathcal{V}'$ of any l.s.c. (resp. u.s.c.) open set in $\mathcal{F}_u(\mathcal{A})$ is an u.s.c. (resp. l.s.c.) open set in $\mathcal{F}_u(\mathcal{A}')$, and conversely.*

Proof. We know from Proposition 8-3-1 that $\mathcal{V} \to \mathcal{V}'$ is a one-to-one mapping from $\mathcal{F}_u(\mathcal{A})$ onto $\mathcal{F}_u(\mathcal{A}')$, and the reciprocal formula holds. On the other hand, if $\mathcal{V} \in \mathcal{F}_u(\mathcal{A})$ and $A \in \mathcal{A}$, $\mathcal{V} \not\subset \mathcal{A}_{A^c}$ if and only if there exists $V \in \mathcal{V}$ such that $V \subset A$, that is, if and only if $A \in \mathcal{V}$, as $\mathcal{V}$ is $\cup$-hereditary. But $A \in \mathcal{V} = \cap\{\mathcal{A}_{A'}, A' \in \mathcal{V}'\}$ is equivalent to $\mathcal{V}' \cap \mathcal{A}'^{A} = \varnothing$. Thus:

$$\begin{aligned} &\mathcal{V} \subset \mathcal{A}_{A^c} \Leftrightarrow \mathcal{V}' \cap \mathcal{A}'^{A} \neq \varnothing \\ &\mathcal{V}' \subset \mathcal{A}'_{A'^c} \Leftrightarrow \mathcal{V} \cap \mathcal{A}^{A'} \neq \varnothing \end{aligned} \tag{8-3-1}$$

This achieves the proof when taking Proposition 8-3-2 into account.

8-4. POINT OF VIEW OF THE INVERSE MAPPINGS

In the preceding sections, we investigated the increasing mappings from $\mathcal{P}(E)$ (or a subset of it) into itself. More generally, now let E and E' be two LCS spaces (no longer necessarily identical), $\mathcal{A} \subset \mathcal{P}(E)$ and ψ an increasing mapping from $\mathcal{A}$ into $\mathcal{P}(E')$, that is, $A, B \in \mathcal{A}$ and $A \subset B$ in E imply $\psi(A) \subset \psi(B)$ in E'. For any $x \in E'$, let us put

$$\mathcal{V}(x) = \psi^{-1}(\mathcal{P}_x(E')) = \{A, A \in \mathcal{A}, x \in \psi(A)\} \tag{8-4-1}$$

The mapping $x \to \mathcal{V}(x)$ from E' into $\mathcal{P}(\mathcal{P}(E))$ is said to be the inverse mapping of ψ. For any $x \in E'$, $\mathcal{V}(x)$ is $\cup$-hereditary in $\mathcal{A}$. Conversely, if $\mathcal{V}$ is a mapping from E' into the space $\mathcal{P}_u(\mathcal{A})$ of the $\cup$-hereditary families in $\mathcal{A}$, $\mathcal{V}$ is the inverse mapping of ψ: $A \to \psi(A) = \{x, A \in \mathcal{V}(x)\}$.

In what follows, the space $\mathcal{A} \subset \mathcal{P}(E)$ will always be either $\mathcal{A} = \mathcal{F}(E)$ or $\mathcal{A} = \mathcal{K}(E)$. If $\mathcal{A} = \mathcal{K}$, we shall put $\mathcal{B} = \mathcal{G}$ and $\mathcal{A}' = \mathcal{F}$; if $\mathcal{A} = \mathcal{F}$, $\mathcal{A}' = \mathcal{K}$ and $\mathcal{B} = \{K^c, K \in \mathcal{K}\}$, that is, in both cases $\mathcal{B} = \{A'^c, A' \in \mathcal{A}'\}$. The space $\mathcal{B}$ will always be provided with the topology induced from $\mathcal{A}'$ by the mapping $A' \to A'^c$. We shall also use the notations defined in Proposition 8-1-4, that is, if ψ is an increasing mapping from $\mathcal{A}$ into $\mathcal{P}(E')$, ψ_b is its smallest extension on $\mathcal{B}$ and ψ_a is the largest extension of ψ_b on $\mathcal{A}$. Explicitly:

$$\begin{aligned} \psi_b(B) &= \cup \{\psi(A), A \in \mathcal{A}, A \subset B\} \qquad (B \in \mathcal{B}) \\ \psi_a(A) &= \cap \{\psi(B), B \in \mathcal{B}, B \supset A\} \qquad (A \in \mathcal{A}) \end{aligned} \tag{8-4-2}$$

With these notations, it is possible to state simultaneously analogous results concerning the spaces $\mathcal{F}(E)$ and $\mathcal{K}(E)$.

First, let us investigate under what condition $\psi = \psi_a$ on $\mathcal{A}$. From (8-4-2), $\psi = \psi_a$ if and only if, for any $A \in \mathcal{A}$ and $x \notin \psi(A)$, there exists $B \in \mathcal{B}$ such that $B \supset A$ and $x \notin \psi(A')$ for any $A' \in \mathcal{A}$, $A' \subset B$ (see Proposition 8-1-4). This is equivalent to $\{A \notin \mathcal{V}(x)$ implies that there exists $B \in \mathcal{B}$ such that $B \supset A$ and $\mathcal{V}(x) \subset \mathcal{A}_{B^c}\}$. But $B^c \in \mathcal{A}'$ (by the definition of $\mathcal{B}$), and thus $\mathcal{A}^{B^c}$ is an open neighborhood of A disjoint of $\mathcal{V}(x)$. Hence $\mathcal{V}(x)$ is closed, that is, $\mathcal{V}(x) \in \mathcal{F}_u(\mathcal{A})$. Conversely, if $\mathcal{V}(x)$ is closed and $\cup$-hereditary for any $x \in E'$, $\complement \mathcal{V}(x)$ is $\cap$-hereditary, and for any $A \notin \mathcal{V}(x)$ there exists a $B \in \mathcal{B}$ such that $\mathcal{A}^{B^c}$ is an open neighborhood of A disjoint of $\mathcal{V}(x)$. Hence $\psi = \psi_a$. More precisely:

Proposition 8-4-1. *Let ψ be an increasing mapping from $\mathcal{A}$ into $\mathcal{P}(E')$ and $\mathcal{V}$: $E' \to \mathcal{P}(\mathcal{A})$ defined by $\mathcal{V}(x) = \{A, x \in \psi(A)\}$. Then $\psi = \psi_a$ if and only if $\mathcal{V}$ maps E' into $\mathcal{F}_u(\mathcal{A})$. Moreover, the four following conditions are equivalent*:

1. *ψ is a u.s.c. mapping from $\mathcal{A}$ into $\mathcal{F}(E')$.*
2. *$\mathcal{V}$ is a u.s.c. mapping from E' into $\mathcal{F}_u(\mathcal{A})$.*
3. *For any $A \in \mathcal{A}$ and $x \in E'$, $\psi(A)$ and $\mathcal{V}(x)$ are closed.*
4. *ψ maps $\mathcal{A}$ into $\mathcal{F}(E')$ and $\psi = \psi_a$.*

Proof. We have just proved $\psi = \psi_a$ if and only if $\mathcal{V}(E') \subset \mathcal{F}_u(\mathcal{A})$. If so, $\psi(A)$ is closed for any $A \in \mathcal{A}$ if and only if $A \in \mathcal{V}(x_n)$ and $\lim x_n = x$ in E' implies $A \in \mathcal{V}(x)$, that is, $\mathcal{V}(x_n) \not\subset \mathcal{A}_{A^c}$ and $x = \lim x_n$ imply $\mathcal{V}(x) \not\subset \mathcal{A}_{A^c}$. It follows from Proposition 8-3-2 that this condition is satisfied if and only if